Probiotics and Prebiotics: Bioactive Foods

Probiotics and Prebiotics: Bioactive Foods

Editor: Myla Dickinson

MURPHY & MOORE
www.murphy-moorepublishing.com

Murphy & Moore Publishing,
1 Rockefeller Plaza,
New York City, NY 10020, USA

Visit us on the World Wide Web at:
www.murphy-moorepublishing.com

ISBN: 978-1-63987-460-6 (Hardback)

Cataloging-in-Publication Data

Probiotics and prebiotics : bioactive foods / edited by Myla Dickinson.
p. cm.
Includes bibliographical references and index.
ISBN 978-1-63987-460-6
1. Probiotics. 2. Prebiotics. 3. Functional foods. 4. Bioactive compounds.
5. Dietary supplements. I. Dickinson, Myla.

RM666.P835 P76 2022
615.329--dc23

Table of Contents

Preface ... VII

Chapter 1 **Bacterial Cell-Free Probiotics Using Effective Substances Produced by Probiotic Bacteria, for Application in the Oral Cavity** ... 1
Tomoko Ohshima, Tomomi Kawai and Nobuko Maeda

Chapter 2 **Probiotics and Prebiotics in Infant Formulae** ... 11
José Maldonado

Chapter 3 **Prebiotics, Probiotics, and Bacterial Infections** ... 23
Christina C. Tam, Kirkwood M. Land and Luisa W. Cheng

Chapter 4 **Cholesterol Uptake and Survival of *Lactococcus lactis* Strains in Fluids Simulating the Human Stomach and Duodenum** ... 39
Małgorzata Ziarno

Chapter 5 **Could Gut Modulation through Probiotic Supplementation be Beneficial in Autism Spectrum Disorder?** ... 60
Rouba El Khatib, Dolla Karam-Sarkis, Anne-Judith Waligora-Dupriet and Marie-José Butel

Chapter 6 **Role of the Microbiome as the First Metal Detoxification Mechanism** ... 84
Rebeca Monroy-Torres, Marco Antonio Hernández-Luna, Xochitl Sofía Ramírez-Gómez and Sergio López-Briones

Chapter 7 **Prebiotics and Probiotics - Potential Benefits in Human Nutrition and Health** ... 100
Maria Inês Sucupira Maciel and Michelle Maria Barreto de Souza

Chapter 8 **Comparision of Antioxidant Activity of Cow and Goat Milk During Fermentation with *Lactobacillus acidophilus* LA-5** ... 113
Jessica Dayara Álvarez-Rosales, César Ozuna, Rubén Salcedo-Hernández and Gabriela Rodríguez-Hernández

Chapter 9 **The Use of Probiotics in Poultry Production for the Control of Bacterial Infections and Aflatoxins** ... 129
Daniel Hernandez-Patlan, Bruno Solis-Cruz, Billy M. Hargis and Guillermo Tellez

Chapter 10 **Prebiotics and the Modulation on the Microbiota-GALT-Brain Axis** ... 148
Elena Franco-Robles, Joel Ramírez-Emiliano, José Sergio López-Briones and Cristina Doriany Balcón-Pacheco

Chapter 11 **Probiotics in Health and Immunity: A First Step toward Understanding the Importance of Microbiota System in Translational Medicine** ... 166
Ciro Gargiulo Isacco, Andrea Ballini, Danila De Vito, Angelo Michele Inchingolo, Stefania Cantore, Gregorio Paduanelli, Kieu Cao Diem Nguyen, Alessio Danilo Inchingolo, Gianna Dipalma and Francesco Inchingolo

Chapter 12 **Use of Prebiotics as an Alternative to Antibiotic Growth Promoters in the Poultry Industry** ... 181
Bruno Solis-Cruz, Daniel Hernandez-Patlan, Billy M. Hargis and Guillermo Tellez

Chapter 13 **Functional Attributes and Health Benefits of Novel Prebiotic Oligosaccharides Derived from Xylan, Arabinan, and Mannan** ... 202
Bradley A. Saville and Sandra H. Saville

Permissions

List of Contributors

Index

Preface

Probiotics and prebiotics are important for human health. Probiotics are live bacteria that can provide various health benefits, and are found in certain foods or supplements. Prebiotics are substances that come from certain types of carbohydrates that humans cannot digest, but are food to the beneficial bacteria in the gut. The gut bacteria are jointly called gut microbiota and perform many vital functions in the body. Eating balanced amounts of both probiotics and prebiotics can help maintain the right balance of these bacteria. Gut microbiota gives protection from harmful bacteria and fungi. Studies have confirmed that these bacteria can improve the immune system functions, depression, obesity, etc. They also provide important nutrients to the cells lining the digestive tract. This book contains some path-breaking studies in the field of bioactive foods. It presents researches and studies performed by experts across the globe. As this field is emerging at a rapid pace, the contents of this book will help the readers understand the different aspects of probiotics and prebiotics.

Significant researches are present in this book. Intensive efforts have been employed by authors to make this book an outstanding discourse. This book contains the enlightening chapters which have been written on the basis of significant researches done by the experts.

Finally, I would also like to thank all the members involved in this book for being a team and meeting all the deadlines for the submission of their respective works. I would also like to thank my friends and family for being supportive in my efforts.

Editor

1

Bacterial Cell-Free Probiotics Using Effective Substances Produced by Probiotic Bacteria, for Application in the Oral Cavity

Tomoko Ohshima, Tomomi Kawai and Nobuko Maeda

Abstract

To avoid side effects of conventional antibiotics and disinfectants used for prevention of oral diseases such as dental caries, periodontitis, and oral candidiasis, application of probiotics has attracted attention recently. However, difficulties arise when applying those probiotics in the oral cavity, because exogenous probiotic bacteria do not colonize easily in the established oral microbiota. Even, if we are able to overcome the restriction of colonization by probiotic bacteria in the oral cavity, it comes with the risk of dental caries due to the potential acidic environment generated by probiotic bacteria. To solve these problems, "biogenics," bacterial cell-free probiotics using only the effective substances metabolically produced by probiotic bacteria, is recommended for application in the oral cavity. The concept and frontline of biogenic research will be introduced and discussed.

Keywords: biogenics, probiotics, oral diseases, dental caries, periodontitis, candidiasis

Introduction

More than 700 bacterial species live in the oral cavity [1, 2]. These bacteria form their own indigenous flora in their habitats, such as teeth, gingival sulcus, and tongue dorsum, making the oral environment considerably complicated. Oral indigenous bacteria coexist with humans and are vital for preventing colonization by foreign pathogenic microorganisms in the oral cavity. Such oral indigenous bacteria proliferate with time, and together with the extracellular polymeric substance (EPS) that they produce, form a biofilm visible even to the naked eye known as dental plaque [3–5].

It has recently been clarified that the formation of biofilms is controlled by quorum-sensing (QS) signals in a communication system between microorganisms that sense each other's abundance [6–8]. It has further been shown that microorganisms constituting a biofilm activate the expression of pathogenic factors when QS signaling molecules, so-called "autoinducers (AI)," act as transcription factors [8].

The potential pathogenicity of dental plaque as oral biofilm

Oral plaque contains also dental caries and periodontal disease causing organ- isms [9–12], and when these exert their potential pathogenicity, they are considered to accelerate biofilm formation even more. Generally it can be said that with increasing thickness of the biofilm, the bacterial metabolites build up at the bottom of the biofilm, and the caries and periodontitis occurrence proliferate.

Caries and periodontal diseases are called the two major dental diseases, and both of them occur as oral infectious diseases which are caused by specific bacteria known as cariogenic bacteria (such as Streptococcus mutans) and periodontal pathogens (such as *Porphyromonas gingivalis*) growing in plaque. This status is interpreted as dysbiosis of the oral flora. In addition, a small number of fungal genus *Candida* is also present in the indigenous oral resident microflora, and some factors also cause its growth in the plaque, resulting in dysbiosis, which causes a major oral mucosal disease, oral candidiasis. However, there is currently no technique to selectively eliminate only those causative microorganisms from the flora.

Oral application of probiotics and problems

The method of preventing caries and periodontal disease is basically the mechanical removal of the entire plaque by brushing, etc. However in the case of onset of the disease, antimicrobial drugs are administered for the treatment of the acute phase of periodontal disease, and antifungal administration is the first choice for treatment of oral candidiasis. However, the use of antimicrobial agents has shown problems regarding adverse effects such as drug-resistant bacteria and allergies, indicating the limitation of chemotherapy [13]. Therefore, attention has recently focused on probiotic bacteria such as bifidobacteria and lactobacilli recognizing the usefulness for improving dysbiosis [14]. Although probiotics were originally intended to improve dysbiosis of the intestinal flora [15], their usefulness is also assumed in the dental field. Attempts have been made for direct use in the oral cavity to prevent diseases such as caries and periodontal disease, and several results have been reported [16–18]. Ishikawa et al. reported that 4 weeks of oral administration of *Lactobacillus salivarius* TI 2711 (LS1) significantly reduced the major periodontal pathogens of P. gingivalis, Prevotella intermedia, and *Prevotella nigrescens* [19].

However, in these reports, there are few basic facts on the effects of probiotics on the oral flora and the antibacterial substances produced by them, so progress and establishment of practical applications based on the underlying mechanism has not been accomplished. In addition, genera *Lactobacillus* and *Bifidobacterium*, which are representative probiotic bacteria, exist though in minority in the oral microbiota, but because they metabolize sugar and produce large amounts of organic acid, the general understanding is that they work cooperatively with cariogenic bacteria or induce hypersensitivity.

As another fundamental issue, previous studies have highlighted the limitation of colonization and fixation of nonnatural probiotic bacteria in the intestinal tract [20, 21]. This phenomenon of transiency, but not permanency in colonization, is also relevant for probiotic applications in the oral cavity [16, 22, 23]. Even if we are able to address the restriction of colonization of probiotic bacteria in the oral cavity, it comes with the risk of dental caries due to the potential acidic environment generated by probiotic bacteria.

The concept of biogenics

To overcome the above problems, "biogenics" as a new idea has been introduced.

Biogenics is a kind of functional food, using only the ingredients, which have a positive effect on the host with regard to immunostimulating or immunosuppressing mutagenesis, tumorigenesis, peroxidation, hypercholesterolemia, or intestinal putrefaction [24]. Achieving a probiotic effect by the intake of nonviable probiotic bacteria has been proposed in previous reports. For example, the life span of mice increased, when they were fed with pasteurized fermented milk [25, 26]. A significant reduction of the Ehrlich ascites tumor growth in mice was also reported [26]. In addition, it was shown that heat-inactivated *Enterococcus faecalis* [27] or L. *gasseri* [28] showed a beneficial regulatory effect in the gut. Moreover, Nakamura et al. [29] found an angiotensin-I-converting enzyme (ACE) inhibitor in a Japanese sterilized milk beverage fermented by L. *helveticus and Saccharomyces cerevisiae.* The active substance in this fermented beverage was identified as lactotripeptide metabolically generated in the fermentation pathway. Follow-up studies were able to determine the bioactive metabolites of probiotic bacteria in addition to the antimicrobial sub- stances, such as bacteriocin [30, 31], and other beneficial active substances, such as conjugated linoleic acid (CLA) [30–32], proteins or peptides [33, 34], and polyphe-nols [35, 36]. Taking all these observations into account, biogenics, which makes use of the bioactive metabolites as foods or medicine, was recently advocated as a new concept [24, 37]. The biogenic effect is independent of the colonization and viability of probiotic bacteria. Hence, biogenics is the direct delivery of an isolated and purified active ingredient of probiotics to the local environment. This strategy may also be useful for oral disease prevention. It may be possible to purify the effective ingredients against oral pathogenic activity of probiotic bacteria for use in the biogenics process. However, this idea requires further study prior to clinical use.

Antibacterial substances produced by lactic acid bacteria

Research of probiotics for intestinal health has revealed several antibacterial substances produced by lactic acid bacteria in addition to organic acids such as lactic and acetic acids [38]. These are (1) hydrogen peroxide [39], (2) bacteriocins [40], and (3) low-molecular-weight antimicrobial substances.

Hydrogen peroxide (H_2O_2)

Hydrogen peroxide is produced by most *lactobacilli* in the presence of oxygen. *lactobacilli* possess oxidases that reduce oxygen to hydrogen peroxide, oxidizing substrates such as pyruvate or NADH [41]. Since they do not produce catalases, H_2O_2 does not suffer auto-degradation. H_2O_2 has a broad-spectrum planktonic bac- teria, but the effect decreases dramatically on biofilm. It appears that Lactobacilli do not produce effective concentrations of H_2O_2 against fungi [42], unlike other bacteria [39].

Bacteriocins

Lactic acid bacteria produce bacteriocins, proteinaceous antimicrobial sub- stances with molecular weights of several thousand daltons or more. Bacteriocins can be divided into five classes according to their primary structure, molecular composition, and physical and functional properties [43, 44]. However, bacteriocins produced by lactic acid bacteria against S. *mutans* and P. *gingivalis* are not yet known. Bacteriocin L23 produced by *Lactobacillus fermentum* L23 [44], plantaricin produced by L. *plantarum* [45], and pentocin TV35b produced by L. pentosus [46] appear to be effective against the yeast form of *Candida*. Bacteriocins effective for the hyphal forms of *Candida* have not yet been identified [47, 48].

Low-molecular-weight antimicrobial substances

Reuterin, an antibacterial substance (also known as 3-hydroxypropionaldehyde; molecular weight, 74 Da; composition formula, C3H6O2), is a product of glycerol fermentation, which has been seen in several probiotic bacteria. These probiotic bacteria include not only *L. reuteri* [49] but also *L. brevis*, *L. buchneri* [50], and *L. collinoides* [51]. Under anaerobic conditions *L. coryniformis* [52] also produces a low-molecular-weight antimicrobial substance that does not contain amino acids [53]. Reuterin was found to exert its antibacterial effects by causing oxidative stress within the bacterial cell [54]. In addition to reuterin, the low-molecular-weight substances of lactobacilli, reutericyclin [55] and diacetyl [56] also showed effective- ness against the yeast forms of *Candida* [57].

As the smallest peptides, diketopiperazines (DKPs, cyclic dipeptides) are known to possess several physiological activities, including an antimicrobial effect. diketopiperazines are a group of cyclic organic compounds where two amino acids are connected by a peptide bond, forming a lactam, and it is the first peptide whose three-dimensional structure has been completely solved by Robert Corey in the 1930s [58]. Corey determined the structure of the cyclic anhydride of the dipeptide glycylglycine. Diketopiperazines are also biosynthesized from amino acids in diverse organisms including mammals and are considered to be secondary metabolites [59]. Although some protease enzymes, such as dipeptidyl peptidase, produce a dipeptide by cleavage from the protein terminus, it is known that the

Cyclic dipeptide	Origin	Biological function	References
Cyclo(Leu-Pro)	*Lactobacillus casei* AST18	Antifungal activity	[64]
Cyclo(Phe-Pro) Cyclo (Phe-4-OH-Pro)	*L. plantarum* MiLAB393	Antifungal activity	[65]
Cyclo(Gly-Leu)	*L. plantarum* VTT E-78076	Antimicrobial activity	[66]
Cyclo(Phe-Pro) Cyclo(Tyr-Pro)	*L. reuteri* RC-14	Antimicrobial activity	[67]
Cyclo(Pro-Pro) Cyclo(Leu-Pro) Cyclo(Met-Pro) Cyclo(His-Pro)	L. amylovorus DSM 19280	Antifungal activity	[68]
Cyclo(Leu-Leu)	L. plantarum AF1	Antifungal activity	[69]
Cyclo(4-OHPro-Leu)	L. fermentum ALAL020	Antimicrobial activity	[70]

Table 1. *Diketopiperazines (cyclic dipeptide) produced by probiotic bacteria.*

resulting dipeptide cyclizes spontaneously to form a diketopiperazine. In addition, diketopiperazines are attractive scaffolds for drug design due to their structural properties such as a rigid structure, optical activity, and various side-chain structures [59]. Both natural diketopiperazines and synthetic diketopiperazines have been reported to possess various physiological activities including antitumor activity [60], antiviral activity [61], antibacterial activity [62], and antimicrobial activity [63]. However, there are only few reports on DKP produced by probiotic bacteria (**Table 1**). In addition, the antimicrobial mechanism is also poorly understood.

Anti-inflammatory substances produced by lactic acid bacteria

Periodontitis and candidiasis are both inflammatory diseases; therefore, inflam- mation symptoms are desired to be cured by biogenics, but there are few candidates for that.

CLA is a general term for regioisomers and structural isomers of linoleic acid having a conjugated diene structure.

Diene structure means there are two double bonds with a single bond in between. Rumenic acid, for example, is one of the 28 isomers of CLAs and exists in the fat and dairy products of ruminants [71]. It is a trans fat; however, CLAs can also appear as cis-fats. CLAs are known to reduce the production level of IgE and a chemical mediator leukotriene in a rat inflammatory model [72]. However, the opposite effect of increasing serum C-reactive protein value and reducing serum adiponectin level in human by CLA supplementation was observed recently [73].

Understanding the property of biofilm

Most bacteria and fungi have the potential to grow in a biofilm, in an environment with liquid flow and solid surfaces. Biofilm formation, which has been experimentally observed in single bacteria, is now known not only to cross species but also to cross the kingdom of microbes. In human bodies, such situations particularly exist in the resident microbiota. Microorganisms including oral pathogens have the potential to express pathogenic properties in biofilms, contrary to the planktonic type. In other words, the so-called biofilm phenotypes upregulate the production of EPS that block the stimuli or stress from outside the biofilm, such as antibiotics and disinfectants. The EPS also provides sticky intercellular binding material and extracellular energy storage compounds [74, 75] to promote interaction among contacting microbial cells [76], resulting in complex and dynamic interplay.

Disruption of the quorum-sensing signals

Recently, a QS inhibitor (QSI) and QS signal quencher (QQ) molecule attracted attention in regard to understanding biofilm infections. Biofilm formation is triggered and controlled by a cell-to-cell communication process in harmony with the bacterial population density known as quorum-sensing system, which is based on small molecules termed autoinducers [77]. Some reports revealed that bacteriocins produced by probiotic lactobacilli such as *L. acidophilus*, *L. plantarum*, and *L. reuteri* functioned as QSI or QQ molecules [78]. It may be possible to purify the effective ingredients of probiotic bacteria against oral pathogenic activity in biofilms for use in the biogenics process. Recently, some instance of QC disruption by cyclic dipeptides has been reported. *L. reuteri*, a human vaginal isolate, was capable of producing the cyclic dipeptides cyclo(L-Phe-L-Pro) and cyclo(L-Tyr-L- Pro), inhibiting the staphylococcal quorum-sensing system driven by the AI named agr, to suppress the expression of toxic shock syndrome toxin-1 in *S. aureus* [79].

The report is useful for a better understanding of interspecies cell-to-cell communication between *Lactobacillus* and *Staphylococcus* and provides a hint to attenuate virulence factor production by bacterial pathogens. However, this idea requires further study before clinical application.

Conclusion

Biogenics is based on the concept of using the active ingredients which were revealed by the mechanism of oral probiotics. Biogenics is expected to be a prevention method for oral diseases that can be implemented without the problems associated with the use of probiotic bacteria, namely the involvement of acids harmful to teeth. The emergence of resistant bacteria against naturally occurring substances of biogenic candidates is not yet known. Furthermore, it is possible to combine substances which contribute to the health of the oral cavity, with those contributing to systemic health, such as control substances for blood sugar level, blood pressure, neutral fat, antioxidants, anti-stress, immune enhancement, anti-inflammation, antianxiety, and antidepressants. Therefore, the progress of practical application is expected. However, the elucidation of the mechanism of action is still in the beginning, and further study is needed.

Acknowledgements

A part of this study was supported by JSPS KAKENHI Grant Number JP18K17057 and MEXT-Supported Program for the Strategic Research Foundation at Private Universities, 2015–2019.

Conflict of interest

There are no conflicts of interest.

Author details

Tomoko Ohshima*, Tomomi Kawai and Nobuko Maeda

Department of Oral Microbiology, School of Dental Medicine, Tsurumi University, Yokohama, Japan

*Address all correspondence to: ohshima-t@fs.tsurumi-u.ac.jp

References

[1] Paster BJ, Olsen I, Aas JA, Dewhirst FE. The breadth of bacterial diversity in the human periodontal pocket and other oral sites. Periodontology 2000. 2006;**42**(1):80-87

[2] Marsh PD, Devine DA. How is the development of dental biofilms influenced by the host? Journal of Clinical Periodontology. 2011;**38**:28-35

[3] Sutherland IW. Biofilm exopolysaccharides: A strong and sticky framework. Microbiology. 2001;**147**:3-9

[4] Bales PM. Purification and characterization of biofilm-associated EPS exopolysaccharides from ESKAPE organisms and other pathogens. PLoS One. 2013;**8**(6):e67950

[5] Koo H. The exopolysaccharide matrix: A virulence determinant of cariogenic biofilm. Journal of Dental Research. 2013;**92**(12):1065-1073

[6] Melissa B. Quorum sensing in bacteria. Annual Review of Microbiology. 2001;**55**:165-199

[7] de Kievit TR et al. Quorum sensing in *Pseudomonas aeruginosa* biofilms. Environmental Microbiology. 2001;**11**(2):279-288

[8] Li YH, Tian X. Quorum sensing and bacterial social interactions in biofilms. Sensors (Basel). 2012;**12**(3):2519-2538

[9] Marsh PD, Bradshaw DJ. Dental plaque as a biofilm. Journal of Industrial Microbiology. 1995;**15**(3):169-175

[10] Stralfors A. Studies of the microbiology of caries; the buffer capacity of the dental plaques. Journal of Dental Research. 1948;**27**(5):587-592

[11] Poole DF, Newman HN. Dental plaque and oral health. Nature. 1971;**234**(5328):329-331

[12] Axelsson P, Lindhe J. The effect of a plaque control program on gingivitis and dental caries in schoolchildren. Journal of Dental Research. 1971;**56**(C):142-148

[13] Watanabe T. Infectious drug resistance in enteric bacteria. The New England Journal of Medicine. 1966;**275**:888-894

[14] Gupta G. Probiotics and periodontal health. Journal of Medicine and Life. 2011;**4**:387-394

[15] Fuller R. Probiotics in man and animals. The Journal of Applied Bacteriology. 1989;**66**:365-378

[16] Krasse P, Carlsson B, Dahl C, Paulsson A, Nilsson A, Sinkiewicz G. Decreased gum bleeding and reduced gingivitis by the probiotic *Lactobacillus reuteri*. Swedish Dental Journal. 2006;**30**:55-60

[17] Vivekananda MR, Vandana KL, Bhat KG. Effect of the probiotic *Lactobacilli reuteri* (Prodentis) in the management of periodontal disease: A preliminary randomized clinical trial. Journal of Oral Microbiology. 2010;**2**:2. DOI: 10.3402/ jom.v2i0.5344

[18] Riccia DN, Bizzini F, Perilli MG, Polimeni A, Trinchieri V, Amicosante G, et al. Anti-inflammatory effects of *Lactobacillus brevis* (CD2) on periodontal disease. Oral Diseases. 2007;**13**:376-385

[19] Ishikawa H, Aiba Y, Nakanishi M, Oh-Hashi Y, Koga Y. Suppression of periodontal pathogenic bacteria by the administration of *Lactobacillus salivarius* TI2711. Journal of the Japanese Society of Periodontology. 2003;**45**:105-112

[20] Haenel H. Aspekte der mikroökologischen beziehungen des makroorganismus. Mikroorganismen im menschlichen und tierischen darm und in anderen organen. Zentralblatt für Bakteriologie. 1960;**176**:305-426

[21] Mitsuoka T, Kaneuchi C. Ecology of the bifidobacteria. The American Journal of Clinical Nutrition. 1977;**30**:1799-1810

[22] Meurman JH, Antila H, Salminen S. Recovery of *Lactobacillus* strain GG (ATCC 53103) from saliva of healthy volunteers after consumption of yoghurt prepared with the bacterium. Microbial Ecology in Health and Disease. 1994;**7**(6):295-298

[23] Caglar E, Topcuoglu N, Cildir SK, Sandalli N, Kulekci G. Oral colonization by *Lactobacillus reuteri* ATCC 55730 after exposure to probiotics. International Journal of Paediatric Dentistry. 2009;**19**(5):377-381

[24] Mitsuoka T. Significance of dietary modulation of intestinal flora and intestinal environment. Bioscience and Microflora. 2000;**19**(1):15-25

[25] Arai K, Murota I, Hayakawa K, Kataoka M, Mitsuoka T. Effects of administration of pasteurized fermented milk to mice on the life-span and intestinal flora. Journal of Japan Society of Nutrition and Food Sciences. 1980;**33**:219-223 [Japanese]

[26] Takano T, Arai K, Murota I, Hayakawa K, Mizutani T, Mitsuoka T. Effects of feeding sour milk on longevity and tumorigenesis in mice and rats. Bifidobacteria and Microflora. 1985;**4**(1):31-37

[27] Terada A, Bukawa W, Kan T, Mitsuoka T. Effects of the consumption of heat-killed enterococcus faecalis EC-12 preparation on microbiota and metabolic activity of the faeces in healthy adults. Microbial Ecology in Health and Diseases. 2004;**16**: 188-194

[28] Sawada D, Sugawara T, Ishida Y, Aihara K, Aoki Y, Takehara I, et al. Effect of continuous ingestion of a beverage prepared with *Lactobacillus gasseri* CP2305 inactivated by heat treatment on the regulation of intestinal function. Food Research International. 2016;**79**:33-39

[29] Nakamura Y, Yamamoto N, Sakai K, Okubo A, Yamazaki S, Takano T. Purification and characterization of angiotensin I-converting enzyme inhibitors from sour milk. Journal of Dairy Science. 1995;**78**:777-783

[30] Ross RP, Mills S, Hill C, Fitzgerald GF, Stanton C. Specific metabolite production by gut microbiota as a basis for probiotic function. International Dairy Journal. 2010;**20**:269-276

[31] O'Shea EF, Cotter PD, Stanton C, Ross RP, Hill C. Production of bioactive substances by intestinal bacteria as a basis for explaining probiotic mechanisms: Bacteriocins and conjugated linoleic acid. International Journal of Food Microbiology. 2012;**152**:189-205

[32] Hayes M, Coakley M, O'sullivan L, Stanton C. Cheese as a delivery vehicle for probiotics and biogenic substances. Australian Journal of Dairy Technology. 2006;**61**:132

[33] Möller NP, Scholz-Ahrens KE, Roos N, Schrezenmeir J. Bioactive peptides and proteins from foods: Indication for health effects. European Journal of Nutrition. 2008;**47**:171-182

[34] Bogsan CS, Florence ACR, Perina N, Hirota C, Soares FASM, Silva RC, et al. Survival of *Bifidobacterium lactis* HN019 and release of biogenic compounds in unfermented and fermented milk is affected by chilled storage at 4°C. Journal of Probiotics and Health. 2013;**4**:114. DOI: 10.4172/2329-8901.1000114

[35] Monagas M, Urpi-Sarda M, Sánchez-Patán F, Llorach R, Garrido I, Gómez-Cordovés C, et al. Insights into the metabolism and microbial biotransformation of dietary flavan- 3-ols and the bioactivity of their metabolites. Food & Function. 2010;**1**: 233-253

[36] Dharmaraj S. Marine *Streptomyces* as a novel source of bioactive substances. World Journal of Microbiology and Biotechnology. 2010;**26**:2123-2139. DOI: 10.1139/cjm-2013-0785

[37] Mitsuoka T. Development of functional foods. Bioscience of Microbiota, Food and Health. 2014;**33**(3):117-128

[38] Taniguchi M, Nakazawa H, Takeda O, Kaneko T, Hoshino K, Tanaka T. Production of a mixture of antimicrobial organic acids from lactose by co-culture of *Bifidobacterium longum* and *Propionibacterium freudenreichii*. Bioscience, Biotechnology, and Biochemistry. 1998;**62**:1522-1527

[39] Piard JC, Desmazeaud M. Inhibiting factors produced by lactic acid bacteria. 1. Oxygen metabolites and catabolism end-products. Le Lait. 1991;**71**(5):525-541

[40] Klaenhammer TR. Bacteriocins of lactic acid bacteria. Biochimie. 1988;**70**:337-349

[41] Marty-Teysset C, De La Torre F, Garel JR. Increased production of hydrogen peroxide by *Lactobacillus delbrueckii* subsp. bulgaricus upon aeration: Involvement of an NADH oxidase in oxidative stress. Applied and Environmental Microbiology. 2000;**66**:262-267

[42] Shokryazdan P, Sieo CC, Kalavathy R, Liang JB, Alitheen NB, Jahromi MF, et al. Probiotic potential of *Lactobacillus* strains with antimicrobial activity against some human pathogenic strains. BioMed Research International. 2014. DOI: 10.1155/2014/927268

[43] Chen H, Hoover DG. Bacteriocins and their food applications. Comprehensive Reviews in Food Science and Food Safety. 2003;**2**:82-100

[44] Pascual LM, Daniele MB, GiordanoW, Pajaro MC, Barberis IL. Purification and partial characterization of novel bacteriocin L23 produced by *Lactobacillus fermentum* L23. Current Microbiology. 2008;**56**:397-402

[45] Sharma A, Srivastava S. Anti- Candida activity of two-peptide bacteriocins, plantaricins (Pln E/F and J/K) and their mode of action. Fungal Biology. 2014;**118**:264-275

[46] Okkers DJ, Dicks LMT, Silvester M, Joubert JJ, Odendaal HJ. Characterization of pentocin TV35b, a bacteriocin-like peptide isolated from *Lactobacillus pentosus* with a fungistatic effect on Candida albicans. Journal of Applied Microbiology. 1999;**87**:726-734

[47] Calderone RA, Fonzi WA. Virulence factors of *Candida albicans*. Trends in Microbiology. 2001;**9**:327-335

[48] Douglas LJ. Candida biofilm and their role in infection. Trends in Microbiology. 2003;**11**:30-36

[49] Talarico TL, Casas IA, Chung TC, Dobrogosz WJ. Production and isolation of reuterin, a growth inhibitor produced by *Lactobacillus reuteri*. Antimicrobial Agents and Chemotherapy. 1988;**32**(12):1854-1858

[50] Schütz H, Radler F. Anaerobic reduction of glycerol to propanediol-1.3 by *Lactobacillus brevis* and *Lactobacillus buchneri*. Systematic and Applied Microbiology. 1984;**5**(2):169-178

[51] Claisse O, Lonvaud-Funel A. Assimilation of glycerol by a strain of *Lactobacillus collinoides* isolated from cider. Food Microbiology. 2000;**17**(5):513-519

[52] Magnusson J, Ström K, Roos S, Sjögren J, Schnürer J. Broad and complex antifungal activity among environmental isolates of lactic acid bacteria. FEMS Microbiology Letters. 2003;**219**(1):129-135

[53] TalaricoTL, DobrogoszWJ. Chemical characterization of an antimicrobial substance produced by *Lactobacillus reuteri*. Antimicrobial Agents and Chemotherapy. 1989;**33**:674-679

[54] Schaefer L, Auchtung TA, Hermans KE, Whitehead D, Borhan B, Britton RA. Theantimicrobial compound reuterin (3-hydroxypropionaldehyde) induces oxidative stress via interaction with thiol groups. Microbiology. 2010;**156**(6):1589-1599

[55] Ganzle MG. Characterization of reutericyclin produced by *Lactobacillus reuteri* LTH2584. Applied and Environmental Microbiology. 2000;**66**:4325-4333

[56] Jay JM. Antimicrobial properties of diacetyl. Applied and Environmental Microbiology. 1982;**44**(3):525-532

[57] Chung TC, Axelsson L, Lindgren SE, Dobrogosz WJ. In vitro studies on reuterin synthesis by *Lactobacillus reuteri*. Microbial Ecology in Health and Diseases. 1989;**2**:137-144

[58] Corey RB. Crystal structure of diketopiperazine. Journal of the American Chemical Society. 1938;**60**:1598-1604. DOI: 10.1021/ ja01274a023

[59] Martins MB, Carvalho I. Diketopiperazines: Biological activity and synthesis. Tetrahedron. 2007;**63**:9923-9932. DOI: 10.1016/j. tet.2007.04.105

[60] Nicholson B. NPI-2358 is a tubulin-depolymerizing agent: In-vitro evidence for activity as a tumor vascular-disrupting agent. Anti-Cancer Drugs. 2006;**17**:25. DOI: 10.1097/01. cad.0000182745.01612.8a

[61] Sinha S, Srivastava R, Clercq D, Erik, Singh RK. Synthesis and antiviral properties of arabino and ribonucleosides of 1,3-dideazaadenine, 4-nitro-1,3-dideazaadenine and diketopiperazine. Nucleosides, Nucleotides & Nucleic Acids. 2004;**23**(12):1815-1824. DOI: 10.1081/ NCN-200040614

[62] Houston DR, Synstad B, Eijsink VGH, Stark MJR, Eggleston IM, van Aalten DMF. Structure-based exploration of cyclic dipeptide chitinase inhibitors. Journal of Medicinal Chemistry. 2004;**47**(23):5713-5720. DOI: 10.1021/jm049940a

[63] Kwon OS, Park SH, Yun B-S, Pyun YR, Kim C-J. Cyclo(dehydroala- L-Leu), an a-glucosidase inhibitor from *Penicillium* sp. F70614. The Journal of Antibiotics. 2000;**53**(9):954-958

[64] Li H, Liu L, Zhang S, Cui W, Lv J. Identification of antifungal compounds produced by *Lactobacillus casei* AST18. Current Microbiology. 2012;**65**(2):156-161

[65] Ström K, Sjögren J, Broberg A, Schnürer J. *Lactobacillus plantarum* MiLAB 393 produces the antifungal cyclic dipeptides cyclo (L-Phe-L-Pro) and cyclo (L-Phe-trans-4-OH-L-Pro) and 3-phenyllactic acid. Applied and Environmental Microbiology. 2002;**68**(9):4322-4327

[66] Niku-Paavola ML, Laitila A, Mattila-Sandholm T, Haikara A. New types of antimicrobial compounds produced by *Lactobacillus plantarum*. Journal of Applied Microbiology. 1999;**86**(1):29-35

[67] Li J, Wang W, Xu SX, Magarvey NA, McCormick JK. *Lactobacillus reuteri*- produced cyclic dipeptides quench agr-mediated expression of toxic shock syndrome toxin-1 in *Staphylococci*. Proceedings of the National Academy of Sciences. 2011;**108**(8):3360-3365

[68] Ryan LA, Zannini E, Dal Bello F, Pawlowska A, Koehler P, Arendt EK. *Lactobacillus amylovorus* DSM 19280 as a novel food-grade antifungal agent for bakery products. International Journal of Food Microbiology. 2011;**146**(3):276-283

[69] Yang EJ, Chang HC. Purification of a new antifungal compound produced by *Lactobacillus plantarum* AF1 isolated from Kimchi. International Journal of Food Microbiology. 2010;**139**(1-2):56-63

[70] Japan Patent JP2018-070463A. Anti-bacterial agent containing cyclic dipeptide against periodontal pathogens; 2018

[71] Banni S. Conjugated linoleic acid metabolism. Current Opinion in Lipidology. 2002;**13**(3):261-266

[72] Sugano M, Tsujita A, Yamasaki M, Noguchi M, Yamada K. Conjugated linoleic acid modulates tissue levels of chemical mediators and immunoglobulins in rats. Lipids. 1998;**33**(5):521-527

[73] Mazidi M, Karimi E, Rezaie P, Ferns GA. Effects of conjugated linoleic acid supplementation on serum C-reactive protein: A systematic review and meta-analysis of randomized controlled trials. Cardiovascular Therapeutics. 2017;**35**(6):e12275

[74] Flemming HC, Wingender J. The biofilm matrix. Nature Reviews. Microbiology. 2010;**8**:623-633

[75] Allison DG. The biofilm matrix. Biofouling. 2003;**19**:139-150

[76] Wimpenny J. An overview of biofilms as functional communities. In: Allison D, Gilbert P, Lappin-Scott HM, Wilson M, editors. Society for General Microbiology Symposium. Community Structure and Co-operation in Biofilms, Vol. 59. Cambridge: Cambridge University Press; 2000

[77] Sperandio V, Torres AG, Jarvis B, Nataro JP, Kaper JB. Bacteria–host communication: The language of hormones. Proceedings of the National Academy of Sciences. 2003;**100**(15):8951-8956

[78] Liévin-Le Moal V, Servin AL. Anti- infective activities of *Lactobacillus* strains in the human intestinal microbiota: From probiotics to gastrointestinal anti-infectious biotherapeutic agents. Clinical Microbiology Reviews. 2014;**27**(2): 167-199

[79] Li J, Wang W, Xu SX, Magarvey NA, McCormick JK. *Lactobacillus reuteri*- produced cyclic dipeptides quench agr-mediated expression of toxic shock syndrome toxin-1 in staphylococci. Proceedings of the National Academy of Sciences. 2011;**108**(8):3360-3365

2

Probiotics and Prebiotics in Infant Formulae

José Maldonado

Abstract

Human breast milk provides all necessary nutrients for the development of term infants. In addition to its universally recognized nutrients, human breast milk contains a number of non-nutritive components that play a potential role in supporting infant growth. Human breast milk also contains bioactive compounds exerting a wide range of beneficial effects, such as promoting immune system maturation and exerting protection against infections. Supplementation of infant formulae with oligosaccharides and bacteria with proven beneficial health effects seems to be well-founded. The purpose of supplementation is to mimic the functional effects of oligosaccharides and bacteria found in human breast milk. Oligosaccharides with prebiotic functions and bacteria strains with probiotic functions have recently been added to infant formulae in the European Union and other countries. However, a systematic review conducted by the Committee on Nutrition of the European Society for Pediatric Gastroenterology, Hepatology, and Nutrition revealed that there is no conclusive evidence supporting the routine use of probiotic- and/or prebiotic-fortified infant formulae. The aim of this review is to analyze the scientific basis for supplementation of infant formula with these compounds.

Keywords: intestinal microbiota, infant formulae, probiotic, prebiotic, symbiotic

Introduction

Human breast milk (HBM) is a complex physiological fluid uniquely suited to nourish infants. Its composition is specifically adapted to the digestive system and nutritional and growth needs of infants. HBM does not only contain essential nutrients, but also a vast array of non-nutritional bioactive components and microbes (microbiota) that confer benefits to the health of infants in the short and long terms. The microbiota mediates bacterial colonization of the newborn gut and supports immune system maturation and metabolic and cognitive development. Protective constituents such as cytokines, oligosaccharides, and bacteria facilitate newborn's adaptation to the extrauterine environment [1, 2]. HBM has been long known to protect neonates and infants from infections. It has been suggested that this protective role could be regulated by the action of a group of components that might inactivate bacteria and viruses individually, additively, and synergistically [3].

Gut microbiota has effects on health, and HBM contributes decisively to its com- position via its bacteria and oligosaccharides supply. In order to achieve the same health effects, infant formulae are supplemented with live bacteria (probiotics), which favor bifidobacteria and lactobacilli growth (prebiotics) or a combination of these components (symbiotics).

Gut microbiota

Our body hosts a vast, diverse community of stable and varying microorganisms that are referred to as microbiota. The gut is the niche with the highest number and diversity of micro-organisms, containing over 10^{14} microbial cells, 10 times the amount of somatic and germinal cells in our body. The microbes that inhabit our gut are known as gut microbiota [4].

Gut microbiota is an open ecosystem that contains a broad diversity of meta- bolically active microbes that coexist in space and time and play a relevant role in the health of their host. The gut microbiota is considered a metabolic organ that is adaptable and rapidly renewable. There is a mutually beneficial interplay between the host and gut microbiota [4, 5].

Gut microbiota and immunity

The relationship between the lymphatic system and gut microbiota in early stages of life is crucial to the appropriate development of interactions between mucosal cell communities and systemic immunomodulation [5]. Animals with a sterile gut have been proven to be highly vulnerable to infections, which demonstrate the important role that gut microbiota plays in the immune system [4].

Bacterial colonization of the newborn's gut is influenced by a variety of factors such as gestation and delivery and breast-feeding mode [6]. HBM is an excellent continuous source of commensal bacteria for the infant gut. Evidence has been provided of a vertical transfer of bacteria from mother to child via breast milk [7, 8]. The fact that facultative anaerobic bacteria in newborn's gut are the predominant bacterial community in HBM microbiota is not a chance. These bacteria play a key role in the prevention of infections in the newborn [9]. Gut microbiota disorders (dysbiosis) in the first stages of life reportedly precede the development of atopy [10].

During the first week of life, the total bacterial count and, more specifically, anaerobic bacteria count progressively increase. The feeding mode of the newborn has a decisive impact on bacterial gut colonization. Bifidobacteria, lactobacilli, and Gram-positive cocci predominate in the feces of breastfed infants, whereas the bifidobacteria count is lower in the feces of formula-fed infants, with the predominance of bacteroides, clostridia, and coliforms [11–13]. Differences in the composition of newborn's gut microbiota based on the type of feeding could be the clue to identifying the bacteria that exert protective effects to breastfed infants [4].

Probiotics

Bacterial concentrations in HBM range between 10^2 and 10^4 ufc/mL. This means that an infant ingesting over 800 ml of milk a day would receive 10^5 to 10^7 ufc [14]. Therefore, HBM is a primary source of commensal and probiotic bacteria to the infant and plays a key role in the initial colonization of the gut. Some bacteria isolated from HBM have proven to have immunomodulatory and anti-infective effects. Therefore, the protective effects of HBM may be conferred by these bacteria. Supplementation of infant formulae with probiotic bacteria isolated from HBM could help improve gut microbial balance in formula-fed infants, thereby mimicking the beneficial effects of HBM.

Evidence has been published that probiotics modulate mucosal and systemic immune function, improve intestinal barrier function, and exert metabolic effects on the host [4]. Some of the lactobacillus strains isolated from HBM [15] have been reported to compete with enteropathogenic bacteria for nutrients and epithelium adhesion and improve gut barrier functions. The ability of lactobacillus and bifido-bacteria strains to stabilize the integrity of gut barrier has been demonstrated [16]. These types of bacteria potentially reduce antigen systemic load and influence immune function via enterocytes, antigen-presenting cells (monocytes and den- dritic cells), regulatory T cells, and effector T and B cells [17, 18].

Infant formula supplemented with probiotics

European Society for Pediatric Gastroenterology, Hepatology, and Nutrition (ESPGHAN) Committee on Nutrition [19] published a systematic review of studies assessing the safety and health effects of probiotic-supplemented infant formulae. No conclusive data were obtained from ESPGHAN's analysis of infant and follow-on formulae due to considerable variability in the type and dose of probiotics used and supplementation periods.

Safety

Formulae supplemented with probiotics do not raise safety concerns with regard to growth and adverse effects [19]. There are sufficient data supporting the safety of probiotics for infants older than 6 months. However, data on the use of probiotic supplementation in infants younger than 4 months are more limited. Studies in breastfed infants younger than 6 months who received a formula supplemented with either *Lactobacillus fermentum* CECT5716 or *Lactobacillus rhamnosus* GG revealed that formulae were well tolerated and had no adverse effects on growth either during the study period or at 3–5 years of age [20–23]. A recent study revealed that growth and food tolerance improved in premature infants >30 weeks of gestational age fed with a formula supplemented with *Saccharomyces boulardii,* and no adverse effects were detected [24].

Prevention and treatment of infant disorders

Conflicting results have been obtained regarding the effects of probiotics on the composition of fecal microbiota. A decrease in bifidobacteria and enterobac-teria concentrations has been reported with respect to controls [25, 26]. Also, no differences have been observed in lactobacillus and bacteroides. By contrast, Maldonado et al. [27] reported an increase in fecal bifidobacteria and lactobacilli concentrations in infants fed with a formula supplemented with *Lactobacillus fermentum* CECT 5716. Also, no differences were found in other bacteria strains. Evidence has been provided that a formula containing *Bifidobacterium lactis* can influence the composition, stability, and function of gut microbiota in low-weight newborns [28].

The literature reports that probiotic supplementation of formula beyond early infancy can produce a decrease in the use of antibiotics and incidence of diarrhea, colic, and/or irritability. Yet, the variety of methods, type and dose of probiotics, and duration of interventions hinders that conclusive data can be obtained on clear clinical effects of probiotic-supplemented formulae in infants younger than 4 months [19].

In general, there is no consistent evidence supporting that supplementation

of follow-on formula with probiotics has protective effects against infectious diarrhea [19]. Yet, a reduction has been reported in the duration and number of episodes of diarrhea associated with the use of probiotic-supplemented formulae [27, 29–31].

A systematic review conducted by Mugambi et al. [32] of controlled, randomized trials did not reveal that supplementation had any effects on infectious diarrhea, colic, crying/irritability, regurgitation, or vomiting. No beneficial effects were documented on either crying or irritability in the review by ESPGHAN Committee on Nutrition. A study that was not included in ESPGHAN study showed that colic symptoms substantially improved with the administration of *Lactobacillus reuteri* DSM 17938 in breastfed infants [33]. There is no sufficient evidence, how- ever, supporting routine supplementation with probiotics for the treatment or prevention of colic, especially in formula-fed infants [34].

In a review on the effects of a variety of immunonutrients in the prevention of necrotizing enterocolitis [35], the authors gathered sufficient data supporting supplementation of infant formulae with probiotics. Several meta-analyses combined these randomized controlled trials and observational studies demonstrated that the use of probiotics was beneficial for the prevention of severe necrotizing enterocolitis, late-onset sepsis, and all-cause mortality in very-low-birth-weight infants, as well as the time to achieve full enteral feeding in preterm infants [36–38]. By contrast, no differences were observed in a multicenter study involving 1315 preterm newborns fed with a hydrolyzed formula supplemented and non- supplemented with the probiotic *Bifidobacterium breve* BBG-001 [39]; the results of this trial provide no evidence of benefit of this probiotic intervention in reducing late-onset sepsis and necrotizing enterocolitis or death.

In relation to respiratory infections, limited available evidence from randomized controlled trials showed that formula supplementation with the probiotics studied is not associated with a reduction in the duration or risk of respiratory infections [19, 32]. A number of studies on formulae supplemented with different probiotic bacteria [30, 27, 40, 41] have shown a significant reduction in the number of upper airway tract infections in infants fed with these formulae. A study on *Lactobacillus fermentum* CECT 5716 [27] reports a significant 30% reduction in the total number of infections.

Significant reductions have been documented in the incidence of influenza and respiratory symptoms in several studies, where *Lactobacillus fermentum* CECT 5716 was administered in combination with anti-influenza vaccine [42]. This effect is explained by increased levels of NK cells and T-helper and T-cytotoxic lymphocytes.

Sufficient evidence has not been published supporting the beneficial effects of supplementation of infant formulae with probiotics on allergies. Several meta- analyses, however, have shown that the use of probiotics reduces the incidence of atopic dermatitis in infants but not of other types of allergies [43–45].

Evidence has been published that dietary treatment with a extensively hydrolyzed formula containing *Lactobacillus rhamnosus* GG is associated with a higher rate of acquisition of tolerance in infants allergic to cow's milk proteins, as com- pared to infants treated with a non-supplemented hydrolyzed formula [46, 47]. A relationship has been documented between dysbiosis in gut microbiota composition and the pathogenesis of cow's milk allergy [48, 49]. In addition, the administration of a hydrolyzed formula supplemented with probiotics reduces the incidence of other allergies and favors tolerance, as it changes the composition of infant's gut microbiota [23, 50].

Some studies suggest that gut microbiota alterations precede the development of the allergic phenotype. Therefore, probiotics could exert preventive and thera-peutical effects [51]. The potential of some strains to favor Th1 and Th3 immune response against Th2 activity in patients with atopy can create the optimal conditions to redirect immune memory and reduce the risk of atopic disease. The Work Allergy Organization (WAO) [52] determined that probiotics confer health benefits in the prevention of eczema. Thus, WAO recommends the use of probiotics in pregnant or breastfeeding women whose infants have a high risk of developing allergies and in infants with a high risk of allergy.

Conclusion

ESPGHAN Committee on Nutrition does not recommend the routine use of infant formulae supplemented with probiotics. However, the evidence obtained in recent studies suggests that infant formulae containing some specific bacteria strains can confer beneficial health effects. A large number of infant formulae currently available on the market contain probiotics, and several panels support their use provided that their safety and benefits for the health and development of the infant have been demonstrated [19, 53, 54]. The European Food Safety Authority (EFSA) supports the safety of formula supplementation with probiotic bacteria.

Yet, EFSA recommends that further studies are conducted to obtain the highest quality evidence on their efficacy [55].

Prebiotics

Prebiotics are defined as oligosaccharides refractory to the human digestive process with ability to stimulate and promote the growth and/or metabolism of bifidobacteria and lactobacilli in human gut [56]. More than 200 oligosaccharide complexes (neutral and cyclical oligosaccharides) have been identified in human breast milk [57]. Neutral oligosaccharides account for 70% of the total count and include the isomers lacto-N-tetraose, lacto-N-neotetraose, lacto-N-hexaose, monofucosyl-lacto-N-hexaose, and difucosyl-lacto-N-hexaose. Low levels of acidic oligosaccharides containing sialic acid or sulfate groups are present in HBM, and they primarily contain 5-N-acetyl-neuraminic acid [58].

Colostrum is composed of higher oligosaccharide concentrations (15–23 g/L), whereas mature HBM contents range from 1 to 10 g/L [59]. Oligosaccharides account for 8% of the total nutrient contents of HBM and are the third prevalent component following lactose and lipids.

Most of these oligosaccharides are non-absorbable and reach the colon, where they have different functions. Thus, they compete for membrane receptors with pathogenic bacteria and viruses in intestinal epithelium; they contribute to acidifi- cation via fermentation by colon bacteria; inhibit the growth of bacteroides, clos- tridia, and coliforms; promote lactobacilli and bifidobacteria growth; and stimulate the development of infant's immune system. A direct relationship has been documented between oligosaccharides and selectins, integrins, and other receptors, and they mediate leukocyte-endothelial cell interactions [59]. Fermentation of prebiotics by gut bacteria produces short-chain fatty acids, which exert a direct anti-inflammatory effect and promote intestinal barrier integrity by stimulating the proliferation and differentiation of gut mucosal cells.

Cow milk oligosaccharide content is substantially lower than that of HBM, and infant formula supplementation with prebiotics with the purpose of obtaining their health benefits is well founded. At present, GOS and FOS combinations are used, and other HBM oligosaccharides have been recently incorporated to infant formulae.

Infant formulae supplemented with prebiotics

The European Scientific Committee on Food approved prebiotic supplemen- tation in infant and follow-on formulate up to a maximum of 0.8 g/100 ml to a GOS:FOS ratio of 9:1. By contrast, a systematic review on the safety and health effects of prebiotic-supplemented infant formulae conducted by ESPGHAN Committee on Nutrition [19] did not provide conclusive evidence due to variability in the type and dose of the prebiotic used and period of intervention.

2′-flucosyllactose, a HBM oligosaccharide, was recently synthesized and has been incorporated to some infant formula [60].

Safety

Formulae fortified with prebiotics do not raise safety concerns with regard to growth and adverse effects. [19]. Infant formulae containing HBM oligosaccharides have proven to be safe and well tolerated, and synthetic oligosaccharides have demonstrated to have similar effects to those of HBM oligosaccharides [60].

Prevention and treatment of infant disorders

There is solid evidence that infant formula containing some prebiotics is associ- ated with less-consistent feces and a higher frequency of defecation [61]. However, inconsistent evidence has been obtained on the association between prebiotics and the frequency of defecations [32, 62].

The use of prebiotic-fortified formulae has been associated with a lower risk for intestinal and respiratory infections [63, 64] and an increase in fecal secretory IgA levels [65]. By contrast, they have not been proven to exert any effects on humoral and cellular immunity [66]. In general terms, there is no conclusive evidence supporting that supplementation of infant formulae with prebiotics exerts any protective effects against infections, colic, crying/irritability, regurgitation, or vomiting [19, 32]. Fortification with 2′-flucosyllactose does seem to improve infant immunity, as it has been reported to be related to a lower incidence of infections, especially respiratory infections [60].

GOS:FOS mixtures favor the growth of bifidobacteria and lactobacilli in the feces of infants receiving fortified formulae. However, they have a limited effect on the reduction of pathogenic bacteria [19]. Yet, some studies suggest that prebiotics reduce pathogenic micro-organism concentrations, while the infant is receiving a formula supplemented with oligosaccharides [67]. A number of studies [25, 32, 62] have failed to demonstrate that bifidobacteria, lactobacilli, or pathogen count decreases with prebiotics.

Other studies have shown similarities between the bifidogenic effect of prebiotic-fortified formulae and HBM, as compared to non-fortified formulae [59, 68, 69]. Indeed, prebiotics have been reported to have special effects on some bifidobacteria species such as *Bifidobacterium breve*. Thus, fecal *Bifidobacterium breve* concentrations in infants fed with a fortified formula have been documented to be similar to those found in breastfed infants.

Although prebiotic-supplemented formulae are thought to prevent eczema in infants at high risk of developing allergies [43, 63, 70], there is no sufficient evidence on the role that prebiotics play in the prevention of eczema, atopic dermatitis, or food hypersensitivity [71, 72]. A partially hydrolyzed formula containing specific prebiotics has been reported to generate a gut microbiota similar to that of breastfed infants. A potential link between microbial activity and eczema onset was identified, which could suggest a suboptimal implementation of gut microbiota in specific developmental stages of infants at high risk of developing allergy [73].

Conclusion

ESPGHAN Committee on Nutrition does not recommend routine use of infant formulae supplemented with prebiotic. In agreement with the American Academy of Pediatrics, they recommend that further studies are conducted to assess the safety and efficacy of prebiotic supplementation.

Symbiotics

Symbiotics are mixtures of probiotics and prebiotics that beneficially affect the host by improving the survival and implantation of the probiotic bacteria and stimulate the activity of the host's endogenous bacteria [56]. Symbiotics are believed to act synergistically to increase the overall gut health by offering more benefits than the use of either a probiotic or prebiotic agent alone. Considering a huge number of possible combinations, the application of symbiotics for the modulation of intestinal microbiota in humans seems promising [74]. A disadvantage to using symbiotics is that it is difficult to predict the selectivity and specificity of each of the components and what the resulting mechanisms of action will be.

Limited data have been provided on concomitant prebiotic and probiotic supplementation of infant formulae. The few studies carried out with symbiotics [19, 32, 75] revealed that symbiotics: (a) do not exert effects on growth; (b) do not reduce the incidence of digestive disorders (colic, regurgitation, crying, vomiting, to name a few) or infections; (c) increase the frequency of daily defecations but do not influence fecal consistency; and (d) no data are available on their effects on the composition of gut microbiota or on immune response.

There is no conclusive evidence on the effects of supplementation of infant formula with symbiotics. Therefore, ESPGHAN Committee on Nutrition does not recommend routine use of infant formula fortified with symbiotics.

Author details

José Maldonado

Unit of Pediatric Gastroenterology, Hepatology and Nutrition, Department of Pediatrics, Virgen de las Nieves University Hospital, University of Granada, Spain

*Address all correspondence to: jmaldon@ugr.es

References

[1] Ojo-Okunola A, Nicol M, du Toit E. Human breast milk bacteriome in health and disease. Nutrients. 2018;**10**:1643. DOI: 10.3390/nu10111643

[2] Aakko J, Kumar H, Rautava S, Wise A, Austran C, Bode L, et al. Human milk oligosaccharide categories define the microbiota composition in human calostrum. Beneficial Microbes. 2017;**8**:563-567. DOI: 10.3920/BM2016.0185

[3] Isaacs CE. Human milk inactivates pathogen individually, additively, and synergistically. The Journal of Nutrition. 2005;**51**:1286-1288. DOI: 10.1093/ jn/135.5.1286

[4] Wallace TC, Guarner F, Madsen K, Cobano MD, Gibson G, Hentges E, et al. Human gut microbiota and its relationship to health and disease. Nutrition Reviews. 2011;**69**:392-403. DOI: 10.1111/j.1753-4887.2011.00402

[5] Guarner F, Melagelada JR. Gut flora in health and disease. Lancet. 2003;**361**:512-519. DOI: 10.1016/ S0140-6736(03)12489-0

[6] López Moriana C, Mach N. Influencia de la gestación, el parto y el tipo de lactancia sobre la microbiota intestinal del neonato. Acta Pediátrica Española. 2014;**72**:37-44

[7] Martín R, Heilig GH, Zoetendal EG, Smidt H, Rodríguez JM. Diversity of the Lactobacillus group in breast milk and vagina of healthy women and potential role in colonization of the infant gut. Journal of Applied Microbiology. 2007;**103**:2638-2644. DOI: 10.1111/j.1365-2672.2007.03497.x

[8] Milani C, Mancabelli L, Lugli GA, Duranti S, Turroni F, Ferrario C, et al. Exploring vertical transmission of bifidobacteria from mother to child. Applied and Environmental Microbiology. 2015;**81**:7078-7087. DOI: 10.1128/ AEM.02037-15

[9] Lara-Villoslada F, Olivares M, Sierra S, Rodríguez JM, Boza J, Xaus J. Beneficial effects of probiotic bacteria isolated from breast milk. The British Journal of Nutrition. 2007;**98**(Suppl. 1):S96-S100. DOI: 10.1017/ S0007114507832910

[10] Rachid R, Chatila TA. The role of the gut microbiota in food allergy. Current Opinion in Pediatrics. 2016;**28**:748-753. DOI: 10.1097/ MOP.0000000000000427

[11] Fallani M, Young D, Scott J, Norin E, Amarri S, Adam R, et al. Intestinal microbiota of 6-week-old infants across Europe: Geographic influence beyond delivery mode, breast-feeding, and antibiotics. Journal of Pediatric Gastroenterology and Nutrition. 2010;**51**:77-84. DOI: 10.1097/MPG. obo13e3181d1b11e

[12] Bezirtzoglou E, Tsiotsias A, Welling GW. Microbiota profile in feces of breast- and formula-fed newborns by using fluorescence in situ hybridization (FISH). Anaerobe. 2011;**17**:478-482. DOI: 10.1016/j.anaerobe.2011.03.009

[13] O'Sullivan A, Farver M, Smilowitz JT. The influence of early infant-feeding practices on the intestinal microbiome and body composition in infants. Nutrition and Metabolic Insights. 2015;**8**:1-9. DOI: 10.4137/NMI. S29530

[14] Rodríguez JM, Jiménez E, Merino V, Maldonado A, Marín ML, Fernández L, et al. Microbiota de la leche humana en condiciones fisiológicas. Acta Pediátrica Española. 2008;**66**:77-82

[15] Olivares M, Díaz-Ropero MP, Martin R, Rodríguez JM, Xaus J. Antimicrobial potential of four lactobacillus strains isolated from breast milk. Journal of Applied Microbiology. 2006;**101**:72-79. DOI: 10.1111/j.1365-2672.2006.02981.x

[16] Rosenfeldt V, Benfeldt E, Valerius NH, Paerregaard A, Michaelsen KF. Effects of probiotics on gastrointestinal symptoms and small intestinal permeability in children with atopic dermatitis. The Journal of Pediatrics. 2004;**145**:612-616. DOI: 10.1016/j.jpeds.2004.06.068

[17] Prescott SL, Björkstén B. Probiotics for the prevention or treatment of allergy diseases. The Journal of Allergy and Clinical Immunology. 2007;**120**:255-262. DOI: 10.1016/j.jaci.2007.04.027

[18] Bermúdez-Brito M, PlaZA-Díaz J, Muñoz-Quezada S, Gómez-Llorente C, Gil A. Probiotics mechanism of action. Annals of Nutrition & Metabolism. 2012;**61**:160-174. DOI: 10.1159/000342079

[19] Braegger C, Chmielewika A, Decsi T, Kolacek SMihatsch W, Morfeno L, ESPGHAN Committee on Nutrition, et al. Supplementation of infant formula with probiotics and/or prebiotics: A systematic review and comment by the ESPGHAN Committee on Nutrition. Journal of Pediatric Gastroenterology and Nutrition. 2011;**52**:238-250. DOI: 10.1097/MPG.0b013e31817b9e80

[20] Gil-Campos M, López MA, Rodríguez-Benítez MV, Romero J, Roncero I, Linares MD, et al. *Lactobacillus fermentum* CECT 5716 is safe and well tolered in infants of 1-6 months of age: A randomized controlled trial. Pharmacological Research. 2012;**65**:231-238. DOI: 10.1016/j.phrs.2011.11.016

[21] Maldonado-Lobón JA, Gil-Campos M, Maldonado J, López-Huertas E, Flores-Rojas K, Valero AD, et al. Long-term safety of early consumption of *Lactobacillus fermentum* CECT5716: A 3-year follow-up of a randomized controlled trial. Pharmacological Research. 2015;**95-96**:12-19. DOI: 10.1016/j. phrs.2015.01.006

[22] Scalabrin DMF, Harris C, Johnston WH, Bersth CL. Long-term safety assessment in children who received hydrolyzed protein formulas with *Lactobacillus rhamnosus* GG: A year follow-up. European Journal of Pediatrics. 2017;**176**:217-224. DOI: 10.1007/soo431-016-2825-4

[23] Berni Canani R, Di Costanzo M, Bedogni G, Amoroso A, Cosenza L, Di Scala C, et al. Extensively hydrolyzed casein formula containing *Lactobacillus rhamnosus* GG reduces the occurrence of other allergic manifestations in children with cow's milk allergy: 3-year randomized controlled trial. The Journal of Allergy and Clinical Immunology. 2017;**139**:1906-1913. DOI: 10.1111/pai.12687

[24] Xu L, Wang Y, Wang Y, Fu J, Sun M, Mao Z, et al. A double-blinded randomized trial on growth and feeding tolerance with *Saccharomyces boulardii* CNCMI-745 in formula-fed preterm infants. Jornal de Pediatria. 2016;**92**:296-301. DOI: 10.1016/j. jped.2015.08.013

[25] Brunser O, Figueroa G, Gatteland M, Haschke-Becher F, Magliola C, Rochat F, et al. Effects of probiotic or prebiotic supplemented milk formulas on fecal microbiota composition in infants. Asia Pacific Journal of Clinical Nutrition. 2006;**15**:368-376

[26] Langhendries JP, Detry J, van Hees J, Lambray JM, Darimont J, Mozin M, et al. Effect of a fermented infant formula containing viable bifidobacteria on the faecal flora composition and pH of healthy full-term infants. Journal of Pediatric Gastroenterology and Nutrition. 1995;**21**:177-181

[27] Maldonado J, Cañabate F, Sempere L, Vela F, Sánchez AR, Narbona E, et al. Human milk probiotic *Lactobacillus fermentum* CECT 5716 reduces the incidence of gastrointestinal and upper respiratory tract infections in infants. Journal of Pediatric Gastroenterology and Nutrition. 2012;**54**:55-61. DOI: 10.1097/MPG. pbo13e318233f18

[28] Chi C, Xue Y, Liu R, Wang Y, Lv N, Zeng H, et al. Effects of a formula with a probiotic *Bifidobacterium lactis* supplement on the gut microbiota of low birth weight infants. European Journal of Nutrition. 13 Jun 2019. DOI: 10.1007/s00394-019-02006-4

[29] Weizman Z, Asli G, Alsheikh A. Effects of a probiotic infant formula on infections in child care centers; comparison of two probiotic agents. Pediatrics. 2005;**115**:5-9. DOI: 10.1542/ peds.2004-1815

[30] Maldonado J, Lara-Villoslada F, Sierra S, Sempere L, Gómez M, Rodríguez JM, et al. Safety and tolerance of the human milk probiotic strain *Lactobacillus salivarius* CECT 5713 in month-old children. Nutrition. 2010;**26**:1082-1087. DOI: 10.1016/j. nut.2009.08.023

[31] Corrêa NB, Péret Filho LA, Penna FJ, Lima FM, Nicoli JR. A randomized formula controlled trial of *Bifidobacterium lactis* and *Streptococcus thermophilus* for prevention of antibiotic-associated diarrhea in infants. Journal of Clinical Gastroenterology. 2005;**39**:385-389. DOI: 10.1097/01. mcg.0000159217.47419.56

[32] Mugambi MN, Musekiwa A, Lombard M, Young T, Blaauw R. Symbiotics, probiotics or prebiotics in infant formula for full term infants: A systematic review. Nutrition Journal. 2012;**11**:81. DOI: 10.1186/1475-2891-11-81

[33] Savino F, Cordisco L, Tarasco V, Palumeri E, Calabrese R, Oggero R, et al. *Lactobacillus reuteri* DSM 17938 in infantile colic: A randomized, double- blind, placebo-controlled trial. Pediatrics. 2010;**126**:e526-e533. DOI: 10.1542/peds.2010-0433

[34] Sung V, Collet S, de Gooyer T, Hiscock H, Tang M, Wake M. Probiotics to prevent or treat excessive infant crying. JAMA Pediatrics. 2013;**167**:1150-1157. DOI: 10.1001/ jamapediatrics.2013.2572

[35] Zhou P, Li Y, Ma LY, Lin HC. The role of immunonutrients in the prevention of necrotizing enterocolitis in preterm very low birth weight infants. Nutrients. 2015;**7**:7256-7270. DOI: 10.3390/nu7095334

[36] Dermyshi E, Wang Y, Yan C, Hong W, Qiu G, Gong X, et al. The "golden age" of probiotics: A systematic review and meta-analysis of randomized and observational studies in preterm infants. Neonatology. 2017;**112**:9-23. DOI: 10.1159/000454668

[37] Sun J, Marwah G, Westgarth M, Buys N, Ellwood D, Gray PH. Effects of probiotics on necrotizing enterocolitis, sepsis, intraventricular hemorrhage, mortality, length of hospital stay, and weight gain in very preterm infants: A meta-analysis. Advances in Nutrition. 2017;**8**:749-763. DOI: 10.3945/ an.116.014605

[38] Aceti A, Gori D, Barone G, Callegari ML, Fantini MP, Indrio F, et al. Probiotics and time to achieve full enteral feeding in human milk-fed and formula-fed preterm infants: Systematic review and meta-analysis. Nutrients. 2016;**8**:471. DOI: 10.3390/nu8080471

[39] Costeloe K, Bowler U, Brocklehurst P, Hardy P, Heal P, Juszczak E, et al. A randomised controlled trial of the probiotic *Bifidobacterium breve* BBG-001 in preterm babies to prevent sepsis, necrotising enterocolitis and death: The Probiotics in Preterm infantS (PiPS) trial. Health Technology Assessment. 2016;**20**:1-194. DOI: 10.3310/hta20660

[40] Taipale T, Pienihäkkinen K, Isolauri E, Larsen C, Brockmann E, Alanen P, et al. *Bifidobacterium animalis* subsp. *lactis* BB-12 in reducing the risk of infections in infancy. The British Journal of Nutrition. 2011;**105**:409-416. DOI: 10.1017/S0007114510003685

[41] Rautava S, Salminen S, Isolauri E. Specific probiotics in reducing the risk of acute infections in infant: A randomised, double-blind, placebo- controlled study. The British Journal of Nutrition. 2009;**101**:1722-1726. DOI: 10.1017/S0007114508116282

[42] Olivares M, Díaz-Ropero MP, Sierra S, Lara-Villoslada F, Fonolla J, Navas M, et al. Oral intakeof *Lactobacillus fermentum* CECT 5716 enhances the effect of influenza vaccination. Nutrition. 2007;**23**:254-260. DOI: 10.1016/j.nut.2007.01.004

[43] Tang ML, Lahtinen SJ, Boyle RJ. Probiotics and prebiotics: Clinical effects in allergic disease. Current Opinion in Pediatrics. 2010;**22**:626-634. DOI: 10.1097/ MOP.0b013e32833d9728

[44] Osborn DA, Sinn JK. Probiotics in infants for prevention of allergic disease and food hypersensitivity. Cochrane Database of Systematic Reviews. 2007;**17**(4):CD006475. DOI: 10.1002/14651858. CD006475.pub2

[45] Pelucchi C, Chatenoud L, Turati F, Galeone C, Moja L, Bach JF, et al. Probiotics supplementation during pregnancy or infancy for the prevention of atopic dermatitis: A meta-analysis. Epidemiology. 2012;**23**:402-414. DOI: 10.1097/EDE.0b013e31824d5da2

[46] Berni Canani R, Nocerino R, Terrin G, Coruzzo A, Cosenza L, Leone L, et al. Effect of *Lactobacillus* GG on tolerance acquisition in infants with cow's milk allergy: A randomized trial. The Journal of Allergy and Clinical Immunology. 2012;**129**:580-582. DOI: 10.1016/j. jaci.2011.10.004

[47] Berni Canani R, Nocerino R, Terrin G, Frediani T, Lucarelli S, Cosenza L, et al. Formula selection for management of children with cow milk allergy influences the rate of acquisition of tolerance: A prospective multicenter study. The Journal of Pediatrics. 2013;**163**:771-777. DOI: 10.1016/j. jpeds.2013.03.008

[48] Thompson-Chagoyan OC, Vieites JM, Maldonado J, Edwards C, Gil A. Changes in faecal microbiota of infants with cow milk protein allergy – a Spanish prospective case-control six-month follow-up study. Pediatric Allergy and Immunology. 2010;**21**:e394-e400. DOI: 10.1111/j.1399-3038.2009.00961.x

[49] Thompson-Chagoyan OC, Fallani M, Maldonado J, Vieites JM, Khanna S, Edwards C, et al. Faecal microbiota and short-chain fatty acid levels in faeces from infants with cow's milk protein allergy. International Archives of Allergy and Immunology. 2011;**156**:325-332. DOI: 10.1159/000323893

[50] Berni Canani R, Sangwan N, Stefka AT, Nocerino R, Paparo L, Aitoro R, et al. *Lactobacillus rhamnosus* GG-supplemented formula expands butyrate-producing bacterial strains in food allergic infants. The ISME Journal. 2016;**10**:742-750. DOI: 10.1038/ ismej.2015.151

[51] Isolauri E, Rautava S, Salminen S. Probiotics in the development and treatment of allergic disease. Gastroenterology Clinics of North America. 2012;**41**:747-762. DOI: 10.1016/j.gtc.2012.08.007

[52] Fiocchi A, Pawankar R, Cuello-García C, Ahn K, Al-Hammadi S, Agarwal A, et al. World Allergy Organization-McMaster University Guidelines for Allergic Disease Prevention (GLAD-P): Probiotics. World Allergy Organization Journal. 2015;**8**:4. DOI: 10.1186/ s40413-015-0055-2

[53] Maldonado LJ. Nuevos ingredientes en las fórmulas para lactantes nacidos a término (I): Probióticos, prebióticos y simbióticos. Acta Pediátrica Española. 2014;**72**:56-62

[54] Bertelsen RJ, Jensen ET, Ringel-Kulka T. Use of probiotics and prebiotics in infant feeding. Best Practice & Research. Clinical Gastroenterology. 2016;**30**:39-48. DOI: 10.1016/j.bpg.2016.01.001

[55] European Food Safety Authority (EFSA). EFSA Panel on dietetic products, nutrition and allergies (NDA). Scientific Opinion on the essential composition of infant and follow-on formulae. EFSA Journal. 2014;**12**(7):3760. DOI: 10.29013/j. efsa.2014.3760

[56] de Vrese M, Schrezenmeir J. Probiotics, prebiotics, and synbiotics. Advances in Biochemical Engineering/ Biotechnology. 2008;**111**:1-66. DOI: 10.1007/10_2008_097

[57] Garrido D, Kim JH, German JB, Raybould HE, Mills DA. Oligosaccharide binding proteins from *Bifidobacterium longum* subsp. *infantis* reveal a preference for host glycans. PLoS One. 2011;**6**:e17315. DOI: 10.1371/journal. pone.0017315

[58] Thurl S, Munzert M, Boehm G, Matthews C, Stahl B. Systematic review of the concentrations of oligosaccharides in human milk. Nutrition Reviews. 2017;**75**:920-933. DOI: 10.1093/nutrit/nux044

[59] Donovan SM, Comstock SS. Human milk oligosaccharides influence neonatal mucosal and systemic immunity. Annals of Nutrition & Metabolism. 2016;**69**(Suppl. 2):42-51. DOI: 10.1159/000452818

[60] Reverri EJ, Devit AA, Kajzer JA, Baggs GE, Borschel MW. Review of the clinical experience of feeding infant formula containing the human milk oligosaccharide 2′-fucosyllactose. Nutrients. 2018;**10**:1436. DOI: 10.3390/ nu10101346

[61] Vandenplas Y, De Greef E, Veeremen G. Prebiotics in infant formula. Gut Microbes. 2014;**5**:681-687. DOI: 10.4161/19490976.2014.972237

[62] Veereman-Wauters G, Staelens S, Van de Broeck H, Plaskic K, Wesling F, Roger LC, et al. Physiological and bifidogenic effects of prebiotic supplements in infancy formulae. Journal of Pediatric Gastroenterology and Nutrition. 2011;**52**:763-771. DOI: 10.1097/ MPG.0b013e3182139f39

[63] Arslonoglu S, Moro GE, Schmitt J, Tandoi L, Rizzardi S, Boehm G. Early dietary intervention with a mixture of prebiotic oligosaccharides reduces the incidence of allergic manifestations and infections during the first two years of life. The Journal of Nutrition. 2008;**138**:1091-1095. DOI: 10.1093/ jn/138.6.1091

[64] Bruzzese E, Volpicelli M, Squeglia V, Bruzzese D, Salvini F, Brisceglia M, et al. A formula containing galacto- and fructo-oligosaccharides prevents intestinal and extraintestinal infections: An observational study. Clinical Nutrition. 2009;**28**:156-161. DOI: 10.1016/j. clnu.2009.01.008

[65] Scholtens PA, Alliet P, Raes M, Allu MS, Kroes H, Boehm G, et al. Fecal secretory immunoglobulin A in increased in healthy infants who receive a formula with short-chain galacto-ologosaccharides and long-chain fructo-oligosaccharides. The Journal of Nutrition. 2008;**138**:1141-1147. DOI: 10.1093/jn/138.6.1141

[66] Raes M, Scholtens PA, Allet P, Hemsen K, Jougen H, Boehm G, et al. Exploration of basal immune parameters in healthy infants receiving an infant milk formula supplemented with prebiotics. Pediatric Allergy and Immunology. 2010;**21**(2 Pt 2):e377-e385. DOI: 10.1111/j.1399-3038.2009.00957.x

[67] Knol J, Boehm G, Lidestri M, Negretti F, Jelinek J, Agosti M, et al. Increase of faecal bifidobacteria due to dietary oligosaccharides induces a reduction of clinically relevant pathogen germs in the faeces of formula-fed infants. Acta Paediatrica. 2005;**94**:S31-S33. DOI: 10.1111/ j.1651-2227.2005.tb02152.x

[68] Sierra C, Bernal MJ, Blasco J, Martínez R, Dalmau J, Ortuño I, et al. Prebiotic effect during the first year of life in healthy infants fed formula containing GOS as the only prebiotic: A multicenter, randomized, double-blind and placebo-controlled trial. European Journal of Nutrition. 2015;**54**:89-99. DOI: 10.1007/s00384-014-689-9

[69] Borewicz K, Suarez-Diez M, Hechler C, Beijers R, de Weerth C, Arts I, et al. The effect of prebiotic fortified infant formulas on microbiota composition and dynamics in early life. Scientific Reports. 2434;**2019**:9. DOI: 10.1038/s41598-018-38268.x

[70] Moro G, Arslonoglu S, Stahl B, Jalinek WH, Boehm G. A mixture of prebiotic oligosaccharides during the first six months of age. Archives of Disease in Childhood. 2006;**91**:814-819. DOI: 10.1136/adc.2006.098251

[71] Osborn DA, Sinn JK. Prebiotics in infants for prevention of allergy disease and food hypersensitivity. Cochrane Database of Systematic Reviews. 2007;**17**(4):CD006474. DOI: 10.1002/14651858.CD006474.pub2

[72] Bozensky J, Hill M, Zelenka R, Skyba T. Prebiotics do not influence the severity of atopic dermatitis in infants: A randomized controlled trial. PLoS One. 2015;**10**(11):e0142897. DOI: 10.1371/journal.pone.0142697. eCollection2015

[73] Wopereis H, Sim K, Shaw A, Warner JO, Knol J, Kroll JS. Intestinal microbiota in infants at high risk for allergy: Effects of prebiotics and role in eczema development. The Journal of Allergy and Clinical Immunology. 2018;**141**:1334-1342. DOI: 10.1016/j. jaci.2017.05.054

[74] Markowiak P, Śliżewska K. Effects of probiotics, prebiotics, and synbiotics on human health. Nutrients. 2017;**9**(9):1021. DOI: 10.3390/ nu9091021

[75] Radke M, Picaud JC, Loui A, Cambonie G, Faas D, Lafeber HN, et al. Starter formula enriched in prebiotics and probiotics ensures normal growth of infants and promotes gut health: A randomized clinical trial. Pediatric Research. 2017;**81**:622-631. DOI: 10.1036/pr.2016.270 **13**

3

Prebiotics, Probiotics, and Bacterial Infections

Christina C. Tam, Kirkwood M. Land and Luisa W. Cheng

Abstract

Bacterial pathogens have developed exquisite virulence mechanisms to survive in the host cells. These virulence mechanisms help them bind and internalize into host cells, replicate, and evade the host immune response. The mammalian host itself has developed its own repertoire of weapons to prevent this from happening. One important component of host response in preventing infections in the gut lumen is the diverse commensal microbiota present. Dysbiosis of the gut microbiota has been implicated in the development of many gastrointestinal diseases. A potential therapeutic pathway to solve these diseases would be by providing probiotics and/or prebiotics to help stimulate growth of the beneficial commensal bacteria.

Here, we will present evidence of commensal microbiota imbalance in the develop- ment of disease as well as potential therapies to restore gut harmony.

Keywords: probiotics, bacterial infections, prebiotics, microbiota, therapeutics

Introduction

Probiotic microorganisms have been extensively studied for their beneficial effects in not only maintaining the normal gut mucosa but also protection from allergens, pathogens, and toxins [1, 2]. The gastrointestinal tract (GI) and its associated microbiota is a complex system that allows for the digestion and absorption of critical nutrients. Additionally, the presence of the commensal bacteria leads to the development and regulation of the mucosal immune system [3]. It is believed that 60% of all fecal matter mass in humans consists of bacteria and that there are between 1010 and 1012 colony-forming units per gram of intestinal content in the colon [4]. The intestinal epithelium is a physical and biochemical barrier that seeks to protect mammalian cells from infection and injury from contaminants such as toxins, pathogenic bacteria, commensal bacteria, and even other luminal contents. Specialized intestinal epithelial cells (IECs) are able to sense and respond to these stimuli with appropriate responses such as increasing their barrier function to activation of anti-pathogenic immune mechanisms [3].

The International Scientific Association for Probiotics and Prebiotics (ISAPP) in 2014 agreed on a consensus definition of probiotics based on the previous Food and Agriculture Organization of the United Nations and World Health Organization (WHO) definition. ISAPP defines probiotics as 'live microorganisms that, when administered in adequate amounts, confer a health benefit on the host" [5].

Probiotics have been used for the treatment of *Helicobacter pylori* infection, irritable bowl syndrome, and inflammatory bowel disease (ulcerative colitis and Crohn's disease) in addition to enhancing the immune system of healthy individuals [6–12].

Though *in vivo* and *in vitro* studies suggest that probiotics can be beneficial [1, 2, 13], the exact mechanism(s) remains to be fully explained. Four mechanism are believed to be involved: (a) maintenance of the gut epithelial barrier, (b) competitive exclusion of pathogenic organisms, (c) secretion of antimicrobial products, and (d) regulation of the mucosal immune system in favor of the hosts.

The major defensive mechanism of the gut is the intestinal barrier which maintains epithelial integrity and to protect the host from the environment. In defense of this barrier, there exists the mucous layer, antimicrobial peptides, secretory IgA and the epithelial junction adhesion complex [14]. Disruption of these defense mechanisms allows for the bacteria and food antigens to reach the submucosa, which can induce an inflammatory response potentially leading to the intestinal disorders such as inflammatory bowel disease [15, 16].

Probiotics

The most common probiotic strains used are *Lactobacillus*, *Bifidobacteria*, and the yeast strain *Saccharomyces cerevisiae* var. *boulardii*. Lactic acid bacteria and bifidobacteria have been shown to remove heavy metals [17], cyanotoxins [18], and mycotoxin from *in vitro* aqueous solutions [19, 20].

In regards to maintenance of the gut epithelial barrier, one can upregulate the genes important for this process [21]. Lactobacilli treatment has been shown to affect several genes including E-cadherin and β-catenin that affect adherence cell junctions in a cell culture model. The phosphorylation and abundance of adherence junction proteins including PKCδ [22] has been seen with Lactobacilli treatment.

The probiotic *Escherichia coli* Nissle 1917 strain (EcN1917) can initiate repair of the intestinal barrier after damage by enteropathogenic *E. coli* by enhancing the expres- sion and redistribution of tight junction proteins of the zonula occludens (ZO-2) and PKC [23, 24]. Treatment with *Lactobacillus casei* DN-114001 [25] and VSL#3 (an eight combination probiotic strain mixture) [26] also affect the gut barrier.

Another method to promote epithelial barrier function may be to increase mucin production thereby leading to increased barrier function as well as exclusion of pathogens and toxins. There have been contradictory data for both *in vitro* and *in vivo* experiments as to whether mucin production occurs in response to probiotic treatment. Some studies have suggested *Lactobacillus* adhesion is required to increase mucin production, which may not occur *in vivo* [27, 28]. However, a *Lactobacillus acidophilus* A4 cell extract has been shown to increase *MUC2* expres- sion in HT29 cells independent of attachment [29]. VSL#3 has also been shown to increase expression of *MUC2*, *MUC3*, and *MUC5AC* in HT-29 cells [26]. *In vivo* studies of mucin production have also been inconsistent. VSL#3 given to mice for 14 days did not show any increase in mucin production or thickness [30] whereas rats given VSL#3 for 7 days have a 60-fold increase in *MUC2* expression and secretion [31].

Prebiotics

Prebiotics and their beneficial effects on human health have been of interest in recent years because of their perceived safety since they are derived from dietary products. The definition of prebiotics has changed somewhat from their initial description in 1995 by Glenn Gibson and Marcel Roberfroid [32]. Today, the general consensus is that "dietary prebiotics" are "selectively fermented ingredients that results in specific changes in the composition and/or activity of the gastrointestinal microbiota, thus conferring benefit(s) upon host health" [33]. There are many types of prebiotics but they can be segregated into the following groups [33]: (1) fructans, (2) galacto-oligosaccharides (GOS), (3) starch and glucose-derived oligosaccha- rides, (4) other oligosaccharides, and (5) non-carbohydrate oligosaccharides.

Fructans, such as inulin and fructose-oligosaccharides (FOS)/oligofructose, generally have a linear chain of fructose with a β(2 + 1) linkage usually with termi- nal glucose units with a β(2 + 1) linkage with variable degrees of polymerization (DP) [33–36]. GOS is the product of lactose extension that can be classified into two subgroups: (i) excess galactose at C3, C4 or C6 and (ii) derived from enzymatic trans-glycosylation [33]. The product of the enzymatic trans-glycosylation is a mixture of tri-to pentasaccharides with galactose known as trans-galacto-oligo- saccharides (TOS) [37, 38]. In addition, there are GOSs derived from lactulose, an isomer of lactose, as well as raffinose family of oligosaccharides (RFO) [33, 37]. Starch that is resistant to the upper gut digestion is known as resistant starch (RS) and is considered a prebiotic along with polydextrose (glucose-derived oligosaccharide) [39, 40]. Pectin derived oligosaccharides (POS) are derived from an extension of galacturonic acid or rhamnose [33]. The carboxyl groups of POS can be modified with methyl esterification as well as acetylated at C_2 or C_3.

Additionally, many different types of sugars (i.e. arabinose, galactose, and xylose) or ferulic acid can be linked to the side chains of POS [41, 42]. Though most of the accepted compounds defined as prebiotics are carbohydrates, there are some non-carbohydrate compounds that are recommended to be classified as prebiotics, i.e. cocoa-derived flavanols [33, 43].

How do prebiotics affect human health? What mechanism(s) are involved? Since prebiotics are derived from dietary products, they provide the metabolic energy for the gut microbiota. This means that they can affect the composition and function of these microorganisms. For example, GOSs can stimulate the growth of *Bifidobacteria* and *Lactobacilli* to a high degree while *Enterobacteria*, *Bacteroidetes*, and *Firmicutes* growth levels occurred at a lower level [35]. Cross-feeding, the production of a by-product that can sustain another microorganism, can occur and an example is the degradation of resistant starch by *Ruminococcus bromii* to provide energy for several other species [44].

In addition to feeding the gut microbiota, the fermentation of prebiotics can generate metabolites such as short chain fatty acids (SCFA) (i.e. lactic acid, butyric acid, and propionic acid) that have dramatic effects not only on the intestinal environment but can affect distant organ sites as well as the immune system. SCFAs decrease the pH of the gut that can alter the composition of the microbiota [45, 46]. A pH unit decrease affects acid sensitive species such as *Bacteroides* and increases butyrate production by *Firmicutes* [45]. Butyrate itself has been shown to be important for intestinal cell development [47]. As reviewed in [33], propionate affects the TH2 helper cells, macrophages, and dendritic cells while peptidoglycan stimulates the innate system.

Bacterial infections and the disruption of gut homeostasis

Bacterial pathogens are microorganisms that have the ability to cause disease due to their specialized virulence factors or that can arise from a dysbiosis such as from antibiotic treatment that can eliminate the normal healthy flora of the gut leading to opportunistic infections from commensals or normally non-pathogenic organisms.

Bacterial pathogens, virulence factors, and mechanisms of pathogenesis

Some of the best-known bacterial pathogens are *Salmonella enterica, Listeria monocytogenes, Vibrio cholera, Shigella flexneri,* Shiga toxin producing *E. coli* (STEC) (i.e. *E. coli* 0157:H7), *Clostridium difficile, Clostridium perfringens,* and *Clostridium botulinum.* Both Gram-positive and Gram-negative pathogens must develop mechanisms to outcompete the normal gut microbiota, bind/invade cells, avoid detection and killing from the host immune system. Some important pathogen virulence mechanisms consist of specialized secretion systems that encode factors important for all the above steps in pathogenesis. The type III secretion systems (TTSS/T3SS) encoded by some Gram-negative pathogens such as *Salmonella, Vibrio, Shigella, Escherichia coli,* and *Yersinia* are well-known examples. Other specialized secretion systems are the T4SS and T6SS. T6SSs are prevalent in both pathogens and commensals suggesting their importance in the intestinal environment [48]. For non-intracellular pathogens, bacterial toxins [i.e., Listeriolysin O (LLO), botulinum neurotoxins (BoNTs), alpha toxin-*C. perfringens,* TcdA/TcdB-*C. difficile*] are important virulence factors that can bind to and enter the intestinal epithelium and/or their target cells to effect their func- tions (i.e. cytotoxicity).

One important growth restriction system on the part of hosts/intestinal flora is the sequestration of iron, which is absolutely required for growth. For example, *Salmonella enterica* can evade lipocalin-2-mediated growth restriction by producing modified siderophores that cannot be bound by lipocalin-2 [49]. The T4SS and T6SS systems can be utilized for intra-and-inter bacterial species warfare. *Bacteroides* strains encoding the T6SS have been shown to target sensitive *Bacteroides* spp. suggesting they limit their competition [50–52]. *Salmonella, Vibrio,* and *E. coli* have also been shown to use T6SS against their competition, the intestinal microbiota [53–55].

The mammalian host response to bacterial infections

Probiotic strains have been shown to induce the release of defensins, small peptides/proteins active against bacteria, fungi, and viruses but also are able to stabilize the gut barrier from epithelial cells. Host cells are able to mount as a first line of defense against pathogens increased production of antimicrobial proteins (AMPs) such as α- and β-defensins, cathelicidins, C-type lectins and ribonucleases. Many of these proteins disrupt the cell wall structures of the bacterial membrane either through enzymatic (i.e. lysozyme, phospholipase A2) or non-enzymatic mechanisms (i.e. pore formation by defensins and cathelicidins) [56–58].

The effect of commensal and probiotic bacteria on the host immune system is complex and not fully understood. It is believed that the effect of probiotic bacteria in modulating the immune system lies with its potential interactions with the host innate immune system by activating pattern recognition receptors (PRRs) that recognize common structures called pathogen-associated molecular patterns (PAMPs) shared by the vast majority of pathogens. Of note are the potential interactions with toll-like receptors (TLRs), extracellular C-type lectin receptors (CLRs), and intra- cellular nucleotide-binding oligomerization domain-containing protein (NOD)-like receptors (NLRs) that recognize PAMPs such as lipopolysaccharide (LPS), peptidoglycan, lipoprotein, flagellin, and CpGDNA. Activation of these receptor complexes will activate multiple downstream signaling pathways that may induce a pro- or anti-inflammatory response. Dysregulation of the pro-inflammatory response has been implicated in Crohn's disease with human intestinal inflammation as well as human autoinflammatory disease [59]. However, expression levels of some of these PRRs are low in immune cells therefore the ability to rapidly induce the expression of the PRRs such as NLRP3 in response to PAMP stimuli are absolutely critical in the defense against potential pathogens [60–64].

Probiotic mechanisms of antagonism against bacterial growth and gene expression

In 1969, Greenberg [65] described the phenomena that *Salmonella typhimurium* was completely excluded from maggots of blowflies. The term "competitive exclusion" was used to define the scenario in which one species of bacteria more vigorously competes for the receptor sites in the intestinal tract than another species.

There are a variety of mechanisms used by one bacterial species to exclude or reduce the growth of another species such as creation of a hostile environment, blocking available receptor sites, production and secretion of antimicrobial products and specific metabolites, and competitive depletion of essential nutrients [66].

Lactobacilli and bifidobacteria have been shown to inhibit a broad range of pathogens including *E. coli*, *Salmonella*, *Helicobacter pylori*, *Listeria monocytogenes* and *rotavirus* [6, 67–73]. Competition for host cell surface receptors by some probiotics has been successful against some enteropathogens [74–76]. *L. rhamnosus* can prevent enterohemorrhagic *E. coli* (EHEC) internalization [77]. Probiotic inhibition of pathogen binding to host cells relies heavily on steric hindrance [78].

Lactobacilli have been shown to produce bacteriocins that are active against some foodborne pathogens [79]. Additionally production of various metabolites and low molecular weight products by probiotics have been shown to have antimicrobial and antifungal properties such as low molecular weight species, deconju- gated bile acids, and cyclic dipeptides among others [80–85]. *Enterococcus faecium* BGPAS1-3 has been shown to produce a cell wall product that has an anti-listerial effect, prevents tight junction disruption, as well as modulating the TLR2/TLR4 immune response to *Listeria monocytogenes* ATCC19111 [86]. *L. plantarum* ATCC 8014 has recently been shown to have *in vitro* antimicrobial activity against *C. butryicum* ATCC 860, *C. difficile* ATCC 9689, and *C. perfringens* ATCC 12924 suggesting that this probiotic strain may have therapeutic potential [87].

The production of antimicrobial substances such as lactic and acetic acid is one example of probiotics making the host environment hostile for pathogens.

Lactobacillus co-cultivation with *E. coli* O157:H7 in broth culture produced organic acids which lead to a decrease in both pH and *stx2A* expression [88]. Low pH also prevented the induction of Stx prophage [89]. Mice given *Lactobacillus reuteri* with *E. coli* O157:H7 had decreased intestinal pathogen count, weight gain, and less kidney damage than controls [90]. The presence of probiotics in cattle feed reduced the amount *E. coli* O157:H7 seen in cattle [91]. *Bifidobacterium* strains decrease STEC and *in vivo* expression of Shiga toxin due to low pH and production of acetate [92, 93]. Pretreatment with live Lactobacilli before *Salmonella enterica Javiana* infection in a tissue culture model showed decrease expression of virulence genes, less cytotoxicity, and reduced host production of inflammatory cytokines [94].

Probiotics and inhibition of bacterial toxins

Clostridium botulinum is an ubiquitous, gram-positive, anaerobic spore-forming organism that is the causative agent of botulinum. The botulinum neurotoxins are one of the most lethal toxins known to mankind with a parenteral lethal dosage of 0.1–1 ng/kg and an oral dose of 1 μg/kg. Due to this high toxicity and potential for bioterrorism, botulinum neurotoxins (BoNTs) are considered Tier 1 category Select Agents by the Centers for Disease Control and Prevention (CDC). BoNTs are a pub- lic health and safety threat in the form of foodborne, wound, and infant botulism.

In order to cause disease for foodborne botulism, BoNTs must first be able to survive in the intestinal lumen, bind to and translocate through the intestinal epithelium to reach the bloodstream [95]. Once in the bloodstream, BoNTs bind to peripheral cholinergic neurons to cleave SNAREs and block exocytosis of neu- rotransmitters hence leading to flaccid muscle paralysis. Similar to other classic A-B chain toxins, the heavy chain (B chain) of BoNTs bind to carbohydrate and protein receptors on their target cell while the light chain (A chain) has the enzymatic function. Therefore, there are two potential therapeutic pathways to block

BoNT intoxication: (1) blocking binding/translocation at the intestinal epithelium/ target cells and (2) degradation or inactivation of the toxin It has been shown that pre-treatment with probiotics (*Saccharomyces cerevisiae* var. *boulardii*, *Lactobacillus acidophilus*, *Lactobacillus rhamnosus* LGG, and *Lactobacillus reuteri*) was able to block toxin binding to cells in an *in vitro* Caco-2 cell culture model and that the mechanism most likely used by the probiotics was steric hindrance of binding to host receptors [96].

Another mechanism to inactivate bacterial toxins would to be to subject them to proteolysis thus rendering them inactive. *S. boulardii* has been shown to produce a 54-KDa protease that is able to cleave and inactivate the two main *C. difficile* toxins, TcdA and TcdB [97] in a HT-29 colonic cell model. Whether or not this occurs in any significant degree in human infection is still unclear.

Probiotics and/or prebiotics as therapeutics to combat gastrointestinal diseases and bacterial infections

Studies in using probiotics as a treatment for a diverse set of diseases ranging from colorectal cancer, traditional gastrointestinal diseases (i.e. IBS/IBD/RCDI), as well as non-gastrointestinal diseases such as arthritis, autism, multiple sclerosis, and Parkinson's among others [99] has been undertaken. In this chapter, we will focus mainly on the effect probiotic and/prebiotic treatments on gastrointestinal diseases.

The therapeutic potential of prebiotics has been investigated for some gastrointestinal disorders. Irritable bowel syndrome (IBS) is a gastrointestinal disease characterized by chronic pain and altered bowel movements with no clear cause. Crohn's disease, a chronic relapsing inflammatory bowel disease (IBD), can affect any part of the gastrointestinal tract. For both conditions, it has been speculated that a shift in the gut microbiota population lays at the foundation of these diseases. It has been shown that the *Bifidobacteria, Faecalibacterium prausnitzii, Bacteroides* to *Firmicutes* population ratio were decreased [100, 101].

Therefore, prebiotics were hypothesized as a potential therapeutic because of its known properties to stimulate the growth of beneficial bacteria. In regards to IBS, the results were unclear for 4 clinical trials. Two clinical trials had no improvement [102, 103] whereas two studies using FOS and GOS showed an improvement in IBS symptoms [104, 105]. In the case of Crohn's disease, one study showed improvement [106] while two did not [107, 108]. As reviewed in [109], their analysis of available studies indicated that generally, the conclusions were supportive of probi- otic treatment for IBS, however, the exact beneficial strains to be used were unclear. The caveats from these studies were the variabilities in the type of prebiotic(s) used, the dosage, time of supplementation, and patient disease stage. As has been used in the treatment of recurrent *C. difficile* infection (RCDI), fecal microbiota transplantation (FMT) has also been used with success in the treatment of IBS and IBD (i.e. Crohn's disease) [99].

The recurrent infection in humans with recurrent *Clostridium difficile* infection (RCDI) in hospitalized patients treated with antibiotics is a severe problem. Studies have successfully used the transfer of healthy gut microbiota to the infected individuals as a treatment [110, 111]. As reviewed in [99], an amazing success rate of

≈92% of RCDI patients was found after FMT therapy.

Another area of medical use that prebiotics may impact is on the health of pre-term neonates. These babies are at significant risk of developing the severe gastrointestinal condition necrotizing enterocolitis (NEC), a life-threatening condition. Studies have shown that FOS and GOS prebiotics can help prime the growth of gut bacteria such as *Bifidobacteria* and reduce pathogenic organisms in preterm babies [112–114] thereby preventing NEC. Additionally, SCFAs from prebiotic fermentation enhances both gastric emptying and bowel motility [115, 116]. A systemic analysis of four randomized controlled trials showed elevated concentrations of fecal *Bifidobacteria* if babies were given FOS, GOS, or their mixture, but there was no significant risk reduction or progression to NEC [117]. In a review of several studies regarding probiotics and their effectiveness in preventing necrotizing enterocolitis (NEC) in preterm infants [13, 118], the authors concluded a beneficial effect of using probiotics but this benefit decreased over time. However, since the studies varied in age, doses, and duration of treatment, this observation has prob- ably very little effect on NEC.

The successful use of probiotics in treating acute infectious diarrhea (AID) in children is well documented and accepted treatment therapy [119]. It has beneficial effects for children at risk (i.e. hospital acquired diarrhea) and should be used early after onset of symptoms. Its usage, however, in healthy populations as a preventive measure to prevent diarrhea in day care centers and communities is currently unknown and not advised.

It has been shown that *E. coli Nissle* 1917 can outcompete *Salmonella* for iron leading to reduced *Salmonella* colonization and inflammation [120]. *E. coli Nissle* 1917 also can prevent *L. monocytogenes* entry into cell lines [121]. Probiotic treatment against *Listeria* infections has best been shown in the poultry industry.

Competitive exclusion (CE) cultures have been developed and used successfully. Pre-treatment with CE prevented the expansion of *Listeria monocytogenes* in young chickens [122].

The development of synthetic oligosaccharide-based mimics such as Synsorb (inert silica particles-linked to synthetic oligosaccharides) have been developed against a variety of toxins including: Stx1/2-Gb3, Stx2e-Gb4, Ctx-GM1, LT-GM1, epsilon toxin–GM2, TcdA-Lewis X and Lewis Y, botulinum neurotoxin- GD1a, GT1b, *E. coli* K88 ad fimbriae-nLc4, *E. coli* P pili- Gb3 and Gb4 [98]. However, the results for these synthetic oligosaccharide conjugates have been mixed. Synsorb-PK was designed as a mimic for Gb3, receptor for Stx1/2, to prevent intoxication with Shigella and STEC strains but failed to prevent the progression of children to hemolytic uremic syndrome (HUS) in a clinical trial [123]. However, there could be at least two reasons as to the failure of this compound, (1) treatment given late in the onset of disease and (2) potential steric hindrance of the size of the compound. Synsorb-90 was developed in the treatment of severe colitis due to *C. difficile* infection. This compound was able to bind TcdA *in vitro* as well as decrease toxin- mediated fluid secretion in a rat-ileal loop model [124]. However, phase III clinical trials for Synsorb-90 was abandoned after promising results from both phase I and phase II trials so we still do not know its efficacy [125].

STEC gastroenteritis has not been traditionally treated with probiotics/FMT as has been seen with acute gastroenteritis and RCDI. There has been a plethora of evidence suggesting the role of probiotic strains in having an antimicrobial effect on STEC but the effects were dependent on the strain(s) used as reviewed in [126]. Additionally, recombinant receptor mimics have been targeted against STEC [98].

In the three studies that mimicked human digestive conditions, *S. cerevisiae* CNCM I-3856 was implicated in having antagonistic effects on STEC including downregulating Stx expression and how the resident microbiota regulates infectivity [126–129].

Though there have been many successful and safe uses of probiotics for treatment of multiple conditions, there have been reported side effects linked to their usage especially in vulnerable populations [130]. As reviewed in [131], there has been movement toward using extracellular vesicles (EVs) derived from probiotic strains (both Gram-negative and Gram-positive) to deliver the same beneficial effects as from using the probiotic strains themselves. There are many different pathways that EVs utilize including bacteria-bacteria communication, affecting host microbial interactions, host immune system, increasing tight junction function, and decreasing inflammatory responses from TLR signaling [131].

Future works and perspectives

It has been shown that the development of gastrointestinal disease is due to an imbalance in the host response (physical, commensal microbiota, adaptive/innate immune systems) to bacterial infections. There has been an increasing accumulation of evidence (*in vitro*, *in silico*, and some *in vivo*) supporting the key role that the resident microbiota in the gut plays in mitigating bacterial infections as well as metabolic and physical diseases. There has been development of novel therapies all designed to replace/regenerate the lost beneficial commensal strains in a variety of diseases such as IBS, IBD, acute gastroenteritis, NEC, RCDI, etc. There has been tremendous success in the treatment of RCDI, IBS, and IBD using FMT therapy.

However, it is still unclear from all the evidence that giving probiotics and/or prebiotics will mitigate all gastrointestinal diseases [13]. The beneficial effects of a probiotic(s)/prebiotic mixture is utterly dependent on many factors including: time of dosage in relation to disease, probiotic strains used, prebiotics given, dosage, time of treatment, pre, post, and the preexisting health and/or the microbiota of the host. Additionally, the clinical trials should also be developed with statistical power to clearly answer the question at hand.

Acknowledgements

This work was funded by the United States Department of Agriculture, Agricultural Research Service, National Program project NP108, CRIS 2030-42000-049-00D.

Conflict of interest

The authors have no conflict of interest.

Author details

Christina C. Tam[1], Kirkwood M. Land[2] and Luisa W. Cheng[1*]

1. Foodborne Toxins Detection and Prevention Research Unit, Agricultural Research Service, United States Department of Agriculture, Albany, California, United States

2. Department of Biological Sciences, University of the Pacific, Stockton, California, United States

*Address all correspondence to: luisa.cheng@usda.gov

References

[1] Rao RK, Samak G. Protection and restitution of gut barrier by probiotics: Nutritional and clinical implications. Current Nutrition & Food Science. 2013;**9**(2):99-107

[2] Bermudez-Brito M, Plaza-Diaz J, Munoz-Quezada S, Gomez-Llorente C, Gil A. Probiotic mechanisms of action. Annals of Nutrition & Metabolism. 2012;**61**(2):160-174

[3] Peterson LW, Artis D. Intestinal epithelial cells: Regulators of barrier function and immune homeostasis. Nature Reviews. Immunology. 2014;**14**(3):141-153

[4] Stephen AM, Cummings JH. The microbial contribution to human faecal mass. Journal of Medical Microbiology. 1980;**13**(1):45-56

[5] Hill C, Guarner F, Reid G, Gibson GR, Merenstein DJ, Pot B, et al. Expert consensus document. The international scientific Association for Probiotics and Prebiotics consensus statement on the scope and appropriate use of the term probiotic. Nature Reviews. Gastroenterology & Hepatology. 2014;**11**(8):506-514

[6] Myllyluoma E, Veijola L, Ahlroos T, Tynkkynen S, Kankuri E, Vapaatalo H, et al. Probiotic supplementation improves tolerance to *Helicobacter pylori* eradication therapy--a placebo-controlled, double-blind randomized pilot study. Alimentary Pharmacology & Therapeutics. 2005;**21**(10):1263-1272

[7] Kajander K, Hatakka K, Poussa T, Farkkila M, Korpela R. A probiotic mixture alleviates symptoms in irritable bowel syndrome patients: A controlled 6-month intervention. Alimentary Pharmacology & Therapeutics. 2005;**22**(5):387-394

[8] Olivares M, Diaz-Ropero MA, Gomez N, Lara-Villoslada F, Sierra S, Maldonado JA, et al. Oral administration of two probiotic strains, *Lactobacillus gasseri* CECT5714 and *Lactobacillus coryniformis* CECT5711, enhances the intestinal function of healthy adults. International Journal of Food Microbiology. 2006;**107**(2):104-111

[9] Olivares M, Diaz-Ropero MP, Gomez N, Lara-Villoslada F, Sierra S, Maldonado JA, et al. The consumption of two new probiotic strains, *Lactobacillus gasseri* CECT 5714 and *Lactobacillus coryniformis* CECT 5711, boosts the immune system of healthy humans. International Microbiology. 2006;**9**(1):47-52

[10] Kim HJ, Vazquez Roque MI, Camilleri M, Stephens D, Burton DD, Baxter K, et al. A randomized controlled trial of a probiotic combination VSL# 3 and placebo in irritable bowel syndrome with bloating. Neurogastroenterology and Motility. 2005;**17**(5):687-696

[11] Derikx LA, Dieleman LA, Hoentjen F. Probiotics and prebiotics in ulcerative colitis. Best Practice & Research. Clinical Gastroenterology. 2016;**30**(1):55-71

[12] Bibiloni R, Fedorak RN, Tannock GW, Madsen KL, Gionchetti P, Campieri M, et al. VSL#3 probiotic- mixture induces remission in patients with active ulcerative colitis. The American Journal of Gastroenterology. 2005;**100**(7):1539-1546

[13] Suez J, Zmora N, Segal E, Elinav E. The pros, cons, and many unknowns of probiotics. Nature Medicine. 2019;**25**(5):716-729

[14] Ohland CL, Macnaughton WK. Probiotic bacteria and intestinal epithelial barrier function. American Journal of Physiology. Gastrointestinal and Liver Physiology. 2010;**298**(6):G807-G819

[15] Hooper LV, Wong MH, Thelin A, Hansson L, Falk PG, Gordon JI. Molecular analysis of commensal host-microbial relationships in the intestine. Science. 2001;**291**(5505):881-884

[16] Hooper LV, Stappenbeck TS, Hong CV, Gordon JI. Angiogenins: A new class of microbicidal proteins involved in innate immunity. Nature Immunology. 2003;**4**(3):269-273

[17] Halttunen T, Collado MC, El-Nezami H, Meriluoto J, Salminen S. Combining strains of lactic acid bacteria may reduce their toxin and heavy metal removal efficiency from aqueous solution. Letters in Applied Microbiology. 2008;**46**(2):160-165

[18] Nybom SM, Salminen SJ, Meriluoto JA. Specific strains of probiotic bacteria are efficient in removal of several different cyanobacterial toxins from solution. Toxicon. 2008;**52**(2):214-220

[19] El-Nezami H, Kankaanpaa P, Salminen S, Ahokas J. Ability of dairy strains of lactic acid bacteria to bind a common food carcinogen, aflatoxin B1. Food and Chemical Toxicology. 1998;**36**(4):321-326

[20] Oatley JT, Rarick MD, Ji GE, Linz JE. Binding of aflatoxin B1 to bifidobacteria in vitro. Journal of Food Protection. 2000;**63**(8):1133-1136

[21] Anderson RC, Cookson AL, McNabb WC, Kelly WJ, Roy NC. *Lactobacillus plantarum* DSM 2648 is a potential probiotic that enhances intestinal barrier function. FEMS Microbiology Letters. 2010;**309**(2):184-192

[22] Hummel S, Veltman K, Cichon C, Sonnenborn U, Schmidt MA. Differential targeting of the E-cadherin/beta-catenin complex by gram- positive probiotic lactobacilli improves epithelial barrier function. Applied and Environmental Microbiology. 2012;**78**(4):1140-1147

[23] Zyrek AA, Cichon C, Helms S, Enders C, Sonnenborn U, Schmidt MA. Molecular mechanisms underlying the probiotic effects of *Escherichia coli* Nissle 1917 involve ZO-2 and PKCzeta redistribution resulting in tight junction and epithelial barrier repair. Cellular Microbiology. 2007;**9**(3):804-816

[24] Stetinova V, Smetanova L, Kvetina J, Svoboda Z, Zidek Z, Tlaskalova- Hogenova H. Caco-2 cell monolayer integrity and effect of probiotic *Escherichia coli* Nissle 1917 components. Neuro Endocrinology Letters. 2010;**31**(Suppl 2):51-56

[25] Parassol N, Freitas M, Thoreux K, Dalmasso G, Bourdet-Sicard R, Rampal P. *Lactobacillus casei* DN-114 001 inhibits the increase in paracellular permeability of enteropathogenic *Escherichia coli*-infected T84 cells. Research in Microbiology. 2005;**156**(2):256-262

[26] Otte JM, Podolsky DK. Functional modulation of enterocytes by gram-positive and gram-negative microorganisms. American Journal of Physiology. Gastrointestinal and Liver Physiology. 2004;**286**(4):G613-G626

[27] Mattar AF, Teitelbaum DH, Drongowski RA, Yongyi F, Harmon CM, Coran AG. Probiotics up-regulate MUC-2 mucin gene expression in a Caco-2 cell-culture model. Pediatric Surgery International. 2002;**18**(7):586-590

[28] Mack DR, Ahrne S, Hyde L, Wei S, Hollingsworth MA. Extracellular MUC3 mucin secretion follows adherence of *Lactobacillus* strains to intestinal epithelial cells in vitro. Gut. 2003;**52**(6):827-833

[29] Kim Y, Kim SH, Whang KY, Kim YJ, Oh S. Inhibition of *Escherichia coli* O157:H7 attachment by interactions between lactic acid bacteria and intestinal epithelial cells. Journal of Microbiology and Biotechnology. 2008;**18**(7):1278-1285

[30] Gaudier E, Michel C, Segain JP, Cherbut C, Hoebler C. The VSL# 3 probiotic mixture modifies microflora but does not heal chronic dextran- sodium sulfate-induced colitis or reinforce the mucus barrier in mice. The Journal of Nutrition. 2005;**135**(12):2753-2761

[31] Caballero-Franco C, Keller K, De Simone C, Chadee K. The VSL#3 probiotic formula induces mucin gene expression and secretion in colonic epithelial cells. American Journal of Physiology. Gastrointestinal and Liver Physiology. 2007;**292**(1):G315-G322

[32] Gibson GR, Roberfroid MB. Dietary modulation of the human colonic microbiota: Introducing the concept of prebiotics. The Journal of Nutrition. 1995;**125**(6):1401-1412

[33] Davani-Davari D, Negahdaripour M, Karimzadeh I, Seifan M, Mohkam M, Masoumi SJ, et al. Prebiotics: Definition, types, sources, mechanisms, and clinical applications. Food. 2019;**8**(3). Article ID: 92

[34] Monsan P, Paul F. Enzymatic synthesis of oligosaccharides. FEMS Microbiology Reviews. 1995;**16**(2-3):187-192

[35] Louis P, Flint HJ, Michel C. How to manipulate the microbiota: Prebiotics. Advances in Experimental Medicine and Biology. 2016;**902**:119-142

[36] Scott KP, Martin JC, Duncan SH, Flint HJ. Prebiotic stimulation of human colonic butyrate-producing bacteria and bifidobacteria, in vitro. FEMS Microbiology Ecology. 2014;**87**(1):30-40

[37] Roberfroid M, Gibson GR, Hoyles L, McCartney AL, Rastall R, Rowland I, et al. Prebiotic effects: Metabolic and health benefits. The British Journal of Nutrition. 2010;**104**(Suppl 2):S1-S63

[38] Macfarlane GT, Steed H, Macfarlane S. Bacterial metabolism and health-related effects of galacto- oligosaccharides and other prebiotics. Journal of Applied Microbiology. 2008;**104**(2):305-344

[39] Fuentes-Zaragoza E, Sánchez-Zapata E, Sendra E, Sayas E, Navarro C, Fernández-López J, et al. Resistant starch as prebiotic: A review. Starch - Stärke. 2011;**63**(7):406-415

[40] Costabile A, Fava F, Röytiö H, Forssten SD, Olli K, Klievink J, et al. Impact of polydextrose on the faecal microbiota: A double-blind, crossover, placebo-controlled feeding study in healthy human subjects. The British Journal of Nutrition. 2012;**108**(3):471-481

[41] Yoo H-D, Kim D, Paek S-H. Plant cell wall polysaccharides as potential resources for the development of novel prebiotics. Biomolecules & Therapeutics (Seoul). 2012;**20**(4):371-379

[42] Gullón B, Gómez B, Martínez- Sabajanes M, Yáñez R, Parajó JC, Alonso JL. Pectic oligosaccharides: Manufacture and functional properties. Trends in Food Science and Technology. 2013;**30**(2):153-161

[43] Tzounis X, Rodriguez-Mateos A, Vulevic J, Gibson GR, Kwik-Uribe C, Spencer JPE. Prebiotic evaluation of cocoa-derived flavanols in healthy humans by using a randomized, controlled, double-blind, crossover intervention study. The American Journal of Clinical Nutrition. 2010;**93**(1):62-72

[44] Ze X, Duncan SH, Louis P, Flint HJ. *Ruminococcus bromii* is a keystone species for the degradation of resistant starch in the human colon. The ISME Journal. 2012;**6**(8):1535-1543

[45] Walker AW, Duncan SH, McWilliam Leitch EC, Child MW, Flint HJ. pH and peptide supply can radically alter bacterial populations and short-chain fatty acid ratios within microbial communities from the human colon. Applied and Environmental Microbiology. 2005;**71**(7):3692-3700

[46] Duncan SH, Louis P, Thomson JM, Flint HJ. The role of pH in determining the species composition of the human colonic microbiota. Environmental Microbiology. 2009;**11**(8):2112-2122

[47] Hamer HM, Jonkers D, Venema K, Vanhoutvin S, Troost FJ, Brummer RJ. Review article: The role of butyrate on colonic function. Alimentary Pharmacology & Therapeutics. 2008;**27**(2):104-119

[48] Rangan KJ, Hang HC. Biochemical mechanisms of pathogen restriction by intestinal bacteria. Trends in Biochemical Sciences. 2017;**42**(11):887-898

[49] Raffatellu M, George MD, Akiyama Y, Hornsby MJ, Nuccio SP, Paixao TA, et al. Lipocalin-2 resistance confers an advantage to *Salmonella enterica* serotype *Typhimurium* for growth and survival in the inflamed intestine. Cell Host & Microbe. 2009;**5**(5):476-486

[50] Wexler AG, Bao Y, Whitney JC, Bobay LM, Xavier JB, Schofield WB, et al. Human symbionts inject and neutralize antibacterial toxins to persist in the gut. Proceedings of the National Academy of Sciences of the United States of America. 2016;**113**(13):3639-3644

[51] Russell AB, Wexler AG, Harding BN, Whitney JC, Bohn AJ, Goo YA, et al. A type VI secretion- related pathway in Bacteroidetes mediates interbacterial antagonism. Cell Host & Microbe. 2014;**16**(2):227-236

[52] Chatzidaki-Livanis M, Geva-Zatorsky N, Comstock LE. *Bacteroides fragilis* type VI secretion systems use novel effector and immunity proteins to antagonize human gut *Bacteroidales* species. Proceedings of the National Academy of Sciences of the United States of America. 2016;**113**(13):3627-3632

[53] Sana TG, Flaugnatti N, Lugo KA, Lam LH, Jacobson A, Baylot V, et al. *Salmonella typhimurium* utilizes a T6SS-mediated antibacterial weapon to establish in the host gut. Proceedings of the National Academy of Sciences of the United States of America. 2016;**113**(34):E5044-E5051

[54] MacIntyre DL, Miyata ST, Kitaoka M, Pukatzki S. The *Vibrio* cholerae type VI secretion system displays antimicrobial properties. Proceedings of the National Academy of Sciences of the United States of America. 2010;**107**(45):19520-19524

[55] Fu Y, Waldor MK, Mekalanos JJ. Tn-Seq analysis of *Vibrio cholerae* intestinal colonization reveals a role for T6SS-mediated antibacterial activity in the host. Cell Host & Microbe. 2013;**14**(6):652-663

[56] Muller CA, Autenrieth IB, Peschel A. Innate defenses of the intestinal epithelial barrier. Cellular and Molecular Life Sciences. 2005;**62**(12):1297-1307

[57] Kagan BL, Selsted ME, Ganz T, Lehrer RI. Antimicrobial defensin peptides form voltage-dependent ion-permeable channels in planar lipid bilayer membranes. Proceedings of the National Academy of Sciences of the United States of America. 1990;**87**(1):210-214

[58] Bals R, Wilson JM. Cathelicidins–A family of multifunctional antimicrobial peptides. Cellular and Molecular Life Sciences. 2003;**60**(4):711-720

[59] Hirota SA, Ng J, Lueng A, Khajah M, Parhar K, Li Y, et al. NLRP3 inflammasome plays a key role in the regulation of intestinal homeostasis. Inflammatory Bowel Diseases. 2011;**17**(6):1359-1372

[60] Bauernfeind FG, Horvath G, Stutz A, Alnemri ES, MacDonald K, Speert D, et al. Cutting edge: NF-kappaB activating pattern recognition and cytokine receptors license NLRP3 inflammasome activation by regulating NLRP3 expression. Journal of Immunology. 2009;**183**(2):787-791

[61] Bauernfeind F, Ablasser A, Kim S, Bartok E, Hornung V. An unexpected role for RNA in the recognition of DNA by the innate immune system. RNA Biology. 2010;**7**(2):151-157

[62] Kim S, Bauernfeind F, Ablasser A, Hartmann G, Fitzgerald KA, Latz E, et al. *Listeria monocytogenes* is sensed by the NLRP3 and AIM2 inflammasome. European Journal of Immunology. 2010;**40**(6):1545-1551

[63] Meylan E, Tschopp J, Karin M. Intracellular pattern recognition receptors in the host response. Nature. 2006;**442**(7098):39-44

[64] Martinon F, Mayor A, Tschopp J. The inflammasomes: Guardians of the body. Annual Review of Immunology. 2009;**27**:229-265

[65] Greenberg B. Salmonella suppression by known populations of bacteria in flies. Journal of Bacteriology. 1969;**99**(3):629-635

[66] Rolfe RD. Population dynamics of the intestinal tract. In: Blankenship LC, editor. Colonization control of human bacterial enteropathogens in poultry. San Diego, CA, USA: Academic Press Inc; 1991

[67] Chenoll E, Casinos B, Bataller E, Astals P, Echevarria J, Iglesias JR, et al. Novel probiotic *Bifidobacterium bifidum* CECT 7366 strain active against the pathogenic bacterium *Helicobacter pylori*. Applied and Environmental Microbiology. 2011;**77**(4):1335-1343

[68] Sgouras D, Maragkoudakis P, Petraki K, Martinez-Gonzalez B, Eriotou E, Michopoulos S, et al. In vitro and in vivo inhibition of *Helicobacter pylori* by *Lactobacillus casei* strain Shirota. Applied and Environmental Microbiology. 2004;**70**(1):518-526

[69] Todoriki K, Mukai T, Sato S, Toba T. Inhibition of adhesion of food- borne pathogens to Caco-2 cells by *Lactobacillus* strains. Journal of Applied Microbiology. 2001;**91**(1):154-159

[70] Chu H, Kang S, Ha S, Cho K, Park SM, Han KH, et al. *Lactobacillus acidophilus* expressing recombinant K99 adhesive fimbriae has an inhibitory effect on adhesion of enterotoxigenic *Escherichia coli*. Microbiology and Immunology. 2005;**49**(11):941-948

[71] Tsai CC, Lin PP, Hsieh YM. Three *Lactobacillus* strains from healthy infant stool inhibit enterotoxigenic *Escherichia coli* grown in vitro. Anaerobe. 2008;**14**(2):61-67

[72] Munoz JA, Chenoll E, Casinos B, Bataller E, Ramon D, Genoves S, et al. Novel probiotic *Bifidobacterium longum* subsp. infantis CECT 7210 strain active against rotavirus infections. Applied and Environmental Microbiology. 2011;**77**(24):8775-8783

[73] Nakamura S, Kuda T, An C, Kanno T, Takahashi H, Kimura B. Inhibitory effects of *Leuconostoc mesenteroides* 1RM3 isolated from narezushi, a fermented fish with rice, on *Listeria monocytogenes* infection to Caco-2 cells and A/J mice. Anaerobe. 2012;**18**(1):19-24

[74] Neeser JR, Granato D, Rouvet M, Servin A, Teneberg S, Karlsson KA. *Lactobacillus johnsonii* La1 shares carbohydrate-binding specificities with several enteropathogenic bacteria. Glycobiology. 2000;**10**(11):1193-1199

[75] Fujiwara S, Hashiba H, Hirota T, Forstner JF. Inhibition of the binding of enterotoxigenic *Escherichia coli* Pb176 to human intestinal epithelial cell line HCT-8 by an extracellular protein fraction containing BIF of *Bifidobacterium longum* SBT2928: Suggestive evidence of blocking of the binding receptor gangliotetraosylceramide on the cell surface. International Journal of Food Microbiology. 2001;**67**(1-2):97-106

[76] Mukai T, Asasaka T, Sato E, Mori K, Matsumoto M, Ohori H. Inhibition of binding of *Helicobacter pylori* to the glycolipid receptors by probiotic *Lactobacillus reuteri*. FEMS Immunology and Medical Microbiology. 2002;**32**(2):105-110

[77] Hirano J, Yoshida T, Sugiyama T, Koide N, Mori I, Yokochi T. The effect of *Lactobacillus rhamnosus* on enterohemorrhagic *Escherichia coli* infection of human intestinal cells in vitro. Microbiology and Immunology. 2003;**47**(6):405-409

[78] Coconnier MH, Bernet MF, Chauviere G, Servin AL. Adhering heat- killed human *Lactobacillus acidophilus*, strain LB, inhibits the process of pathogenicity of diarrhoeagenic bacteria in cultured human intestinal cells. Journal of Diarrhoeal Diseases Research. 1993;**11**(4):235-242

[79] Nielsen DS, Cho GS, Hanak A, Huch M, Franz CM, Arneborg N. The effect of bacteriocin-producing *Lactobacillus plantarum* strains on the intracellular pH of sessile and planktonic *Listeria monocytogenes* single cells. International Journal of Food Microbiology. 2010;**141**(Suppl 1):S53-S59

[80] Lievin V, Peiffer I, Hudault S, Rochat F, Brassart D, Neeser JR, et al. Bifidobacterium strains from resident infant human gastrointestinal microflora exert antimicrobial activity. Gut. 2000;**47**(5):646-652

[81] Fujiwara S, Hashiba H, Hirota T, Forstner JF. Proteinaceous factor(s) in culture supernatant fluids of bifidobacteria which prevents the binding of enterotoxigenic *Escherichia coli* to gangliotetraosylceramide. Applied and Environmental Microbiology. 1997;**63**(2):506-512

[82] Magnusson J, Schnurer J. *Lactobacillus coryniformis* subsp. coryniformis strain Si3 produces a broad-spectrum proteinaceous antifungal compound. Applied and Environmental Microbiology. 2001;**67**(1):1-5

[83] Rouse S, Canchaya C, van Sinderen D. *Lactobacillus hordei* sp. nov., a bacteriocinogenic strain isolated from malted barley. International Journal of Systematic and Evolutionary Microbiology. 2008;**58**(Pt 9):2013-2017

[84] Rouse S, van Sinderen D. Bioprotective potential of lactic acid bacteria in malting and brewing. Journal of Food Protection. 2008;**71**(8):1724-1733

[85] Dal Bello F, Clarke CI, Ryan LAM, Ulmer H, Schober TJ, Ström K, et al. Improvement of the quality and shelf life of wheat bread by fermentation with the antifungal strain *Lactobacillus plantarum* FST 1.7. Journal of Cereal Science. 2007;**45**(3):309-318

[86] Popovic N, Djokic J, Brdaric E, Dinic M, Terzic-Vidojevic A, Golic N, et al. The influence of heat-killed *Enterococcus faecium* BGPAS1-3 on the tight junction protein expression and immune function in differentiated Caco-2 cells infected with *Listeria monocytogenes* ATCC 19111. Frontiers in Microbiology. 2019;**10**:412

[87] Monteiro C, do Carmo MS, Melo BO, Alves MS, Dos Santos CI, Monteiro SG, et al. In vitro antimicrobial activity and probiotic potential of bifidobacterium and *Lactobacillus* against species of *Clostridium*. Nutrients. 2019;**11**(2):448

[88] Carey CM, Kostrzynska M, Ojha S, Thompson S. The effect of probiotics and organic acids on Shiga-toxin 2 gene expression in enterohemorrhagic *Escherichia coli* O157:H7. Journal of Microbiological Methods. 2008;**73**(2):125-132

[89] Imamovic L, Muniesa M. Characterizing RecA-independent induction of Shiga toxin2-encoding phages by EDTA treatment. PLoS One. 2012;**7**(2):e32393

[90] Eaton KA, Honkala A, Auchtung TA, Britton RA. Probiotic *Lactobacillus reuteri* ameliorates disease due to enterohemorrhagic *Escherichia coli* in germfree mice. Infection and Immunity. 2011;**79**(1):185-191

[91] LeJeune JT, Wetzel AN. Preharvest control of *Escherichia coli* O157 in cattle. Journal of Animal Science. 2007;**85**(13 Suppl):E73-E80

[92] Asahara T, Shimizu K, Nomoto K, HamabataT,OzawaA,TakedaY.Probiotic bifidobacteria protect mice from lethal infection with Shiga toxin-producing *Escherichia coli* O157:H7. Infection and Immunity. 2004;**72**(4):2240-2247

[93] Fukuda S, Toh H, Hase K, Oshima K, Nakanishi Y, Yoshimura K, et al. Bifidobacteria can protect from enteropathogenic infection through production of acetate. Nature. 2011;**469**(7331):543-547

[94] Burkholder KM, Fletcher DH, Gileau L, Kandolo A. Lactic acid bacteria decrease *Salmonella enterica* Javiana virulence and modulate host inflammation during infection of an intestinal epithelial cell line. Pathogens and Disease. 2019;**77**(3):ftz025

[95] Lam TI, Stanker LH, Lee K, Jin R, Cheng LW. Translocation of botulinum neurotoxin serotype A and associated proteins across the intestinal epithelia. Cellular Microbiology. 2015;**17**(8):1133-1143

[96] Lam TI, Tam CC, Stanker LH, Cheng LW. Probiotic microorganisms inhibit epithelial cell internalization of botulinum neurotoxin serotype A. Toxins (Basel). 2016;**8**(12):377

[97] Castagliuolo I, Riegler MF, Valenick L, LaMont JT, Pothoulakis C. *Saccharomyces boulardii* protease inhibits the effects of *Clostridium difficile* toxins A and B in human colonic mucosa. Infection and Immunity. 1999;**67**(1):302-307

[98] Paton AW, Morona R, Paton JC. Designer probiotics for prevention of enteric infections. Nature Reviews. Microbiology. 2006;**4**(3):193-200

[99] Suchodolski JS, Jergens AE. Recent advances and understanding of using probiotic-based interventions to restore homeostasis of the microbiome for the prevention/therapy of bacterial diseases. Microbiology Spectrum. 2016;**4**(2):VMBF-0025-2015

[100] Whelan K. Mechanisms and effectiveness of prebiotics in modifying the gastrointestinal microbiota for the management of digestive disorders. The Proceedings of the Nutrition Society. 2013;**72**(3):288-298

[101] Wilson B, Whelan K. Prebiotic inulin-type fructans and galacto- oligosaccharides: Definition, specificity, function, and application in gastrointestinal disorders. Journal of Gastroenterology and Hepatology. 2017;**32**(Suppl 1):64-68

[102] Hunter JO, Tuffnell Q , Lee AJ. Controlled trial of oligofructose in the management of irritable bowel syndrome. The Journal of Nutrition. 1999;**129**(7 Suppl):1451S-1453S

[103] Olesen M, Gudmand-Hoyer E. Efficacy, safety, and tolerability of fructooligosaccharides in the treatment of irritable bowel syndrome. The American Journal of Clinical Nutrition. 2000;**72**(6):1570-1575

[104] Paineau D, Payen F, Panserieu S, Coulombier G, Sobaszek A, Lartigau I, et al. The effects of regular consumption of short-chain fructo-oligosaccharides on digestive comfort of subjects with minor functional bowel disorders. The British Journal of Nutrition. 2008;**99**(2):311-318

[105] Silk DB, Davis A, Vulevic J, Tzortzis G, Gibson GR. Clinical trial: The effects of a trans- galactooligosaccharide prebiotic on faecal microbiota and symptoms in irritable bowel syndrome. Alimentary Pharmacology & Therapeutics. 2009;**29**(5):508-518

[106] Lindsay JO, Whelan K, Stagg AJ, Gobin P, Al-Hassi HO, Rayment N, et al. Clinical, microbiological, and immunological effects of fructo- oligosaccharide in patients with Crohn's disease. Gut. 2006;**55**(3):348-355

[107] Benjamin JL, Hedin CR, Koutsoumpas A, Ng SC, McCarthy NE, Hart AL, et al. Randomised, double-blind, placebo-controlled trial of fructo-oligosaccharides in active Crohn's disease. Gut. 2011;**60**(7):923-929

[108] Joossens M, De Preter V, Ballet V, Verbeke K, Rutgeerts P, Vermeire S. Effect of oligofructose-enriched inulin (OF-IN) on bacterial composition and disease activity of patients with Crohn's disease: Results from a double-blinded randomised controlled trial. Gut. 2012;**61**(6):958

[109] Rodino-Janeiro BK, Vicario M, Alonso-Cotoner C, Pascua-Garcia R, Santos J. A review of microbiota and irritable bowel syndrome: Future in therapies. Advances in Therapy. 2018;**35**(3):289-310

[110] van Nood E, Dijkgraaf MG, Keller JJ. Duodenal infusion of feces for recurrent *Clostridium difficile*. The New England Journal of Medicine. 2013;**368**(22):2145

[111] van Nood E, Vrieze A, Nieuwdorp M, Fuentes S, Zoetendal EG, de Vos WM, et al. Duodenal infusion of donor feces for recurrent *Clostridium difficile*. The New England Journal of Medicine. 2013;**368**(5):407-415

[112] Knol J, Boehm G, Lidestri M, Negretti F, Jelinek J, Agosti M, et al. Increase of faecal bifidobacteria due to dietary oligosaccharides induces a reduction of clinically relevant pathogen germs in the faeces of formula-fed preterm infants. Acta Paediatrica. Supplement. 2005;**94**(449):31-33

[113] Boehm G, Lidestri M, Casetta P, Jelinek J, Negretti F, Stahl B, et al. Supplementation of a bovine milk formula with an oligosaccharide mixture increases counts of faecal bifidobacteria in preterm infants. Archives of Disease in Childhood. Fetal and Neonatal Edition. 2002;**86**(3):F178-F181

[114] Kapiki A, Costalos C, Oikonomidou C, Triantafyllidou A, Loukatou E, Pertrohilou V. The effect of a fructo-oligosaccharide supplemented formula on gut flora of preterm infants. Early Human Development. 2007;**83**(5):335-339

[115] Indrio F, Riezzo G, Raimondi F, Bisceglia M, Cavallo L, Francavilla R. Effects of probiotic and prebiotic on gastrointestinal motility in newborns. Journal of Physiology and Pharmacology. 2009;**60**(Suppl 6):27-31

[116] Indrio F, Riezzo G, Raimondi F, Francavilla R, Montagna O, Valenzano ML, et al. Prebiotics improve gastric motility and gastric electrical activity in preterm newborns. Journal of Pediatric Gastroenterology and Nutrition. 2009;**49**(2):258-261

[117] Srinivasjois R, Rao S, Patole S. Prebiotic supplementation of formula in preterm neonates: A systematic review and meta-analysis of randomised controlled trials. Clinical Nutrition. 2009;**28**(3):237-242

[118] Patel RM, Underwood MA. Probiotics and necrotizing enterocolitis. Seminars in Pediatric Surgery. 2018;**27**(1):39-46

[119] Lo Vecchio A, Buccigrossi V, Fedele MC, Guarino A. Acute infectious diarrhea. Advances in Experimental Medicine and Biology. 2019;**1125**:109-120

[120] Deriu E, Liu JZ, Pezeshki M, Edwards RA, Ochoa RJ, Contreras H, et al. Probioticbacteriareduce *Salmonella typhimurium* intestinal colonization by competing for iron. Cell Host & Microbe. 2013;**14**(1):26-37

[121] Altenhoefer A, Oswald S, Sonnenborn U, Enders C, Schulze J, Hacker J, et al. The probiotic *Escherichia coli* strain Nissle 1917 interferes with invasion of human intestinal epithelial cells by different enteroinvasive bacterial pathogens. FEMS Immunology and Medical Microbiology. 2004;**40**(3):223-229

[122] Hume ME, Byrd JA, Stanker LH, Ziprin RL. Reduction of caecal *Listeria monocytogenes* in Leghorn chicks following treatment with a competitive exclusion culture (PREEMPT). Letters in Applied Microbiology. 1998;**26**(6):432-436

[123] Trachtman H, Cnaan A, Christen E, Gibbs K, Zhao S, Acheson DW, et al. Effect of an oral Shiga toxin-binding agent on diarrhea- associated hemolytic uremic syndrome in children: A randomized controlled trial. Journal of the American Medical Association. 2003;**290**(10):1337-1344

[124] Castagliuolo I, LaMont JT, Qiu B, Nikulasson ST, Pothoulakis C. A receptor decoy inhibits the enterotoxic effects of *Clostridium difficile* toxin A in rat ileum. Gastroenterology. 1996;**111**(2):433-438

[125] Weiss K. Toxin-binding treatment for *Clostridium difficile*: A review including reports of studies with tolevamer. International Journal of Antimicrobial Agents. 2009;**33**(1):4-7

[126] Giordano M, Baldassarre ME, Palmieri V, Torres DD, Carbone V, Santangelo L, et al. Management of STEC gastroenteritis: Is there a role for probiotics? International Journal of Environmental Research and Public Health. 2019;**16**(9):1649

[127] Etienne-Mesmin L, Livrelli V, Privat M, Denis S, Cardot JM, Alric M, et al. Effect of a new probiotic *Saccharomyces cerevisiae* strain on survival of *Escherichia coli* O157:H7 in a dynamic gastrointestinal model. Applied and Environmental Microbiology. 2011;**77**(3):1127-1131

[128] Thevenot J, Etienne-Mesmin L, Denis S, Chalancon S, Alric M, Livrelli V, et al. Enterohemorrhagic *Escherichia coli* O157:H7 survival in an in vitro model of the human large intestine and interactions with probiotic yeasts and resident microbiota. Applied and Environmental Microbiology. 2013;**79**(3):1058-1064

[129] Thevenot J, Cordonnier C, Rougeron A, Le Goff O, Nguyen HT, Denis S, et al. Enterohemorrhagic *Escherichia coli* infection has donor- dependent effect on human gut microbiota and may be antagonized by probiotic yeast during interaction with Peyer's patches. Applied Microbiology and Biotechnology. 2015;**99**(21):9097-9110

[130] Marteau P, Shanahan F. Basic aspects and pharmacology of probiotics: An overview of pharmacokinetics, mechanisms of action and side-effects. Best Practice & Research. Clinical Gastroenterology. 2003;**17**(5):725-740

[131] Molina-Tijeras JA, Galvez J, Rodriguez-Cabezas ME. The immunomodulatory properties of extracellular vesicles derived from probiotics: A novel approach for the management of gastrointestinal diseases. Nutrients. 2019;**11**(5):1038

4

Cholesterol Uptake and Survival of *Lactococcus lactis* Strains in Fluids Simulating the Human Stomach and Duodenum

Małgorzata Ziarno

Abstract

Scientific evidence exists showing that lactic acid bacteria, including the genus *Lactococcus*, have the capacity to bind and remove cholesterol. However, in many cases, in vivo and in vitro results are not unambiguous or reproducible; thus it appeared valid to conduct a study which would explain what factors determine adhesion and assimilation of cholesterol by *Lactococcus*. The study on *Lactococcus* bacteria under in vitro conditions in model digestive fluids may contribute to the explanation of the observed ambiguities. In vitro research has proven that *Lactococcus* is capable of removing free cholesterol under in vitro conditions in broths without bile salts, as well as in a simulated gastric fluid and in the conditions of simulated intestinal fluid. The amount of cholesterol removed by live cells of *Lactococcus* is directly proportionately dependent on the concentration of this sub- stance, incubation temperature, count, and viability of cells. However, oftentimes these relationships are not linear. Under the conditions of model gastric fluid or intestinal fluid, the cultures of *Lactococcus* release portion of the previously bound cholesterol, independent of cell viability. The survival rate of *Lactococcus* cells in simulated gastric fluid or simulated intestinal fluid depends on the tested bacterial culture and does not depend on the presence of cholesterol.

Keywords: cholesterol, *Lactococcus*, survival, gastrointestinal tract, duodenum, gastric acid

Introduction

Coronary ischemia, known as the coronary disease, is one of the modern civilization diseases, whose cause is coronary atherosclerosis (so-called atherosclerotic coronary plaque) in over 90% of cases, leading to their stenosis. One of the risk factors for the formation of atherosclerotic coronary plaques is hypercholesterolemia, in particular elevated concentration of LDL cholesterol. Scientific data exist indicating that consumption of fermented milk products reduces the level of cholesterol in humans. Some of the studies (on animals and human volunteers) indicate that the reduction of the cholesterol level in blood serum is caused by lactic acid bacteria present in fermented milk drinks. Numerous in vitro studies demonstrate that the capacity to reduce the cholesterol level may be exhibited not only by the strains with documented probiotic traits but also some "traditional" lactic acid bacteria used in the production of cheese, cream, or fermented milk products. The role of *Lactococcus* in dairy fermentation is mostly down to the production of lactic acid; however these bacteria utilize less than 0.5% of lactose from milk. Only *Lactococcus lactis* is appli- cable in the dairy industry, with its two subspecies: *Lactococcus lactis* subsp. *lactis* and *Lactococcus lactis* subsp. *cremoris*. These subspecies comprise the basic component of dairy mesophilic starter cultures, used for the production of cream, buttermilk, cottage cheeses, cheeses, and fermented milk [1–3]. In the process and functional terms, *Lactococcus* possess all the traits required for starter cultures: the capacity to ferment lactose, resistance to low pH, low temperature, and high concentrations of cooking salt. They are characterized by stability and suitable survival time during lyophilization and freezing and in the storage process of starter cultures [1, 2, 4].

Furthermore, lactic acid bacteria have the capacity to reduce the level of cholesterol in simulated conditions, i.e., culture media. It is known that lactic acid bacteria are not capable of metabolizing cholesterol, meaning its transformation into other compounds. It has been noted that bacterial cells are capable of binding cholesterol, consisting in adhesion of substances by the cell wall or its assimilation into cell wall. It has also been suspected that lactic acid bacteria are capable of deconjugating bile salts being the component of bile, followed by coprecipitation of cholesterol with deconjugated bile acids. Furthermore, tests on gnotobiotic animals demonstrated that hydrolysis of bile enhances its secretion and thus may contribute to reduction of the cholesterol level in blood serum. Moreover, the cholesterol level in the human organism may be also influenced by exopolysaccharides (EPS) produced by numerous lactic acid bacteria species. It is believed that these bacteria, similar to fiber, can bind cholesterol and bile acid molecules present in intestines and remove them from the human organism.

In many cases, results of in vitro studies are not unambiguous, or lack of their reproducibility has been determined. It turns out that also in vivo tests conducted on human volunteers or experimental animals do not produce unambiguous results or that their results are divergent. Considering that it is difficult to explain as to why this happens, such studies are frequently criticized for methodological and technical errors and lack of reproducibility.

Influence of lactic acid bacteria, including Lactococcus, on the cholesterol level in humans

The extensive collection of scientific publications devoted to health-promoting properties of lactic acid bacteria includes articles presenting studies on the possibility of reducing the cholesterol level in human and animal organisms through consumption of fermented milk products including traditional and probiotic strains of lactic acid bacteria.

As early as in 1974, Mann and Spoerry [5, 6] determined the reduced level of cholesterol in the blood serum of men from the African Maasai, which stemmed from the consumption of high amounts of fermented milk containing wild lactic acid bacteria strains. This research enabled researchers to look for the methods of reducing the cholesterol level in the human organism, although the first reports on the positive impact of fermented milk drinks on the reduction of the cholesterol level in live organisms were criticized due to their methodological and technical errors. However, these studies opened a new route for researchers in terms of the search of methods of cholesterol level reduction in the human organism, increasing the chances of the modern human populations in the combat with cardiovascular disorders [7–13].

The interpretation of study results concerning the influence of lactic acid bacteria on the cholesterol level obtained under in vivo conditions on living organ- isms is not easy. Organisms of animals and humans differ in terms of mechanisms of regulation of lipid metabolism, including cholesterol. It should be taken into account that introduction of lactic acid bacteria to the gastrointestinal tract does not only have a direct influence on the cholesterol metabolism but also on the entire intestinal microflora, which is capable of metabolizing cholesterol and other lipids, as indicated by the study results obtained by Hosono et al. [14]. This might be the cause for the difficulties in proving the positive influence of lactic acid bacteria on the cholesterol level in the human organism.

Certain in vitro studies from this field conducted within the last dozen or so years enabled assumption that it is lactic acid bacteria that produce the effect of cholesterol level decrease in humans and animals consuming fermented milk products. Numerous study results are available in the literature concerning cholesterol level reduction under laboratory conditions in model media. Decrease of the cholesterol level in culture media has been determined for numerous species and strains of lactic acid bacteria.

The majority of research concerns thermophilic bacilli of the genus *Lactobacillus* [14–22]. Other genera of bacteria exhibiting similar property include *Streptococcus*, *Enterococcus*, *Lactococcus*, and *Leuconostoc* [20, 21, 23–26]. According to these studies, the cholesterol binding capacity can be exhibited not only by strains with probiotic characters documented by research but also certain lactic acid bacteria species that are traditionally used to manufacture dairy products and included in dairy starter cultures.

It should be borne in mind that despite the results of in vitro and in vivo studies on animals and humans, it is impossible to unambiguously confirm or negate the capacity of lactic acid bacteria to reduce the cholesterol level in the blood serum due to the possibility of methodological and technical errors and the lack of reproducibility [27]. The more so that the level of cholesterol in blood serum is positively correlated not only with the amount of cholesterol taken with food but also depends on the intake of saturated fatty acids and refined carbohydrates. Therefore, the definite confirmation of the manner in which lactic acid bacteria exercise a beneficial influence on the level of cholesterol in humans is still missing [28–30].

Mechanism of cholesterol level reduction by lactic acid bacteria including Lactococcus in humans

The assumption that lactic acid bacteria may cause reduction of the cholesterol level directly in fermented milk products or live organisms was made on the basis of numerous in vivo and in vitro studies demonstrating that certain lactic acid bacteria produced a reduction of the cholesterol level in the blood serum of experimental animals or human volunteers or in model culture media. This type of research has been conducted since the 1970s [5, 7, 8, 14, 31–36]. The majority of these studies concern the influence of consumption of fermented products or products containing lactic acid bacteria strains, including primarily probiotic strains. In that time, several scientific hypotheses were formed on the mechanisms through which the phenomenon of cholesterol level reduction performed by lactic acid bacteria may occur. Literature data lists here primarily cholesterol binding, enzymatic deconjuga-tion of bile salts, production of exopolysaccharides, and synthesis of short-chain fatty acids (SCFAs) [15, 21, 23–25, 28, 29, 36–45].

Cholesterol binding by the bacterial cell wall and its incorporation into the cell wall or cytoplasmic membrane of bacterial cells are listed among the major mechanisms [9, 19, 23–25, 43]. It is known that cholesterol binding may have different paths. Certain bacteria incorporate cholesterol into the cell wall, as exhibited by such genera as *Micrococcus*, *Bacillus*, *Proteus*, or *Mycoplasma*. In the case of lactic acid bacteria, it has been thus far believed that cholesterol is solely attached by the cell through physical adhesion and it is not subject to subsequent metabolism. However, in vitro tests demonstrated that lactic acid bacteria are also capable of incorporating cholesterol into the cell wall [19, 23, 28, 30]. Many scientists have determined that the amount of cholesterol bound by lactic acid bacteria cells depends on, among others, genus, species, and culture of bacteria, growth phase, viability, and cell count [5, 10, 23–25, 38–40, 43]. Research results demonstrate that the strains commercially used in fermented food production are less efficient in binding cholesterol in comparison with the strains isolated from the alimentary tract of humans and animals [16, 46]. According to the majority of literature data, the phenomenon of cholesterol binding by lactic acid bacteria occurs primarily in anaerobic conditions and with the presence of bile salts [5, 17, 35, 47]. However, scientific reports have been published indicating a lack of or poor correlation between tolerance of bile salts and the capacity to bind cholesterol [22, 46, 48].

Another proposed mechanism for cholesterol level reduction in the human organism by lactic acid bacteria is the deconjugation of bile salts, associated with the activity of bile salt hydrolase (BSH) enzyme [16, 17, 21, 30, 39, 40, 45, 49, 50]. Bile salt hydrolase also referred to as cholylglycine hydrolase EC.3.5.1.24 catalyzes hydrolysis (also known as deconjugation) of the amide bond in bile acids conjugated with taurine or glycine, with the release of primary bile acids and amino acids, taurine or glycine [30, 45]. Hydrolysis of bile salts conjugated with taurine or glycine is one of the best known microbiological biotransformations of bile salts. BSH activity is observed for certain bacteria species isolated from the alimentary tract of humans and animals, i.e., strains from genera *Lactobacillus*, *Enterococcus*, *Peptostreptococcus*, *Bifidobacterium*, *Clostridium*, and *Bacteroides*, that is, microflora from environments rich in conjugated bile acids [14, 30, 35, 39, 40, 45, 51, 52]. A study conducted by Tanaka et al. [53] demonstrated BSH activity also in *Lactococcus lactis*, *Leuconostoc mesenteroides*, and *S. thermophilus* strains. The details on the function of BSH are unknown. It is believed that a relationship exists between BSH activity and the natural environment of bacteria. It is likely that hydrolysis of bile salts catalyzed by BSH constitutes a protection mechanism against the toxic effect of these salts, present in the natural environment of the bacteria. As demonstrated in the literature data, the influence of bile salts on the surface of bacterial cells may result in changes in the metabolism and structure of the cell wall and membrane [23, 54–56]. Such changes have been observed in *Lactobacillus* bacteria, among others. However, some researchers believe that bile acids released by BSH are even more toxic toward bacterial cells than their forms conjugated with taurine or glycine [48, 53, 57–61]. Recently, a mechanism has been proposed, according to which BSH facilitates incorporation of cholesterol or bile salts into bacterial cell membrane [62]. The positive effects stemming from the capacity of lactic acid bacteria for bile salt hydrolysis are sometimes understated in the literature. It is believed that deconjugated bile salts may return to the liver and then to the intestines, where the intestinal microflora transforms them into secondary bile salts (SBS), which are considered cytotoxic [63]. Deoxycholate and lithocholate are examples of such secondary bile salts and are formed by removing the 7α-hydroxyl group from primary bile salts, cholane and chenodeoxycholate, respectively [28, 30, 64, 65]. Removal of the 7α-hydroxyl group from primary bile salts is catalyzed by an enzyme known as 7α-dehydroxylase. It is suspected that BSH along with 7α-dehydroxylase plays a significant role in the gallstone formation [53]. However, no 7α-dehydroxylase activity could be found for *Lactobacillus* strains isolated from humans or dairy products [64, 65]. This debunks the myth that lactic acid bacteria and bifidobacteria contribute to the formation of secondary bile acids and gallstones.

Another probable mechanism of cholesterol level reduction in the blood serum is associated with the capacity of numerous lactic acid bacteria to synthesize exopoly- saccharides. However, this mechanism remains among the group of hypotheses that have been poorly understood and studied [28, 29, 42, 66, 67]. It is suspected that EPS influence the absorption of cholesterol, free bile acids, or salts from the intestines through binding and removing them from the organism via the same principle as it is performed by nutritional fiber or plant polysaccharides [42, 68]. Nakajima

et al. [67] demonstrated that the level of cholesterol in the blood serum was lowest in the rats fed with milk containing EPS-producing streptococci. Similarly, the HDL cholesterol fraction ratio to its total content was highest in the rats fed with diet including these streptococci. This shows that EPS produced by *Lactococcus lactis* subsp. *cremoris* SBT 0495 had a positive impact on the metabolism of cholesterol in rats. Moreover, results of in vitro tests carried out by Pigeon et al. [42] suggested that bile acid binding by EPS could influence reduction of the cholesterol level via its usage in the synthesis of new bile acids in the place of those associated with EPS and thus removed from the system. Moreover, they formed a hypothesis that the full EPS efficiency in terms of cholesterol or bile acid removal requires the activity of BSH-type enzymes in lactic acid bacteria and bifidobacteria. However, these researchers did not verify whether this relationship is also present with regard to conjugated bile acids (e.g., glycocholic acid, taurocholic acid), as then it would be possible that the phenomenon of bile salt binding by EPS does not require activity of bile salt hydrolase and it may occur in the conditions prevailing in the intestine. Perhaps the cholesterol removal by EPS-producing bacteria is even more complex than in the case of bacteria that do not produce these substances. Moreover, it is unknown whether the cholesterol bound by EPS is biologically available to human organism, as the literature lacks information as to whether this research has been conducted in vivo.

Another mechanism, associated with production of short-chain fatty acids, has been mentioned among other possible mechanisms of cholesterol level reduction in the human organism by lactic acid bacteria [28, 29, 36, 69]. In the human organism, propionic acid penetrates to the liver and inhibits the hypercholesterolemic effect of acetate, the precursor of cholesterol and a product of fermentation activity of lactic acid bacteria. Thus far no in vivo tests have been conducted to confirm this phenomenon. St-Onge et al. [36] further point out to the fact that synthesis of acetate by lactic acid bacteria predominates synthesis of other SCFAs.

The aforementioned supposed mechanisms concern reduction of the level of cholesterol in the blood serum by lactic acid bacteria. It is presumed that considering lactic acid bacteria and bifidobacteria do not metabolize cholesterol, then it is possible that only binding (adhesion and/or assimilation) of cholesterol by cell wall or cytoplasmic membrane occurs in food products. Thus far, it has not been demonstrated that lactic acid bacteria are capable of metabolizing cholesterol, although the literature provides examples of studies on the introduction of genes encoding activity of such genes to the cells of lactic acid bacteria [70–73]. It is known that many other microorganisms produce enzymes that decompose cholesterol to other compounds, e.g., cholesterol reductase or cholesterol oxidase [70, 71, 74, 75].

Worth noting are intestinal microorganisms producing the enzyme of cholesterol reductase that transforms cholesterol into coprostanol (5β-cholestan-3β-ol). In the human organism, the anaerobic intestinal microflora transforms cholesterol primarily to 5β-coprostanol [76]. It should be noted that coprostanol is poorly absorbed in the gastrointestinal tract and it is easily eliminated from the organism [14]. *Eubacterium coprostanoligenes* is a bacteria species that includes cholesterol reductase. These bacteria could be used for production of probiotic foods with a naturally reduced cholesterol level, and such attempts have been made, yet thus far with poor effects [77]. In vivo tests on animals demonstrated that administration of *Eubacterium coprostanoligenes* has a positive influence on reduction of cholesterol in the blood serum [29, 78–80]. This is an indication that providing lactic acid bacteria cells with the capacity for cholesterol transformation into coprostanol may enable reduction of the cholesterol level already at the stage of fermented product formation.

By examining the hypocholesterolemic influence of lactic acid bacteria on the level of cholesterol in the blood serum of volunteering humans or experimental animals, it should be borne in mind that introduction of additional microflora to the intestines may significantly alter the quantitative and qualitative composition of the entire intestine ecosystem and its function. As shown in the results of the study of Hosono et al. [14], despite the fact that lactic acid bacteria cells do not possess the capacity to transform cholesterol into coprostanol, they are capable of influencing the amount at which it is excreted from the organism. This forms the effect of the influence of lactic acid bacteria on the remaining microorganisms present in the intestinal microflora.

Cholesterol binding sites by bacterial cells

As stated above, binding (adhesion or assimilation) by bacteria cells is one of the major mechanisms for the removal of cholesterol by bacteria from the environment. Hosono and Tono-Oka [24] have suggested that it is the chemical nature and structure of peptidoglycan present in bacterial cell wall that fulfill a major role in cholesterol binding. This hypothesis was confirmed by Usman and Hosono [43]. They further suggested that a portion of cholesterol could be embedded into the bacterial cell walls. The possibility for incorporating cholesterol into the cell membrane of lactic acid bacteria was demonstrated in the study of Noh et al. [19].

The phenomenon of cholesterol binding by the cell wall has been indicated by similar research conducted on the binding of aflatoxin B1 (AFB1) by lactic acid bacteria cells [81–86]. Many researchers point out to the phenomenon of AFB1 aflatoxin binding by live and dead lactic acid bacteria cultures, which do not possess the capacity to metabolize this compound [81, 83–85, 87, 88].

As it is known, Gram-positive bacteria are characterized by a thick cell wall.

The cell wall of Gram-positive bacteria comprises of peptidoglycan (murein) and its associated teichoic and/or teichuronic and lipoteichoic acids and proteins [89]. The wall thickness ranges between 15 and 50 nm, corresponding to 20–30 individual murein layers. Murein is built of saccharide chains comprising of alternately arranged N-acetylglucosamine and N-acetylmuramic acid, joined with a β-(1 → 4)-glycoside bond [89]. Apart from the saccharide chains, murein contains short peptides. The free carbonyl group of muramic acid forms the acceptor for the first peptides amino acid. Typically, l-alanine is the first amino acid. The protein portion of murein exhibits considerably greater diversity than its saccharide part, as its composition depends on the bacteria species, environmental conditions, and even the cell age. In Gram-positive bacteria, the cell wall further contains proteins active in various physiological and biochemical processes—energy transfer, electron and proton transport, cell casing synthesis, etc. [89]. Moreover, various types of polymers are associated with murein, such as teichoic acids (teichoic and lipotei- choic acid) and teichuronic acid. Considering the manner in which these acids are attached, they are sometimes referred to as secondary (after murein) polymers of bacterial cell wall. The importance of these acids has not been fully explained, although numerous assumptions have been made explaining the presence of these compounds in the bacterial cell wall. It is possible that these acids play a certain role in bacteria adhesion, biofilm formation, tolerance to environmental acidity, resistance to antibiotics, bacteriophages, or UV radiation. Lactic bacteria synthe- size teichoic and lipoteichoic acids simultaneously or lipoteichoic acids only. The characteristic feature of lipoteichoic acids is the presence (at the end of the chain) of a glycolipid anchored into the cytoplasmic membrane, and the structure of this connection depends on bacteria species. The qualitative composition of cytoplasmic membrane phospholipids depends on environmental factors, such as availability of nutrients, temperature, pH, and presence of toxic materials. The fatty acids profile changes also depending on the genus and species of bacteria and their growth phase, which has been used for microorganism grouping and classification attempts [69, 74, 90]. C16 fatty acids are the most common, while C12, C14, and C18 fatty acids are less frequently found. Methylated, hydroxylated, and branched fatty acids or those containing cyclopropane ring are common. Lactobacillic acid—a fatty acid containing cyclopropane ring—was first detected in the cytoplasmic membrane of lactic bacilli [50, 69].

Literature data indicate that the cell wall or cytoplasmic membrane can form the cholesterol binding site. In the case of the cell wall, the bond may have physical (adhesion) or chemical character (analogous to incorporation of teichoic, lipotei- choic, and teichuronic acid incorporation). In the case of a chemical bond, we deal with cholesterol assimilation, that is, its incorporation into the cell wall. Cholesterol molecules are oriented in the cell membrane in the same manner as phospholipid molecules. The polar portion of cholesterol molecule adheres to the polar portion of phospholipid. Perhaps, in cytoplasmic membranes of bacterial cells, cholesterol molecules are located in the same manner as in membranes of eukaryotic organisms. The bacterial cytoplasmic membrane contains compounds with a structure similar to steroids, which further indicates the possibility for cholesterol incorporation into the cytoplasmic membrane of bacterial cells. However, in order to be incorporated into the cytoplasmic membrane, cholesterol molecules must be transported through the cell wall. As shown in the studies of Kurdi et al. [91], Pigeon et al. [42], and Kurdi et al. [61] on bile acids, transport of such large molecules through the cell wall is possible even if it threatens the survival of the bacteria. Cholesterol binding by the cell membrane is not neutral to the bacterial cell itself. The pres- ence of such substances as cholesterol in the environment influences the ratio of saturated acids to unsaturated acids in the cytoplasmic wall, as well as the structure and properties of this membrane. Goldberg and Eschar [92] noted that addition of Tween 80 to culture medium increases the concentration of certain fatty acids with the concomitant influence on the ratio of saturated acids to unsaturated acids. The same happens when the cholesterol molecules are being bound. Dambekodi and Gilliland [23] proved that incorporation of cholesterol into the cell membrane of bifidobacteria was manifested by changes in its composition and resulted in an increase of the resistance of cells growing in the presence of cholesterol to ultrasonic lysis. In turn, Taranto et al. [50] demonstrated that bacterial cells growing in the presence of cholesterol or bile salts are more resistant to lysis than those growing in their absence, contrary to the cells growing in the absence of cholesterol. The cited authors observed that addition of cholesterol to culture broth resulted in an increase of saturated fatty acid content in lactic bacilli biomass from 44.3% to 56.5% of total acids and unsaturated acids from 1.26% to 43.5% of the total amount of fatty acids.

Furthermore, Kimoto et al. [25] reached a conclusion that the change in the distribution of fatty acids by *Lactococcus lactis* cells growing in the presence of cholesterol is an effect of its removal from the culture medium and incorporation into the cell membrane. Liong and Shah [38] examined the influence of cholesterol on the profile of fatty acids of lactic bacilli and determined that the strains growing in the medium without addition of cholesterol demonstrated a stronger percentage content of

unsaturated acids (oleic and linoleic acids) than the samples, to which cholesterol was added. According to Boggs [93] cholesterol forms hydrogen bonds with the amide group N–H of bile acids and oxygen molecules of hydroxyl groups of sac- charides in fatty acids. It is likely that the same bonds connect the cholesterol with phospholipids and glycolipids of bacterial cell membrane [50]. However, according to other literature data, no strict relationship exists between lactic acid bacteria resistance to bile salts and their capacity to bind cholesterol [46, 48].

Influence of selected factors on cholesterol removal by Lactococcus cells

It can be stated that the phenomenon of cholesterol binding and removal by bacterial cells is complex. It can be concluded that the contribution of the phenomenon of cholesterol molecule assimilation or adhesion by lactic acid bacteria cells depends on a wide range of factors, which are not always possible to reproduce or replicate in subsequent experiments. Perhaps this depends on the different chemical structure of the cell wall, particularly peptidoglycan, as well as lipid profile of phospholipids of the cytoplasmic membrane in bacterial cells.

Influence of cholesterol concentration on cholesterol removal by Lactococcus

The capacity of lactic streptococci to reduce the cholesterol level under in vitro conditions was also tested by Hosono and Tono-Oka [24] and Kimoto et al. [25].

The cited researchers carried out cultures at 37°C for 24 h. In the study of Hosono and Tono-Oka [24], the percentage of cholesterol bound by *Lactococcus lactis* subsp. *lactis* 12007 and 12546 strains was 25.1 and 30.3%, respectively. Four strains of *Lactococcus lactis* subsp. *cremoris* bound from 14.2 to 20.9% of cholesterol and two strains of *Lactococcus lactis* subsp. *lactis* biovar. *diacetilactis*—29.7 and 33.9%, respectively. The capacity to remove cholesterol from culture broth was demon- strated also in the case of *Leuconostoc mesenteroides* subsp. *cremoris*, and it amounted to between 11.4% and 14.9%, depending on the strain. In turn, in the experiments of Kimoto et al. [25], bacterial cells from *Lactococcus lactis* subsp. *lactis* and *Lactococcus lactis* subsp. *lactis* biovar. *diacetilactis* strains removed 53.9–86.7% and 31.0–97.3% of cholesterol, respectively, from GM17-THIO broth, containing addi- tion of 0.2% sodium taurocholate and 0.070 g cholesterol per 1 dm3 of medium. Moreover, Ziarno [4] examined the capacity of isolates from the genus *Lactococcus* originating from fermented dairy products to remove cholesterol depending on the concentration of cholesterol in culture broth (in a range from slightly above 0 g/dm3 to close to 2 g/dm3). Considering it is known that lactic acid bacteria do not metabolize cholesterol, its loss from post-culture liquid can be seen as the amount of cholesterol removed and bound by bacterial cells. Ziarno [4] demonstrated that the amount of cholesterol removed by bacterial cells is determined by the preliminary concentration of this substance in the culture medium. In general, the more cholesterol was introduced to the culture broth, the more of it was removed by bacterial cells. However, the above statement is true only for low cholesterol concentrations in culture broth. With higher concentration of cholesterol in the culture broth, amounting to over 1–1.5 g/dm3, its removal by bacterial cells was still observed; however the dynamics of this removal was far less pronounced than in broths with lower cholesterol concentration. The earlier research indicates a different capacity of lactic acid bacteria cultures to remove cholesterol from culture media [6, 94, 95]. The differences were observed between individually tested cultures and between individual replications for the same culture. This is a confirmation of observations of other researchers [5, 6, 10, 16, 17, 21–26, 38, 41, 46, 48, 96]. A significant effect on the diversity of the results obtained not only within the strains but also repeti- tions seems to be also held by the fact that the mechanism of cholesterol binding by bacterial cells can occur via adhesion of cholesterol molecules through the cell wall or by embedding it into the cell wall or membrane [9, 19, 23–25, 43]. It appears to be obvious that cholesterol adhesion does not produce strong binding, and this substance is very easily washed back to the culture broth. In turn, embedding cholesterol into the cell wall or cytoplasmic membrane is more durable. This may explain the observed considerable dispersion of results and the lack of experiment reproducibility.

Influence of culture temperature on cholesterol removal by Lactococcus

Usman and Hosono [43] demonstrated that *Lactococcus lactis* subsp. *lactis* biovar. *diacetilactis* bacteria are capable of binding and removing cholesterol already after culture is started, independent of its temperature in the range from 10 to 70°C. After addition of salts of such metals as Mg2+, Na+, Ca2+, or K+, the cholesterol binding was inhibited. The bacteria bound the highest amount of cholesterol when the pH value was about 7.0. The applied culture temperature range indicates that dead bacterial cells are also capable of binding cholesterol, which comprised the subject of further tests of this study. In turn, Noh et al. [19] demonstrated that lactic bacilli bind cholesterol in a culture with constant pH of 6.0, as well as during growth without pH value control. Ziarno [4] examined the capacity of *Lactococcus* cells isolated from industrial dairy starters to remove cholesterol in M17 culture broth with application of several temperature variants of culture (4, 25, and 30°C). The temperature of 4°C aimed at stimulating refrigeration conditions and ensuring inhibition of bioactivity of bacterial cells [97, 98]. The temperature of 30°C was utilized as the optimum conditions for the development of mesophilic bacteria. In turn, the temperature of 25°C was used to simulate room temperature conditions. It was proven that bacterial cells from all tested lactic acid bacteria cultures reduced the level of cholesterol in culture medium in the applied experimental conditions. As it could be expected, the degree at which cholesterol is removed depended on the applied temperature of lactic acid bacteria incubation. The initial cholesterol concentration in culture broths was on average 0.606 g/dm3. When the cultures were kept at the temperature of 4°C, mesophilic cultures of *Lactococcus* removed low amounts of cholesterol (from 0.005 to 0.021 g/dm3) [4]. When the cultures were carried out at 25°C, the discussed cultures bound from 0.065 to 0.085 g/dm3. In turn, at the temperature of 30°C, which is optimum for the development of mesophilic cultures, the obtained values of removed cholesterol ranged from 0.068 to 0.104 g/dm3 [4].

Influence of Lactococcus live cell biomass concentration on cholesterol removal

Usman and Hosono [43] determined that cholesterol binding was significantly dependent on the amount of bacterial cell biomass and it increased proportionate to the increase of the cell count. Furthermore, Liong and Shah [38] observed that the amount of cells has a significant impact on the differences in the amount of cholesterol bound by lactic acid bacteria, whereas the growth dynamics for individual strains determines the amount of cell biomass and differences in experimental results. Ziarno [4] verified the manner in which the concentration level of live cell biomass originating from monocultures and multi-species cultures of *Lactococcus* influences the capacity of cultures to remove cholesterol from M17 culture broth. As expected, the highest amount of cholesterol was removed in the cultures containing 10-fold concentrated *Lactococcus* biomass. At this cell biomass concentration, the studied *Lactococcus* cultures removed on average between 0.113 and 0.129 g/dm3 of cholesterol from its initial content of 0.611 g in 1 dm3 of M17 broth. In turn, bacterial biomasses with a 10-fold lower concentration (1×) produced approximately 1.3–1.6- fold reduction of the amount of cholesterol removed. From a culture broth containing a 10-fold diluted bacterial cell biomass culture, from 0.054 g/dm3 to 0.066 g/dm3 of cholesterol was removed after culture maintained for 20h, thus 1.3–1.5 times less than in the case of 1× concentrated biomasses [4]. It is worthy of emphasis that in multi-species commercial mesophilic starter cultures, used in the dairy industry, e.g., cheese and cream production, similar capacities to remove cholesterol were observed as in lactic acid bacteria monocultures. However, it should be expected that with 10-fold decrease of cell biomass concentration, the amount of cholesterol removed from culture broth will be decreased proportionately (by 10-fold). However, minor differences were observed in the amount of cholesterol removed by biomasses with selected live cell concentration levels. This can be explained with two phenomena. Firstly, the applied cultures were live and biologically active. During the experiments, bacteria propagated, significantly altering the amount of biomass capable of binding cholesterol. Microbial analyses demonstrated that the strongest increase of *Lactococcus* population was observed in the culture with the lowest initial biomass concentration (10-fold diluted). Bacteria propagation was poorest in the cultures with the highest initial concentration of biomass (10×). After completion of experiments, in the cultures containing 10-fold diluted biomass of the tested mesophilic cultures, the live cell bacteria count was determined at 7–8 log CFU/cm3. In cultures with 10-fold concentrated biomass, an average of 6–7 log CFU/cm3 was determined [4]. The second explanation for the minor differences in the amount of cholesterol removed by *Lactococcus* biomass with the used live cell concentration levels is the concomitant adhesion and assimilation of cholesterol molecules. Most likely, with poor growth of bacterial cells, the phenomenon of cholesterol removal through its adhesion by the cell wall is predominant. And as it could be expected, this type of cholesterol binding is not durable and cholesterol is easily released. In turn, the high biological activity of bacterial cells may favor permanent embedding of cholesterol into the wall or cytoplasmic membrane of bacteria cells, which likely occurred in the experiments of this stage of research, in cultures with the lowest initial biomass concentration (diluted 10-fold), in which the greatest increase in population was observed.

The obtained study results may find implications for the explanation of hypo- cholesterolemic influence of products containing lactic acid bacteria. A considerable amount of literature data is available on the subject, but these are often contradictory [5–7, 10–13]. Based on the results of this study, a hypothesis can be formed that in this case the count of live bacteria in the product is important. In order for bacterial cells to assimilate cholesterol molecules, their high biological activity is required, as demonstrated by Hosono and Tono-Oka [24] for *Lactococcus lactis* subsp. *lactis* biovar. *diacetilactis* R-43 strain; the course of this phenomenon is most intensive in the logarithmic growth phase. The physical binding of cholesterol by the cell wall does not require cell activity, only a suitably long contact time between the cells and cholesterol molecules. The same team of researchers noted that not only live but also dead cells of the tested strain were capable of binding cholesterol.

Influence of Lactococcus dead cell biomass concentration on cholesterol removal

The sparse literature data on cholesterol removal by inactivated cells prove that lactic acid bacteria monocultures are capable of removing cholesterol from culture media even after their thermal death [21, 24, 25, 43, 95]. The amount of cholesterol removed by inactivated cells is considerably lower than by biologically active cells, which likely stems from the fact that in the case of dead cells cholesterol may not be built into the cell wall or cytoplasmic membrane, but it only undergoes adhesion by the cells. Furthermore, Ziarno [4] demonstrated that biomass of dead (thermally inactivated) cells of *Lactococcus*, originating from industrial monocultures and multi-species cultures, influences cholesterol uptake from the M17 culture broth.

The highest amount of cholesterol was removed in the cultures containing 10x con- centrated biomass of dead bacterial cells, from 0.074 to 0.083 g/dm3 of cholesterol from the M17 broth. Bacterial biomasses with 1x concentration removed 1.4–1.9- fold less cholesterol. Tenfold diluted biomass of dead bacterial cells bound from 0.021 to 0.029 g/dm3 of cholesterol, thus 1.7–2.3-fold less than 1× concentrated biomasses. The fact that cholesterol removal occurs even when the bacterial cells are dead confirms that the physical binding of cholesterol molecules by the cell wall (adhesion) is one of the mechanisms of cholesterol removal by *Lactococcus* cells.

Survival of Lactococcus cells in the human gastrointestinal tract

Literature contains studies confirming the capacity of lactic bacilli to survive under in vivo conditions in the human alimentary tract [28, 30, 35, 52, 99, 100]. The factors with a significant impact on lactic acid bacteria survivability in the alimentary tract include low gastric pH value, intestine peristalsis, presence of bile acids in the pancreatic fluid and various digestive enzymes present in the individual sections of the alimentary tract, presence of nutrients, as well as bacteria passage time through the alimentary tract and their initial count [35, 50, 101–103]. The mentioned factors result in a decrease of lactic acid bacteria survival rate, but at the same time they may constitute a criterion for the selection of probiotic strains [28, 104].

In order to determine the survival rate of lactic acid bacteria, scientists first deter- mine their resistance to low pH present in certain sections of the gastrointestinal tract. Gastric fluid comprises of the secretion of foveolar cells secreting mucus, chief cells secreting digestive enzymes (pepsin), and parietal cells secreting hydrochloric acid.

The pH of gastric fluid is between 1.5 and 3.0. Secretion of gastric fluid is inhibited when the pH drops below 2.0. The temperature inside the stomach is over 37°C, and the alimentary content, depending on the individual physiological and emotional circum- stances, remains in the stomach for average 1–3 h [4]. Results of in vitro tests concern survival rate of different lactic acid bacteria strains under conditions imitating low pH of the gastric fluid [18, 35, 41, 105–109]. Strains traditionally used to manufacture dairy products have also been commonly found to survive the conditions of gastric fluid [101]. Also the study of Lankaputhra and Shah [107] indicated that numerous lactic bacilli strains survived perfectly the conditions simulating the pH of gastric fluid.

Another subject of the study is the capacity of lactic acid bacteria to survive during transport through subsequent sections of the alimentary canal. Here, a particular significance is exhibited by the section of the small intestine [102, 110, 111]. Literature data show that bile salts comprise a serious obstacle for lactic acid bacteria, as they contain toxic bile acids [19, 106, 107, 112]. Ziarno and Bartosz [113] provided evidence for the influence of cell biomass on the survival of lactic bacilli in model intestinal fluid. Cholesterol influences the composition and functioning of the bacterial cell wall and membrane, thus producing change in the relationship with the surrounding environment, such as resistance to bile acids, pH, or tempera- ture [46, 48, 50]. Cholesterol uptake by bacterial cells is not neutral to them and results in a change of, among others, the profile of fatty acids of the cell membrane [23, 25, 38, 50]. Cell responds to stress conditions of the environment with a change of the composition of the cell membrane, and it may result in an increase of the resistance of the cell to stress factors [50]. Doubtlessly, this is significant for the survival of lactic acid bacteria in various environments they inhabit, such as the alimentary tract or food products.

The strains which are not probiotics exhibit lower survival rate of their cells in model gastric fluid as compared with probiotic strains [41, 108]. Ziarno and Margol [109] examined the capacity of bacteria from several mesophilic starter cultures to survive in a simulated gastric fluid. Also in their study, industrial starter cultures containing bacteria from the genus *Lactococcus* were used, which, after propagation, were kept in a simulated gastric fluid with pH of 2.4 for 3 h at 37°C. The study demonstrated that the present streptococci were not resistant to the environment of a simulated gastric fluid [109]. On the other hand, intestinal fluid has a more complex enzymatic and chemical composition than broths used by other researchers, but its influence on lactic acid bacteria cells is typically referred to probiotic strains of thermophilic lactic acid bacteria [38–40, 108, 110, 111]. Ziarno [4] tested the viability of *Lactococcus* in model conditions of the alimentary tract in the presence of cholesterol, separately for the simulated gastric fluid and simulated intestinal fluid. *Lactococcus* isolated from industrial starter cultures were used for the experiments. No influence of addition of cholesterol on the viability of *Lactococcus* cultures in a simulated gastric fluid could be demonstrated, although reduction of live cells in the range from 1 to 3 log CFU/cm3 was observed. *Lactococcus* cells exhibited low tolerance also to the conditions of simulated intestinal fluid, considerably lower than the simulated gastric fluid. From the initial cell population of average 6–7 log CFU/ cm3, only 2–3 log CFU/cm3 remained after 6 h of experiment, with few exceptions surviving at the level of 6 log CFU/cm3, independent of the addition of cholesterol. The lack of influence of cholesterol in simulated intestinal fluid on the survival rate of lactic acid bacteria cells was also demonstrated in earlier research [114, 115]. The study conducted by Ziarno [114, 115] utilized bacteria cultures isolated from commercial pharmaceutical preparations and commercially available dairy products or dairy starter monocultures. Cells of lactobacilli tolerated conditions of simulated intestinal fluid better than bifidobacteria cells and *Lactococcus lactis* cells.

The good tolerance of bacterial cells to the conditions of simulated digestive fluids can be explained by the occurrence of these bacteria in the alimentary tract of humans and animals. Numerous factors determine lactic acid bacteria viability, including pH, temperature, oxygenation, and presence of toxic substances toward bacterial cells [101–103]. Bacteria not forming the natural intestinal microflora do not possess the natural resistance to the conditions of the intestinal fluid [106].

Viability of bacterial cells determines the level of cholesterol removal. It seems obvi- ous that the count of live and dead bacterial cells holds significance for the removal of cholesterol under the conditions of a human alimentary tract. Thus, a hypothesis can be formed that the factors determining survival rate of bacterial cells further influence the cholesterol removal level by lactic acid bacteria and bifidobacteria cells. Such relationships may further impede interpretation of the results of experiments realized under in vitro or in vivo conditions and may prevent interpolation of results obtained in vitro onto the conditions of human or animal organisms.

Cholesterol uptake and release by Lactococcus in the simulated human gastrointestinal tract

Cholesterol uptake by Lactococcus under conditions of simulated gastric fluid

Cholesterol uptake by *Lactococcus* cells in simulated gastric fluid depends on the amount of biomass [4]. Ziarno [4] carried out in vitro experiments with the use of industrial starter cultures of mesophilic lactic bacteria, including *Lactococcus*. The cultures were grown for 5 h at 37°C in a simulated gastric fluid containing addition of 0.511 g/dm3 of cholesterol. The study demonstrated that higher amount of cholesterol was bound by *Lactococcus* cells contained in mixed cultures than cells from *Lactococcus lactis* monocultures. Bacterial cells present in the mixed cultures removed cholesterol in the range from 0.012 to 0.020 g/dm3. In turn, bacterial cells from *Lactococcus lactis* cultures reduced cholesterol concentration in the simulated gastric fluid by an average 0.005 g/dm3. 10× concentrated bacterial biomasses removed 1.4–2.3 times more cholesterol than bacterial cells with 1× cell concentration. In turn, biomasses with 0.1× bacterial cell concentration bound 2.2–4.6 less cholesterol than bacterial cultures with 1× cell concentration. This means that the conditions prevailing in the stomach may favor removal of cholesterol by bacterial cells independent of their viability. However, it remains unknown whether bacterial cells release the bound cholesterol after entering the gastrointestinal tract and whether it may penetrate to the blood. Similar studies concerning aflatoxin B bound by the cell wall of lactic bacilli suggest that such assimilation by the cell wall may be robust [81–84]. This may indicate that cholesterol binding is also robust.

Release of cholesterol bound by Lactococcus in the conditions of simulated gastric fluid

The study of Ziarno [4] indicates that the binding of a portion of cholesterol by lactic acid bacteria cells is robust enough so that it is not released in the conditions of gastric fluid. The study was carried out using isolates of *Lactococcus* originating from industrial monocultures and mixed cultures. Bacterial cells present in the tested cultures released 51–84% of the removed and bound cholesterol independent of bacterial cells' viability. The biomass of dead cells released lower amount of cholesterol than the biomass with viable cells, but it also bound and removed lower amount of cholesterol from the culture medium earlier. Biomass of live *Lactococcus lactis* cells removed an average of 0.063 g cholesterol/dm3, whereas biomass of dead cells removed average of 0.033 g/dm3 [4].

Similar tendencies are observed in the case of studies conducted on aflatoxin B1 binding by lactic acid bacteria [81, 84]. El-Nezami et al. [81] observed that aflatoxin B1 uptake from culture medium by selected lactic acid bacteria cultures depended on their population and culture temperature. The same was demonstrated by Lee et al. [84]. Identical relationships were observed in the present study with regard to binding and release of cholesterol by lactic acid bacteria cells. Moreover, Lee et al. [84] concluded that thermal killing of bacteria resulted in a change of the surface of bacteria cells and uncovering of additional binding sites for aflatoxin B1.

Cholesterol removal by Lactococcus in the conditions of simulated intestinal fluid

As stated by Ziarno and Bartosz [113], cholesterol removal by lactic acid bacteria in intestinal fluid is less pronounced than in culture broth. This is further confirmed by the experiments of Ziarno [4] conducted under in vitro conditions with *Lactococcus* isolates originating from industrial starter cultures. The mentioned cultures were grown at 37°C for 6 h in a simulated intestinal fluid with addition of cholesterol. The tested *Lactococcus* cultures resulted in a reduction of cholesterol from the initial content of 0.543 g/dm3 to the level between 0.011 and 0.087 g/dm3. In the majority of the tested cultures, the influence of biomass concentration on the degree of cholesterol removal was statistically significant; however, 10-fold concen- trated biomass did not remove 10 times more cholesterol than onefold concentrated biomass. Therefore, Ziarno [4] demonstrated that in not all of the tested *Lactococcus* cultures the degree of biomass concentration had a significant influence on the amount of cholesterol removed. This can be explained by the activity of enzymes such as BSH, which caused bile hydrolysis and coprecipitation of cholesterol with released bile acids, independent of the amount of cells in the culture.

The chemical composition of simulated intestinal fluid seems to be of significance for the obtained results [18]. This indicates additional methodological factors influencing the results obtained in laboratory experiments conducted under in vitro conditions. In order to prepare simulated intestinal fluid, cattle bile was also used containing conjugated and deconjugated bile salts; therefore bile salt hydrolase activity (produced by the majority of intestinal lactic acid bacteria strains) was not necessary for cholesterol precipitation with free bile acids to occur [30, 35, 39, 40, 45]. Active BSH enzyme results in hydrolysis of bile salts, whereas cholesterol molecules may coprecipitate with the released bile acids [16, 17, 21, 30, 35, 39, 40, 45, 49, 50]. Such phenomenon has been observed in numerous lactic acid bacteria species, but not in *Lactococcus* thus far [49, 53]. It is known that coprecipitates of cholesterol with bile acids are formed at a low pH below 5.5 [17, 18, 21, 39, 49, 52]. However, with a renewed increase of pH to over 5.5, such coprecipitates were rapidly dissolved [15, 18, 28, 39]. Bile secreted from the liver is introduced to the duodenum, where it neutralizes the acidic food pulp that leaves the stomach and then the pH in the small intestine has a value of over 6.0. Under these conditions, the coprecipitates of bile acids and cholesterol are dissolved. Thus, the hypocholes- terolemic effect caused by cholesterol coprecipitation with deconjugated bile acids is probably impossible to occur under in vivo conditions.

Release of cholesterol bound by Lactococcus in the conditions of simulated intestinal fluid

Ziarno [4] studied whether the cholesterol previously bound by *Lactococcus* cells is released under the conditions of simulated intestinal fluid. It was determined that certain isolates of *Lactococcus lactis* released up to 60–90% of cholesterol, which was earlier bound by these cells. Lower amount of cholesterol under conditions of simulated intestinal fluid is released by *Lactococcus lactis* cells (average of 45%), meaning that in these bacteria cultures cholesterol was bound with sufficient force by the cell wall so that it was not released under the conditions of simulated intestinal fluid.

This may confirm the hypothesis of Lee et al. [84] on structural changes in the wall of dead bacterial cells.

Conclusions

One important conclusion should be drawn from the research results presented above, namely, that lactic acid bacteria may cause a different hypocholesterolemic effect in the human digestive system. They may exhibit a clear capacity for permanent binding and removal of cholesterol or to not bind it at all. It is also possible that they may cause such change of the intestinal microflora. Hosono et al. [14] formed a hypothesis that lactic acid bacteria may influence the amount of cholesterol eliminated from the organism despite the fact that they do not have the capacity to trans- form it into coprostanol. This is an effect of the influence of lactic acid bacteria on other microorganisms present in the intestinal microflora, including microorganisms capable of transforming cholesterol into coprostanol. This is particularly possible in the case of probiotic strains of lactic acid bacteria and bifidobacteria, which are distinguished due to their capacity to produce low-molecular antimicrobial substances. Based on the results of experiments conducted by Ziarno [4], it can be stated that the phenomenon of cholesterol binding depends on such a wide array of factors influencing the cell wall and cytoplasmic membrane of bacteria that it may not be impossible to predict the hypocholesterolemic effect unambiguously.

It can be concluded that lactic acid bacteria are capable of binding cholesterol molecules present in their environment. Cholesterol can be subject to adhesion by the cell wall or assimilation via the cytoplasmic membrane or cell wall of lactic acid bacteria, including *Lactococcus*. However, the degree and force of this bond depend on numerous environmental factors. A change of even one of these parameters results in the hypocholesterolemic effect which is no longer reproducible in the experiments. It is likely that this is the manner in which the results and the discrepancies found between in vitro and perhaps also in vivo tests on human volunteers and experimental animals should be interpreted and explained.

Acknowledgements

This work was supported by a grant from Warsaw University of Life Sciences (WULS-SGGW).

Conflict of interest

The author has declared that she does not have any conflict of interest for publishing this research.

Author details

Małgorzata Ziarno

Division of Milk Biotechnology, Department of Biotechnology, Microbiology and Food Evaluation, Faculty of Food Sciences, Warsaw University of Life Sciences— SGGW (WULS-SGGW), Warsaw, Poland

*Address all correspondence to: malgorzata_ziarno@sggw.pl

References

[1] Ziarno M, Godlewska A. Significance and application of *Lactococcus* species in dairy industry. Medycyna Weterynaryjna. 2008;**64**:35-39

[2] Ziarno M, Zaręba D, Piskorz J. Fortifying buttermilk with calcium, magnesium, and whey proteins. Żywność Nauka Technologia Jakość. 2009;**2**:14-27

[3] Ziarno M. Characteristics of commercial dairy starter cultures. Medycyna Weterynaryjna. 2007;**63**:909-913

[4] Ziarno M. Studies on the binding and removal of cholesterol by bacterial cells of lactic fermentation and bifidobacteria in ex vivo conditions [habilitation dissertation]. Warsaw: Warsaw University of Life Sciences—SGGW; 2008

[5] Gilliland SE, Nelson CR, Maxwell C. Assimilation of cholesterol by *Lactobacillus acidophilus*. Applied and Environmental Microbiology. 1985;**49**:377-381

[6] Ziarno M, Zając A. Fermented dairy products and cholesterol levels. Przemysł Spożywczy. 2007;**3**:44-46

[7] Harrison VC, Peat G. Serum cholesterol and bowel flora in the newborn. American Journal of Clinical Nutrition. 1975;**28**:1351-1355

[8] Anderson JW, Gilliland E. Effect of fermented milk (yogurt) containing *Lactobacillus acidophilus* L1 on serum cholesterol in hypercholesterolemic humans. Journal of the American College of Nutrition. 1999;**18**:43-50

[9] Tabuchi M, Tamura A, Yamada N, Ishida T, Hosoda M, Hosono A. Hypocholesterolemic effects of viable and heat-sterilized cells of *Lactobacillus* GG in rats fed a high-cholesterol diet. Milchwissenschaft. 2004;**58**:249-253

[10] Lin SY, Ayres JW, Winkler W, Sandine WE. *Lactobacillus* effects on cholesterol: *In vitro* and *in vivo* results. Journal of Dairy Science. 1989;**72**:2885-2899. DOI: 10.3168/jds. S0022-0302(89)79439-X

[11] Jaspers DA, Massey LK, Luedecke LO. Effect of consuming yogurt prepared with three culture strains on human serum lipoproteins. Journal of Food Science. 1984;**49**: 1178-1181. DOI: 10.1111/j.1365- 2621.1984.tb10422.x

[12] Mcnamara DJ, Lowell AE, Sabb JE. Effect of yoghurt intake on plasma-lipid and lipoprotein levels in normolipidemic males. Atherosclerosis. 1989;**79**:167-171. DOI: 10.1016/0021-9150(89)90121-4

[13] Thompson LU, Jenkins DJA, Amer MAV, Reichert R, Jenkins A, Kamulsky J. The effect of fermented and unfermented milks on serum cholesterol. American Journal of Clinical Nutrition. 1982;**36**:1106-1111. DOI: 10.1093/ajcn/36.6.1106

[14] Hosono A, Otani H, Yasui H, Watanuki M. Impact of fermented milk on human health: Cholesterol- lowering and immunomodulatory properties of fermented milk. Animal Science Journal. 2002;**73**:241-256. DOI: 10.1046/j.1344-3941.2002.00034.x

[15] Brashears MM, Gilliland SE, Buck LM. Bile salt deconjugation and cholesterol removal from media by *Lactobacillus casei*. Journal of Dairy Science. 1998;**81**:2103-2110. DOI: 10.3168/jds.S0022-0302(98)75785-6

[16] Buck LM, Gilliland SE. Comparisons of freshly isolated strains of *Lactobacillus acidophilus* of human intestinal origin for ability to assimilate cholesterol during growth. Journal of Dairy Science. 1994;**77**:2925-2933. DOI: 10.3168/jds.S0022-0302(94)77233-7

[17] Grill JP, Cayuela C, Antoine JM, Schneider F. Effects of *Lactobacillus* amylovorus and *Bifidobacterium* breve on cholesterol. Letters in Applied Microbiology. 2000;**31**:154-156. DOI: 10.1046/j.1365-2672.2000.00792.x

[18] Lin MY, Chen TW. Reduction of cholesterol by *Lactobacillus acidophilus* in culture broth. Journal of Food and Drug Analysis. 2000;**8**:97-102

[19] Noh DO, Kim SH, Gilliland SE. Incorporation of cholesterol into the cellular membrane of *Lactobacillus acidophilus* ATCC 43121. Journal of Dairy Science. 1997;**80**:3107-3113. DOI: 10.3168/jds.S0022-0302(97)76281-7

[20] Rašić JL, Vujičić IF, Škrinjar M, Vulić M. Assimilation of cholesterol by some cultures of lactic acid bacteria and bifidobacteria. Biotechnology Letters. 1992;**14**:39-44. DOI: 10.1007/ BF01030911

[21] Tahri K, Grill JP, Schneider F. Bifidobacteria strain behavior toward cholesterol: Coprecipitation with bile salts and assimilation. Current Microbiology. 1996;**33**:187-193

[22] Walker DR, Gilliland SE. Relationships among bile tolerance, bile salt deconjugation, and assimilation of cholesterol by *Lactobacillus acidophilus*. Journal of Dairy Science. 1993;**76**:956-961. DOI: 10.3168/jds.S0022-0302(93)77422-6

[23] Dambekodi PC, Gilliland SE. Incorporation of cholesterol into the cellular membrane of *Bifidobacterium* longum. Journal of Dairy Science. 1998;**81**:1818-1824. DOI: 10.3168/jds. S0022-0302(98)75751-0

[24] Hosono A, Tono-Oka T. Binding of cholesterol with lactic acid bacterial cells. Milchwissenschaft. 1995;**50**:556-559

[25] Kimoto H, Ohmono S, Okamoto T. Cholesterol removal from media by lactococci. Journal of Dairy Science. 2002;**85**:3182-3188. DOI: 10.3168/jds.S0022-0302(02)74406-8

[26] Taranto MP, Gonzales De Llano D, Rodriguez A, De Ruiz Holgado PA, De Valdez FG. Bile tolerance and cholesterol reduction by *Enterococcus faecium*, a candidate microorganism for the use as a dietary adjunct in milk products. Milchwissenschaft. 1996;**51**:383-385

[27] Lourens-Hattingh A, Viljoen BC. Yogurt as probiotic carrier food. International Dairy Journal. 2001;**11**:1-17. DOI: 10.1016/ S0958-6946(01)00036-X

[28] Ziarno M. Mechanisms of cholesterol lowering by bacteria of the genus *Lactobacillus*. Żywienie Człowieka i Metabolizm. 2004;**2**:10-18

[29] Ziarno M. Pro-health properties of lactic bacteria. Przegląd Mleczarski. 2004;**11**:4-10

[30] Ziarno M. The significance of bile salts hydrolase activity of bacteria from *Lactobacillus* genus. Postępy Mikrobiologii. 2004;**43**:285-296

[31] De Rodas B, Gilliland SE, Maxwell CV. Hypocholesterolemic action of *Lactobacillus acidophilus* ATCC 43121 and calcium in swine with hypercholesterolemia induced by diet. Journal of Dairy Science. 1996;**79**:2121-2128. DOI: 10.3168/jds. S0022-0302(96)76586-4

[32] Grunewald KK, Mitchell K. Serum cholesterol levels in mice fed fermented and unfermented acidophilus milk. Journal of Food Protection. 1983;**46**:315-318

[33] Grunewald KK. Serum cholesterol level in rats fed skim milk fermented by *Lactobacillus acidophilus*. Journal of Food Science. 1982;**47**:2078-2079. DOI: 10.1111/ j.1365-2621.1982. tb12955.x

[34] Kiesling G, Schneider J, Jahreis G. Long-term consumption of fermented dairy products over 6 months increases HDL cholesterol. European Journal of Clinical Nutrition. 2002;**56**:843-849. DOI: 10.1038/ sj.ejcn.1601399

[35] Lim HJ, Kim SY, Lee WK. Isolation of cholesterol-lowering lactic acid bacteria from human intestine for probiotic use. Journal of Veterinary Science. 2004;**5**:391-395. DOI: 10.4142/ jvs.2004.5.4.391

[36] St-Onge MP, Farnworth ER, Jones PJH. Consumption of fermented and nonfermented dairy products: Effects on cholesterol concentrations and metabolism. American Journal of Clinical Nutrition. 2000;**71**:674-681. DOI: 10.1093/ajcn/71.3.674

[37] Jones ML, Chen H, Ouyang W, Metz T, Prakash S. Microencapsulated genetically engineered *Lactobacillus plantarum* 80 (pCBH1) for bile acid deconjugation and its implication in lowering cholesterol. Journal of Biomedicine and Biotechnology. 2004;**1**:61-69. DOI: 10.1155/ S1110724304307011

[38] Liong MT, Shah NP. Acid and bile tolerance and cholesterol removal ability of lactobacilli strains. Journal of Dairy Science. 2005;**88**:55-66. DOI: 10.3168/ jds. S0022-0302(05)72662-X

[39] Liong MT, Shah NP. Bile salt deconjugation ability, bile salt hydrolase activity and cholesterol co-precipitation ability of lactobacilli strains. International Dairy Journal. 2005;**15**:391-398. DOI: 10.1016/j. idairyj.2004.08.007

[40] Liong MT, Shah NP. Bile salt deconjugation and BSH activity of five bifidobacterial strains and their cholesterol co-precipitating properties. Food Research International. 2005;**38**:135-142

[41] Pereira DIA, Gibson GR. Cholesterol assimilation by lactic acid bacteria and bifidobacteria isolated from the human gut. Applied and Environmental Microbiology. 2002;**68**:4689-4693. DOI: 10.1128/ aem.68.9.4689-4693.2002

[42] Pigeon RM, Cuesta EP, Gilliland SE. Binding of free bile acids by cells of yogurt starter culture bacteria. Journal of Dairy Science. 2002;**85**:2705-2710. DOI: 10.3168/ jds. S0022-0302(02)74357-9

[43] Usman B, Hosono A. Binding of cholesterol to the cells and peptidoglycan of *Lactobacillus* gasseri. Milchwissenschaft. 1999;**54**:495-498

[44] Ziarno M, Sękul E, Makowska M. The assimilation of cholesterol by starter cultures of mesophilic lactococci. Biotechnologia. 2006;**2**:234-246

[45] Ziarno M. The significance of bile salts hydrolase activity of bacteria of *Bifidobacterium* genus. Biotechnologia. 2005;**2**:183-195

[46] Gilliland SE, Walker DK. Factors to consider when selecting a culture of *Lactobacillus acidophilus* as dietary adjunct to produce a hypocholesterolemic effect in humans. Journal of Dairy Science. 1989;**73**:905-911. DOI: 10.3168/jds. S0022-0302(90)78747-4

[47] Aloglu H, Öner Z. Assimilation of cholesterol in broth, cream, and butter by probiotic bacteria. European Journal of Lipid Science and Technology. 2006;**108**:709-713. DOI: 10.1002/ ejlt.200600137

[48] Gopal A, Shah NP, Rogiński H. Bile tolerance, taurocholate deconjugation and cholesterol removal by *Lactobacillus acidophilus* and *Bifidobacterium* spp. Milchwissenschaft. 1996;**51**:619-623

[49] Klaver FAM, Van Der Meer R. The assumed assimilation of cholesterol by lactobacilli and *Bifidobacterium* bifidum is due to their bile salt-deconjugating activity. Applied and Environmental Microbiology. 1993;**59**:1120-1124

[50] Taranto MP, Murga MLF, Lorca G, Valdez GF. Bile salts and cholesterol induce changes in the lipid membrane of *Lactobacillus reuteri*. Journal of Applied Microbiology. 2003;**95**:86-91. DOI: 10.1046/j.1365-2672.2003.01962.x

[51] Gilliland SE, Speck ML. Deconjugation of bile acids by intestinal lactobacilli. Applied and Environmental Microbiology. 1977;**33**:15-18

[52] Taranto MP, Sesma F, De Ruiz Holgado AP, De Valdez GF. Bile salts hydrolase plays a key role on cholesterol removal by *Lactobacillus reuteri*. Biotechnology Letters. 1997;**19**:845-847. DOI: 10.1023/A:1018373217429

[53] Tanaka H, Doesburg K, Iwasaki T, Mierau I. Screening of lactic acid bacteria for bile salt hydrolase activity. Journal of Dairy Science. 1999;**82**:2530-2535. DOI: 10.3168/jds. S0022-0302(99)75506-2

[54] Gomez Zavaglia A, Kociubinski G, Perez P, Disalvo E, De Antoni G. Effect of bile on the lipid composition and surface properties of bifidobacteria. Journal of Applied Microbiology. 2002;**93**:794-799. DOI: 10.1046/j.1365-2672.2002.01747.x

[55] Perez PF, Minnaard Y, Disalvo EA, De Antoni GL. Surface properties of bifidobacterial strains of human origin. Applied and Environmental Microbiology. 1998;**64**:21-26

[56] Perrin S, Grill JP, Schneider F. Effects of fructooligosaccharides and their monomeric components on bile salt resistance in three species of bifidobacteria. Journal of Applied Microbiology. 2000;**88**:968-974. DOI: 10.1046/j.1365-2672.2000.01070.x

[57] Clark PA, Cotton LN, Martin JH. Selection of bifidobacteria for use as dietary adjuncts in cultured dairy foods: II. Tolerance to simulated pH of humans stomachs. Cultured Dairy Products Journal. 1993;**9**:11-14

[58] Clark PA, Martin JH. Selection of bifidobacteria for use as dietary adjuncts in cultured dairy foods: III. Tolerance to simulated bile concentrations of human small intestines. Cultured Dairy Products Journal. 1994;**29**:18-21

[59] Gunn JS. Mechanisms of bacterial resistance and response to bile. Microbes and Infection. 2000;**2**:907-913. DOI: 10.1016/S1286-4579(00)00392-0

[60] Kim GB, Yi SH, Lee BH. Purification and characterization of three different types of bile salt hydrolases from *Bifidobacterium* strains. Journal of Dairy Science. 2004;**87**:258-266. DOI: 10.3168/ jds.S0022-0302(04)73164-1

[61] Kurdi P, Tanaka H, Van Veen HW, Asano K, Tomita F, Yokota A. Cholic acid accumulation and its diminution by short-chain fatty acids in bifidobacteria. Microbiology. 2003;**149**:2031-2037. DOI: 10.1099/mic.0.26376-0

[62] Begley M, Hill C, Gahan CGM. Bile salt hydrolase activity in probiotics. Applied and Environmental Microbiology. 2006;**72**:1729-1738. DOI: 10.1128/ AEM.72.3.1729-1738.2006

[63] Marshall VM. Bioyogurt: How healthy? Dairy Industries International. 1996;**61**:28-29

[64] Ahn YT, Kim GB, Lim KS, Baek YJ, Kim HU. Deconjugation of bile salts by *Lactobacillus acidophilus* isolates. International Dairy Journal. 2003;**13**:303-311. DOI: 10.1016/ S0958-6946(02)00174-7

[65] Takahashi T, Morotomi M. Absence of cholic acid 7-alpha-dehydroxylase activity in the strains of *Lactobacillus* and *Bifidobacterium*. Journal of Dairy Science. 1994;**77**:3275-3286. DOI: 10.3168/jds.S0022-0302(94)77268-4

[66] De Vuyst L, De Vin F, Vaningelgem F, Degeest B. Recent developments in the biosynthesis and applications of heteropolysaccharides from lactic acid bacteria. International Dairy Journal. 2001;**11**:687-707. DOI: 10.1016/ S0958-6946(01)00114-5

[67] Nakajima H, Suzuki Y, Kaizu H, Hirota T. Cholesterol lowering activity of ropy fermented milk. Journal of Food Science. 1992;**57**:1327-1329. DOI: 10.1111/j.1365-2621.1992. tb06848.x

[68] Varki A. Biological roles of oligosaccharides: All of the theories are correct. Glycobiology. 1993;**3**:97-130. DOI: 10.1093/glycob/3.2.97

[69] Zaręba D. The studies on the use of chromatographic identification of lactic acid bacteria in assessing the functionality of probiotic milk drink [doctoral thesis]. Warsaw: Warsaw University of Life Sciences—SGGW; 2012

[70] Fujishiro K, Uchida H, Shimokawa K, Nakano M, Sano F, Ohta T, et al. Purification and properties of a new Brevibacterium sterolicum cholesterol oxidase produced by *E. coli* MM294/pnH10. FEMS Microbiology Letters. 2002;**215**:243-248. DOI: 10.1111/ j.1574-6968.2002. tb11397.x

[71] Smith M, Sullivan C, Goodman N. Reactivity of milk cholesterol with bacterial cholesterol oxidases. Journal of Agricultural and Food Chemistry. 1991;**39**:2158-2162. DOI: 10.1021/jf00012a011

[72] Somkuti GA, Solaiman DKY, Johnson TL, Steinberg DH. Transfer and expression of a Streptomyces cholesterol oxidase gene in *Streptococcus thermophilus*. Biotechnology and Applied Biochemistry. 1991;**13**:238-245. DOI: 10.1111/j.1470-8744.1991.tb00153.x

[73] Somkuti GA, Solaiman DKY, Steinberg DH. Expression of Streptomyces sp. cholesterol oxidase in *Lactobacillus casei*. Applied Microbiology and Biotechnology. 1992;**37**:330-334. DOI: 10.1007/BF00210988

[74] Boudreau A, Arul J. Cholesterol reduction and fat fractionation technologies for milk fat. Journal of Dairy Science. 1993;**76**:1772-1781. DOI: 10.3168/jds.S0022-0302(93)77509-8

[75] Sanders ME. Summary of conclusion from a consensus panel of experts on health attributes of lactic cultures: Significance to fluid milk products containing cultures. Journal of Dairy Science. 1993;**76**:1819-1828. DOI: 10.3168/jds.S0022-0302(93)77514-1

[76] Lund EG, Kerr TA, Sakai J, Li WP, Russell DW. cDNA cloning of mouse and human cholesterol 25-hydroxylases, polytopic membrane proteins that synthesize a potent oxysterol regulator of lipid metabolism. Journal of Biological Chemistry. 1998;**273**: 34316-34327. DOI: 10.1074/ jbc.273.51.34316

[77] Madden UA, Osweiler GD, Knipe L, Beran GW, Beitz DC. Effects of *Eubacterium coprostanoligenes* and *Lactobacillus* on pH, lipid content, and cholesterol of fermented pork and mutton sausage-type mixes. Journal of Food Science. 1999;**64**:903-908. DOI: 10.1111/j.1365-2621.1999.tb15937.x

[78] Cardona ME, De Vanay V, Midtvedt T, Norin E. Probiotics in gnotobiotic mice. Conversion of cholesterol to coprostanol *in vitro* and *in vivo* and bile acid deconjugation *in vitro*. Microbial Ecology in Health and Disease. 2000;**12**:219-224. DOI: 10.1080/08910600050216200

[79] Li L, Batt SM, Wannemuehler M, Dispirito A, Beitz DC. Effect of feeding of a cholesterol-reducing bacterium, *Eubacterium coprostanoligenes*, to germ- free mice. Laboratory Animal Science. 1998;**48**:253-255

[80] Li L, Buhman KK, Hartman PA, Beitz DC. Hypocholesterolemic effect of *Eubacterium coprostanoligenes* ATCC 51222 in rabbits. Letters in Applied Microbiology. 1995;**20**:137-140. DOI: 10.1111/j.1472-765X.1995.tb00410.x

[81] El-Nezami H, Kankaanpaa P, Salminen S, Ahokas J. Ability of dairy strains of lactic acid bacteria to bind a common food carcinogen, aflatoxin B1. Food and Chemical Toxicology. 1998;**36**:321-326. DOI: 10.1016/ S0278-6915(97)00160-9

[82] Gratz S, Mykkanen H, Ouwehand AC, Juvonen R, Salminen S, El-Nezami H. Intestinal mucus alters the ability of probiotic bacteria to bind aflatoxin B1 *in vitro*. Applied and Environmental Microbiology. 2004;**70**:6306-6308. DOI: 10.1128/ AEM.70.10.6306-6308.2004

[83] Lahtinen SJ, Haskard CA, Ouwehand AC, Salminen SJ, Ahokas JT. Binding of aflatoxin B1 to cell wall components of *Lactobacillus rhamnosus* strain GG. Food Additives & Contaminants. 2004;**21**:158-164. DOI: 10.1080/02652030310001639521

[84] Lee YK, El-Nezami H, Haskard CA, Gratz S, Puong KY, Salminen S, et al. Kinetics of adsorption and desorption of aflatoxin B1 by viable and nonviable bacteria. Journal of Food Protection. 2003;**66**:426-430. DOI: 10.4315/0362-028X-66.3.426

[85] Thyagaraja N, Hosono A. Binding propertied of lactic acid bacteria from "idly" towards food-borne mutagens. Food and Chemical Toxicology. 1994;**32**:805-809. DOI: 10.1016/0278-6915(94)90156-2

[86] Peltonen K, El-Nezami H, Haskard C, Ahokas J, Salminen S. Aflatoxin B1 binding by dairy strains of lactic acid bacteria and bifidobacteria. Journal of Dairy Science. 2001;**84**:2152-2156. DOI: 10.3168/jds. S0022-0302(01)74660-7

[87] Haskard CA, El-Nezami HS, Kankaanpaa PE, Salminen S, Ahokas JT. Surface binding of aflatoxin B1 by lactic acid bacteria. Applied and Environmental Microbiology. 2001;**67**:3086-3091. DOI: 10.1128/ AEM.67.7.3086-3091.2001

[88] Oatley JT, Rarick MD, Ji GE, Linz JE. Binding of aflatoxin B1 to bifidobacteria *in vitro*. Journal of Food Protection. 2000;**63**:1133-1136. DOI: 10.4315/0362-028X-63.8.1133

[89] Madigan MT, Martinko JM, Parker J. Brock Biology of Microorganisms. 11th ed. Upper Saddle River: Prentice Hall; 2006. p. 1058

[90] Lonvaud-Funel A, Desens C. Constitution en acides gras des membranes des bacteries lactiques du vin. Sciences des Aliments. 1990;**10**:817-829

[91] Kurdi P, Van Veen Hendrik W, Tanaka H, Mierau I, Konings WN, Tannock GW, et al. Cholic acid is accumulated spontaneously, driven by membrane ΔpH in many lactobacilli. Journal of Bacteriology. 2000;**182**:6525-6528. DOI: 10.1128/ JB.182.22.6525-6528.2000

[92] Goldberg I, Eschar L. Stability of lactic acid bacteria to freezing as related to their fatty acid composition. Applied and Environmental Microbiology. 1977;**33**:489-496

[93] Boggs JM. Lipid intermolecular hydrogen bonding: Influence on structural organization and membrane function. Biochimica et Biophysica Acta. 1987;**906**:353-404

[94] Ziarno M, Sękul E, Lafraya Aguado A. Cholesterol assimilation by commercial yoghurt starter cultures. ACTA Scientiarum Polonorum Technologia Alimentaria. 2007;**6**:83-94

[95] Ziarno M. The influence of cholesterol and biomass concentration on the uptake of cholesterol by *Lactobacillus* from MRS broth. ACTA Scientiarum Polonorum Technologia Alimentaria. 2007;**6**:29-40

[96] Dilmi-Bouras A. Assimilation (*in vitro*) of cholesterol by yogurt bacteria. Annals of Agricultural and Environmental Medicine. 2006;**13**:49-53

[97] Ziarno M, Makowska M. Sensory properties of fermented sour cream containing probiotic strains of lactic acid bacteria. Przemysł Spożywczy. 2005;**10**:46-49

[98] Ziarno M, Makowska M. Viability of technical microflora in yoghurt cream during refrigerated storage. Medycyna Weterynaryjna. 2008;**64**:461-464

[99] Moser A, Savage DC. Bile salt hydrolase activity and resistance to toxicity of conjugated bile salts are unrelated properties in lactobacilli. Applied and Environmental Microbiology. 2001;**67**:3476-3480. DOI: 10.1128/ AEM.67.8.3476-3480.2001

[100] Oozeer R, Leplingard A, Mater D, Mogenet A, Michelin R, Seksek I, et al. Survival of *Lactobacillus casei* in human digestive tract after consumption of fermented milk. Applied and Environmental Microbiology. 2006;**8**:5615-5617. DOI: 10.1128/ AEM.00722-06

[101] Hood S, Zottola E. Effect of low pH on the ability of *Lactobacillus acidophilus* to survive and adhere to human intestinal cells. Journal of Food Science. 1988;**53**:1514-1516. DOI: 10.1111/j. 1365-2621.1988.tb09312.x

[102] Marteau P, Minekus M, Havenaar R, Huis In't veld JH. Survival of lactic acid bacteria in a dynamic model of the stomach and small intestine: Validation and the effects of bile. Journal of Dairy Science. 1997;**80**:1031-1037. DOI: 10.3168/jds. S0022-0302(97)76027-2

[103] Elli M, Callegari ML, Ferrari S, Bessi E, Cattivelli D, Soldi S, et al. Survival of yoghurt bacteria in human gut. Applied and Environmental Microbiology. 2006;**7**:5113-5117. DOI: 10.1128/AEM.02950-05

[104] Hoier E. Use of probiotic starter cultures in dairy products. Food Australia. 1992;**44**:418-420

[105] Corcoran BM, Stanton C, Fitzgerald GF. Survival of probiotic lactobacilli in acidic environments is enhanced in the presence of metabolizable sugars. Applied and Environmental Microbiology. 2005;**71**:3060-3067. DOI: 10.1128/ AEM.71.6.3060-3067.2005

[106] Kailasapathy K, Chin J. Survival and therapeutic potential of probiotic organism with reference to *Lactobacillus acidophilus* and *Bifidobacterium* spp. Immunology & Cell Biology. 2000;**78**:80-88. DOI: 10.1046/j.1440-1711.2000.00886.x

[107] Lankaputhra WEV, Shah NP. Survival of *Lactobacillus acidophilus* and *Bifidobacterium* in the presence of acid and bile salts. Cultured Dairy Products Journal. 1995;**30**:2-6

[108] Vinderola CG, Reinheimer JA. Lactic acid starter and probiotic bacteria: A comparative "*in vitro*" study of probiotic characteristics and biological barrier resistance. Food Research International. 2003;**36**:895-904. DOI: 10.1016/S0963-9969(03)00098-X

[109] Ziarno M, Margol B. Research into the ability of some selected starter lactic acid bacteria to survive in a model gastric juice and cholesterol binding under those these conditions. Żywność Nauka Technologia Jakość. 2007;**6**:304-314

[110] Xanthopoulos V, Hatzikamari M, Adamidis T, Tsakalidou E, Tzanetakis N, Litopoulou-Tzanetaki E. Heterogeneity of *Lactobacillus plantarum* isolates from feta cheese throughout ripening. Journal of Applied Microbiology. 2000;**88**:1056-1064. DOI: 10.1046/j.1365-2672.2000.01056.x

[111] Xanthopoulos V, Litopoulou- Tzanetaki E, Tzanetakis N. Characterization of *Lactobacillus* isolates from infant faeces as dietary adjuncts. Food Microbiology. 2000;**17**:205-215. DOI: 10.1006/ fmic.1999.0300

[112] Bezkorovainy A. Probiotics: Determinants of survival and growth in the gut. American Journal of Clinical Nutrition. 2001;**73**:399-405. DOI: 10.1093/ajcn/73.2.399s

[113] Ziarno M, Bartosz P. The cholesterol binding by yoghurt bacteria in simulated intestinal juice. Żywność Nauka Technologia Jakość. 2007;**4**:126-138

[114] Ziarno M. Survival of lactic acid bacteria in simulated duodenal fluid depending on the cholesterol presence. Polish Journal of Food and Nutrition Sciences. 2007;**57**:625-631

[115] Ziarno M. The survival of lactic acid bacteria in simulated intestinal juice. In: Proceedings of the III International Conference "Quality And Safety In Food Production Chain", Wrocław. 2007. p. 402

5

Could Gut Modulation through Probiotic Supplementation be Beneficial in Autism Spectrum Disorder?

Rouba El Khatib, Dolla Karam-Sarkis, Anne-Judith Waligora-Dupriet and Marie-José Butel

Abstract

Evidence is mounting to a possible link between autism spectrum disorder (ASD) and gut microbiota through the well-known gut-brain axis. Numerous mechanisms have been suggested including bacterial metabolites that could involve in chemokines, antimicrobial peptides, or neuropeptides production. Hence, numerous studies reported dysbiosis in autistic patients. Antibiotic courses are known to more or less improve neurobehavioral symptoms; however, it could lead to side effects. Modulation of the gut microbiota using pro- and/or prebiotics is therefore an appealing way of treatment. Fecal microbiota transfer is suggested to be an alternative new approach that could be promising. The aim of our chapter will be first to briefly review the current data concerning the possible role of the gut microbiota and its mechanisms in ASD and second to review the interest and limits of the pre- and probiotic supplementations in ASD treatment. Lastly, we will discuss on the potential interest of the microbiota transfer in ASD.

Keywords: autism spectrum disorder, gut microbiota, dysbiosis, probiotics, prebiotics, fecal microbiota transplantation

The human microbiota

Definition and functions

The term microbiota describes the entirety of all bacterial, viral, fungal, protozoal, and archaeal microorganisms living on almost every cutaneous and mucosal surface of the body [1]. The gut microbiota (GM) inhabited by several trillion microorganisms that live in a symbiotic relationship with the host represents the most heavily colonized area of the human body. It is mainly dominated by organisms belonging to four major phyla that together account for more than 90% of the total bacterial population: *Bacteroidetes*, *Firmicutes*, *Proteobacteria*, and *Actinobacteria*, followed by the minor phyla *Fusobacteria* and *Verrucomicrobia*.

With a number of microorganisms being around 10 times higher than the number of human cells and a number of genes nearly 150 times greater than the human genome, the GM is now recognized as an environmental factor that affects normal host physiology, metabolism, immunity, brain function, and behavior.

During recent decades, the development of high-throughput sequencing technology has hugely contributed to increased understanding of host-microbe interactions in health and disease [2]. First, GM offers a barrier, protecting against external factors and proliferation of pathogenic microbes, through various mechanisms such as increasing mucus production, reinforcing intestinal epithelium permeability, and production of bacteriocins and antimicrobial peptides. GM is also involved in many fundamental metabolic functions such as synthesis of essential nutrients, hormones, vitamins, supply of energy from dietary sources otherwise unavailable to host and clearance of drugs, and toxins. Furthermore, GM is shown to be involved in the maturation of the host immune system, where specific strains such as *Bifidobacterium* and *Lactobacillus* produce anti-inflammatory cytokines, and others such as *Clostridium* and *Ruminococcus* produce pro-inflammatory cytokines. Hence, GM is essential for the development of innate and acquired immunity through stimulation of local and systemic immune responses.

Lastly, very recently there is mounting evidence of the significant influence of GM in the modulation of brain activity and behavior across the so-called "microbi- ota-gut-brain axis (GBA)" [3]. The GBA consists of bidirectional communication between the central nervous system (CNS), the enteric nervous system, and the gut linking emotional and cognitive centers of the brain with peripheral intestinal functions [4]. The exact mechanisms of signal transmission within this network are not completely elucidated. The CNS asserts its role over the GM through influencing gut motility patterns, altering the equilibrium in the gut permeability, and modulating mucus secretion which are known to exert control over gut microbial composition [5]. Conversely, the GM claims its influence over the CNS by regulating the hypothalamic-pituitary-adrenal axis and the production and turnover of cytokines and neurotransmitters. In addition to their effects on development and maturation of the enteric nervous system, these neuroactive metabolites can signal beyond the local gastrointestinal (GI) tract to the distant CNS potentially through signaling pathways that include the vagus nerve. Catecholamines can modulate important processes, including neurogenesis, myelination, microglia activation, brain plastic- ity, and blood-brain barrier permeability [6]. Indeed, recent studies demonstrated that, under extreme conditions (e.g., in a germ-free environment or during anti- biotic (ATB) treatment), GM absence is associated with several abnormalities in brain gene expression and neurophysiology [7]. Interestingly, these aberrations are reversed after colonization with a conventional GM [8] or even specific bacterial species [7]. Thus, the disruption of neural, endocrine, immune, and metabolic mechanisms that are involved in gut-CNS signaling seems to be involved in neuropsychiatric, neurobehavioral, neurodegenerative, and mental disorders.

Colonization of intestinal ecosystem in early life and its evolution

The composition of the GM varies widely from fetal life to adult age. Until recently, babies were believed to be born sterile and only populated by microbes on exposure to their first postdelivery environments. Recent research suggests that the process of microbial colonization of the GI tract could begin prenatally as acquisition of maternal microbiota might occur during intrauterine life via placenta [9].

However, these findings are questionable, and recent data strongly suggest that bacteria isolated in utero are rather a contamination linked to the sampling methods than a specific microbiota [10]. Neonates show unstable and highly dynamic intes- tinal microbiota with a low microbial diversity. First colonizers in healthy neonates are enterobacteria, *Staphylococcus*, and *Streptococcus*, followed by strict anaerobes such as *Bacteroides*, *Bifidobacterium*, and *Clostridium* [11]. This pattern of microbial diversity provides an efficient means for adaptation to the variable circumstances over a lifetime such as changes in lifestyle, illness, puberty, and others. Over the first few years of life, GM matures and stabilizes to a more balanced "adult-like" composition at around the end of the third year [12].

Interestingly, the brain of neonates grows to approximately 90% of its future adult volume until the age of two, and the formation of new synapses in the brain peaks during this period [13]. Thus, the critical window for establishment of a healthy microbial composition falls into the same critical time window for brain development. Therefore, understanding GM establishment and its critical developmental window in early child- hood is important because any perturbation during this period causes long-lasting effects on the development of the CNS. Being more flexible at infantile in contrast to the subsequent life, this temporal requirement may have important ramifications for potential preventative and therapeutic remediation strategies.

Autism spectrum disorder and the gut microbiota

Gut microbiota involved in the pathogenesis of autism spectrum disorders

Microbiota role in health and disease is as crucial as is complex. Alterations in normal commensal GM (known as dysbiosis) have been widely reported as a key contributor to the etiology and/or pathogenesis of various diseases including several neurobehavioral conditions, such as Parkinson's disease, schizophrenia, Alzheimer's disease, depression, anxiety, and most compellingly autism spectrum disorder (ASD) [14].

ASD refers to a group of heterogeneous and complex neurodevelopmental dis- orders characterized by impaired social interactions and reciprocal communication skills as well as restricted, repetitive, and stereotyped patterns of behavior, interests, and activities [15]. Over the last decades, a steady increase of ASD prevalence has been reported worldwide. ASD is currently estimated to affect about 1 in every 68 children, with greater incidence found among boys (4:1) [14]. To date, the etiology of ASD remains elusive, and it is thought to involve both genetic predisposition and different environmental triggers. Although several genetic factors are known to influence the etiology of different types of ASD, these only apply to a minor part of the autistic population. By estimate, the heritability accounts approximately for only 35–40% of the contributing elements and the remaining 60–65% results from the combination of prenatal, perinatal, and postnatal environmental factors as well as related medical disorders [16]. Besides, along with significant psychiatric symptoms, ASD is often characterized by a number of medical comorbidities, the most prominent of which implicates the GI tract. Children with ASD experience significantly more GI symptoms than children without ASD occurring nearly ata fourfold greater rate [17]. Symptoms include constipation, diarrhea, abdomi- nal pain, bloating, gastroesophageal reflux, and food selectivity and seem to be strongly associated with the severity of ASD behaviors. Clinical abnormalities such as altered GI motility and increased gut permeability have also been reported [18]. The cause of ASD-associated GI problems is difficult to ascertain, but it appears to partly relate to the excessive use of oral ATBs which can alter GM. Indeed, several studies report increased use of oral ATBs in children with ASD compared to neurotypical children. By eliminating beneficial indigenous GM, long-term ATB use destabilizes microbial community and creates favorable environment for colonization by potentially harmful (toxin-producing) microorganisms. Thus, considering the potential interactions between intestinal microbes and the CNS, loss of the protective commensal microbiota along with the overgrowth of pathogenic micro- organisms is hypothesized to cause or contribute not only to GI dysfunction but also to ASD-related behavioral symptoms. All these findings have gained an insight into the influence of GM in ASD as potential mediator of risk factors.

Disruptions of microbial colonization in autism spectrum disorder

Imbalances in GM composition at the first stages of life and concomitant behav- ioral changes have been related to various prenatal and early-life events [8].

Mode of birth, whether through natural birthing process or Cesarean section (C-section), greatly affects the initial microbial settlement. Vaginally delivered babies harbor bacteria similar to their mothers' fecal and vaginal microbiota with dominance of *Bifidobacterium*, whereas babies born via C-section would acquire an altered GM composition resembling to their mother's skin microbiota as well as microbes from the surrounding environment with a delayed colonization by bacteria from maternal origin, mainly enterobacteria, bifidobacteria, and *Bacteroides* [11]. Interestingly, these modifications can persist for several months [2]. Thus, concern has arisen that this reduced microbial diversity and altered bacterial profile in babies delivered via C-section could contribute to ASD. Indeed, studies employing animal models have revealed that, compared with vaginally delivered animals, those delivered by C-section suffer more frequently from behavioral abnormalities that have been associated with ASD [19]. Consistent with this, some clinical studies report that C-section births in comparison to natural vaginal delivery are associated with a significantly 23% higher risk of the child developing ASD [20]. Thus, the importance of C-section in conditioning negative effects on the CNS is debated.

In terms of gestational age at birth, preterm neonates (PT) are characterized by a delayed microbial colonization, missed or reduced acquisition of *Lactobacillus* and *Bifidobacterium*, greater abundance of *Proteobacteria*, and earlier acquisition of *Firmicutes* [11], hence increasing the risk to develop later disease. One study reported a relationship between premature birth and an increase risk of ASD development [21]. Interestingly, each week of shorter gestation was associated with an increased risk of ASD. The early development of infant at birth has also been recognized as a biomarker of future risk of neurodevelopmental disorders. Besides, gestational growth affects the neurodevelopment, and both infants born small or large for gestational age are associated with higher rates of ASD [22].

Early feeding pattern also interferes greatly in the regulation of the intestinal colonization, with breastfed infants harboring a less diverse but more stable and uniform GM. Specific biological markers of healthy GM, including early coloni- zation of *Bifidobacterium* and greater prevalence of *Lactobacillus*, characterize microbial communities of breastfed infants [11]. By contrast, increased species richness accompanied by an overrepresentation of *Clostridium* prevails in bottle- fed ones. In fact, cognitive functioning and neurological development of children have been associated with the duration of exclusive breast-feeding and introduction of formula feed [23]. Remarkably, breast-feeding appears to be less frequent and, when present, occurs for a much shorter duration in children with ASD [24]. Further differentiation occurs after the introduction of solid foods. Feeding pat- terns and strong food preferences of ASD children for nutrient-poor starchy foods while rejecting fruits, vegetables, and proteins may lay the foundation of an abnormal GM. The high prevalence of ASD in some countries has been correlated with typical carbohydrate rich diet and consequently predominance of genera *Prevotella* and *Megasphaera* [25]. Therefore, understanding how these changes affect human GM suggests that early dietary habits have a more complex effect on the metabolic programming of a child than previously anticipated.

Additional modifications are induced by exposure to drugs either directly or indirectly from the mother. ATB-induced shifts in the gut microbial composition can persist several months after cessation of the treatment, inducing long-term dysbiosis [26]. An association between long-term ATB use, hospitalization, abdominal discomfort, and the onset of ASD symptoms has been shown [27]. Likewise, a population-based cohort study revealed the use of various ATBs during pregnancy as a potential risk factor for ASD [28]. Using animal models, periconceptional exposure to nonabsorbable ATB was shown to induce variations in GM composition in offspring associated to reduction in social interactions and increased anxiety [29]. Other drugs lead to similar observations. In mice, modeling maternal exposure to valproic acid, an antiepileptic drug induced lasting changes in the offspring GM composition, which was associated with neuroinflammation, abnormal GI physiology, and ASD-like behavioral abnormalities [30]. All these observations point to the common hypothesis that early exposure to drugs can modify GM composition transiently or permanently, possibly affecting the severity of non-GI-related symptoms of ASD patients.

Apart from exposition to drugs, another emerging explanation for the difference in GM between ASD and healthy individuals is immunological. Maternal infection during pregnancy was shown to alter microbial composition and is a primary environmental risk factor for ASD [31]. GM alterations (especially in the bacterial classes clostridia and bacteroidia) and higher gut permeability are seen in a maternal immune activation model of ASD [18]. Male progenies of pregnant mice injected during pregnancy with a viral mimetic develop abnormal communication and sociability, repetitive behaviors, and increased anxiety. Another well-estab- lished brain-gut connection is the role of stress and its mediators in altering the GM. Maternal separation is a typical model of early-life stress employed in animal studies. In rats, early maternal separation does not only lead to a dysbiotic state of the GM that persists into adulthood but also results in functional GI symptoms and long-term cognitive and behavioral deficits [32]. Likewise, psychological stress was shown to alter colonic mucosa-associated microbiota, with significant decrease in abundance of health-benefit bacteria, such as *Lactobacillus*, while abundance of clostridia increased [33]. Although one cannot extrapolate from mice to human without further evidence, these findings nonetheless suggest that the differences seen in the GM of ASD patients may be a result of immunological changes.

Further, metals and other contaminants have also been identified to increase the risk for ASD knowing their capacity to interfere with the composition and metabolic activity of the GM. In fact, results of a study evaluating the interaction between environmental chemicals and GI microorganisms suggest that alterations in the levels of seven elements (Pb, As, Cu, Zn, Mg, Ca, and Hg) and nine genera of GM (*Bacteroides*, *Parabacteroides*, *Sutterella*, *Lachnospira*, *Bacillus*, *Bilophila*, *Lactococcus*, *Lachnobacterium*, and *Oscillospira*) may be related to ASD [34].

Excessive accumulation of these typical neurotoxic elements leads to abnormally increased abundance of several genera and is reported to be closely related to poorer intellectual function [35].

Taken together, ASD is thought to be a result of a combination of different interacting mechanisms that each contributes a fraction of disease risk.

Microbial and metabolic dysregulations in autism spectrum disorder

Previously, much research effort on ASD focused on genetic, neurological, and behavioral aspects of disease. Recently, the evidence of the impact of dysbio- sis on CNS raised interest in the analysis of the potential link between GM and ASD. So far, various studies demonstrated that children with ASD exhibit different compositions of GM compared to healthy controls [36–43]. Moreover, exciting work with animal models widely deepened the possible role of gut microorganisms in the pathogenesis of such disorders [44, 45]. These evidences have led to the hypothesis that GM alteration is not only associated with ASD but may play a key role in the exacerbation of ASD symptoms and/or its pathogenesis, at least for some ASD subgroups [14]. Overall, most studies agree that GM composition is distinctive in ASD compared to healthy controls, but results are often inconsistent as to the nature and/or extent of GI bacterial community differences, failing to generate a coherent picture. Microbiota analyses reported tenfold higher counts of pathogenic *Clostridium* spp. in children with ASD compared with healthy controls [46, 47].

Clostridium is known to produce neurotoxins and P-cresol, cause higher propionic acid levels, and promote conditions that favor inflammation and exacerbate ASD symptoms. On the other hand, there have been some consistent findings of decrease in certain beneficial bacteria, specifically *Bifidobacterium* [40, 48, 49] known for its health-benefit properties. Then, the composition of GM of ASD individuals has been characterized, showing a reduction of the *Bacteroidetes/Firmicutes* ratio which pointed to elevated numbers of *Firmicutes* in contrast to decreased levels of *Bacteroidetes* [36, 39, 41, 49–51]. *Bacteroidetes* are short-chain fatty acids (SCFA) producing bacteria, and their metabolites, especially propionic acid, may influence the CNS and autism behavior by modulating the GBA [36]. Moreover, species of *Desulfovibrio* were also isolated from the stool of patients with ASD, and, to a lesser extent, in non-affected siblings [52]. *Desulfovibrio* could be an important contributor to GI inflammation, as its major metabolic by-product—hydrogen sulfide—is cytotoxic to colonic epithelial cells. Furthermore, the presence of autistic symptoms in children with ASD has been correlated with a less diverse gut microbiome, with less carbohydrate degrading and fermenting bacteria of the genera *Prevotella*, *Coprococcus*, and the unclassified *Veillonellaceae* in ASD microflora samples as compared to the healthy controls [39]. This decrease in GM diversity can lead to a loss of key signals required for normal brain maturation. Additionally, increased *Sutterella* were found in significant numbers in intestinal biopsies and stools of ASD children [48, 50]. This genus is known to regulate mucosal metabolism and intestinal epithelial integrity. Lastly, there were still some conflicting results about the alterations of *Akkermansia*, *Ruminococcus*, and *Faecalibacterium* in ASD patients. *Akkermansia* and *Ruminococcus* are mucin-degrading bacteria [40], and *Faecalibacterium* is regarded as commensal or even beneficial due to its function of producing anti-inflammation butyrate [45].

Thus, the existence of a GI dysbiosis as an actor in the ASD etiopathogenesis remains a controversial topic. Indeed, other studies comparing children with ASD and their healthy siblings reported no meaningful difference in GM composition [53, 54]. According to the authors, other explanations for the GI dysfunction in this population should be considered, including elevated levels of anxiety and self-restricted diets. Therefore, given the higher incidence of ATB usage and often different diets compared with neurotypical individuals, both of which can alter the composition of the GM, such data should be interpreted with care.

Dysbiosis in ASD involves not only bacterial species but also yeasts, as reported in recent studies [40, 41, 55, 56]. One culture-based study showed significant presence of *Candida* species in the feces of children with ASD, mainly *Candida albicans*. It also identified hyphae formation, suggesting that the dimorphic yeast had switched to its invasive and adhesive form. However, another study did not report such overrepresentation of *Candida* in ASD children compared to control ones [49].

Moreover, correlations between ASD and GI disturbances may not alone be driven by the composition of the GM but also by differences in its functionality, such as the bacterial metabolites that could play a role in the GBA. Indeed, overproduction of bacterial metabolites (e.g., p-cresol) and SCFA (e.g., propionic acid) is frequently described in infants with ASD [30, 49, 57]. These compounds induce intracellular signaling and modulate host gene expression related to neurotransmission systems and behavior [58]. P-cresol seems to negatively affect the homeostasis of colonic epithelial cells in children with ASD. When tested *in vitro*, excessive concentration of p-cresol showed deleterious metabolic and genotoxic effects on colonocytes [59]. In addition, early exposure to p-cresol may contribute to the severity of behavioral symptoms and cognitive impairment in ASD toddlers. On the other hand, propionic acid is known to have a number of direct effects on gut physiology. It increases the contraction of colonic smooth muscle, dilates colonic arteries, stimulates serotonin release, and decreases gas- tric motility, which could be easily related to the GI abnormalities frequently observed in many ASD patients [60]. In fact, lower plasmatic levels of propionic acid have been reported in ASD children as an aspect of metabolic alteration in gut host-microbial co-metabolism. Authors related the occurrence of lower propionic acid in the plasma to elevated levels of propionic acid in the brain [61]. Additionally, administration of neurotoxic dose of propionic acid to animals was effective in inducing autistic features. Orally dispensed propionic acid was reported to induce oxidative stress. Elevated inter- leukin 6 (IL-6), tumor necrosis factor α (TNF-α), and interferon γ (INF-γ) confirmed the neuroinflammatory effect of propionic acid [61]. When propionic acid was injected into the cerebral ventricles of rats, the rats showed biologic, chemical, and pathologic changes that are characteristic of ASD [62]. These chemicals can also alter the intercellular spaces between the cells, resulting in the "leaky gut syndrome" that can lead to detrimental neurologic effects.

Dysbiosis can also affect the functional intestinal barrier that can lead to an alteration in the intestinal permeability referred to as "leaky gut" state, a fundamental factor underlying the relationships between ASD and the GM [63]. Indeed, several reports show increased gut permeability in ASD patients [64]. Disrupted barrier function facilitates the translocation of bacterial metabolites from the gut into the bloodstream to possibly reach the otherwise sterile CNS inducing directly inflammatory reactions. One important bacterial component is the lipopolysaccha- ride (LPS) that was shown to increase the activity of areas deputed to the emotion-alism control such as amygdale [65]. It also leads to the production of inflammatory cytokines that critically alter the physiological brain activity, modulating the neuro- peptides synthesis

[66]. Moreover, it has been demonstrated that the administration of low doses of LPS in healthy subject is associated to increased pro-inflammatory cytokines and plasma norepinephrine, with higher depression rates, fatigue, and apathy [67]. Consistent with this, a study showed LPS serum levels were significantly higher in autistic patients compared to heath individuals and correlated with socialization scores in an inverse and independent manner [68].

Restoring the gut ecosystem: therapeutic outlooks for autism spectrum disorder

Despite increased ASD diagnoses, there remains no US Food and Drug Administration (FDA)-approved pharmaceutical treatment to alleviate core symp- toms of ASD [69]. Currently, recommended management strategies essentially involve rehabilitation, educational interventions, speech therapies, psychiatric medications, and specific treatments for individual comorbidities [70], all with lim- ited success [71]. Considered the emerging role of gut dysbiosis in ASD, interest in rebalancing human GM as a possible therapeutic approach is growing [72]. Indeed, targeting the GM in children with ASD through administration of ATBs, pro- and prebiotics, and nutritional approaches or, more recently, through fecal microbiota transplantation (FMT) has been shown to improve not only GI disturbances but also behavioral and neurophysiological abnormalities associated with ASD [18, 24].

Antibiotics/antifungals

ATB therapy is used for the management of ASD and is routinely prescribed to treat ASD symptoms associated with several underlying disorders like pediatric autoimmune neuropsychiatric disorder associated with streptococcal infections (PANDAS) and pediatric acute-onset neuropsychiatric syndrome (PANS) [72].

Short-term administration of vancomycin was shown to provide significant improvement in both GI and behavioral disturbances in a subset of children with ASD [73]. Unfortunately, this attempt had only partial success since children relapsed, and benefits waned upon treatment termination. As vancomycin is a poorly absorbed ATB known to destroy Gram-positive anaerobes, observed improvement is believed to be the consequence of temporary elimination of toxin- producing clostridia population. Hence, symptomatic relapse was attributed to the spore-forming capacity of these bacteria, and spores resistant to ATB would later germinate into vegetative forms once treatment has stopped [46].

Children with ASD, particularly those with GI disease, are sometimes treated with antifungal agents, as they may have increased incidence of fungal infection. Despite the lack of evidence of fungal overgrowth in children with ASD, parents find that antifungal therapies can often be beneficial and may be a useful adjuvant for the treatments of ASD [74].

Thus, it is possible to speculate that anti-infectives, through modulation of GM, should be able to influence symptoms and expression of psychiatric disorders.

However, ATB resistance (and to some extent antifungal resistance) is a major public health concern, making the safety of ATB/antifungal treatments ethically problematic, if they do prove to be beneficial. Therefore, extensive and prolonged use of ATB/antifungal treatments may not be advisable as a long-term therapy for ASD.

Probiotics

One of the most nutritional popular approaches selectively modulating the GM is probiotic supplementation due to ease of use, wide availability, and good safety profile. Probiotics are defined as live microorganisms that when administered in adequate amounts confer a health benefit to the host [75]. The main species used are one *Escherichia coli* strain, several lactic acid-producing *Lactobacillus*, a number of bifidobacterial strains, and a yeast *Saccharomyces boulardii*. Research on probiotics has shown efficacy in prevention or treatment of a wide variety of diseases associated with GI difficulties. Recently, some evidence has been accumulated regarding the possible role of probiotics in modulating symptoms of certain psychological diseases such as depression and anxiety as well as ASD [76–78]. Hence, a new class of probiotics, termed as psychobiotics, has emerged and refers to living organisms with beneficial effects on mental health [79]. In spite of the well-documented beneficial effects of probiotics, testing for ASD is still in its infancy, and the exact mechanism of their action has not been thoroughly elucidated to date, though there are several hypotheses. Implicating correction of both composition and/or activities of GM is the first mechanism of action of probiotics through several mechanisms including bacteriocins and metabolites, such as lactic, propionic, and acetic acids.

Hence, because ASD patients present GI dysbiosis, which may exacerbate the disease, these patients could benefit from GM modulation through probiotic supplementation. Conversely, other investigations have shown that probiotic administration could act independently from GM alterations by inducing a stabilization of the microbial communities making them less susceptible to perturbations from stressors such as ATBs, poor diet, and psychological stress [80]. Moreover, probiotics may correct the imbalance in the activity of the GM without changing its composition, via their metabolites released in the gut lumen or by potentially correcting the overproduction of harmful and underproduction of beneficial gut bacterial products [81].

The increased intestinal permeability in ASD may also be ameliorated by probiotics which are able to enhance the mucosal barrier. This supports the concept that probi- otics can provide protective effect by preventing the metabolites of exogenous toxic substances from leaving the gut and entering the bloodstream to affect the brain [70]. Probiotics may therefore maintain or improve gut barrier integrity and aid in ASD rehabilitation by promoting "leaky gut" healing. Lastly, given the multiple findings of aberrant immune activation and higher levels of gut inflammation in a subset of individuals with ASD, that a major part of the immune system is concentrated in/around the intestinal mucosa, and that the GM plays an important role in the maturation and the regulation of the immune response, another mechanism of action of probiotics may be on the immune system [82]. Probiotics can downregulate gut and CNS inflammatory pathways in a species- and strain-specific manner [83], by promoting the production of regulatory T cells, diminishing the levels of LPS, providing tolerogenic signals, and boosting the brain-derived neurotrophic factor (BDNF).

BDNF is a protein that promotes the survival of existent neurons in the developing brain, fosters the growth and differentiation of new neurons, and regulates the formation and plasticity of synaptic connections, thereby playing an essential role in the normal neurological development [6]. Impairment in BDNF signaling in early developmental phases is thought to be related to CNS abnormalities, the most severe forms of ASD, as well as intellectual disability [84]. Probiotics have been shown also to modulate the intestinal immune system by the production of secreted factors and metabolites that affect growth and function of intestinal epithelial and immune cells

[85]. Moreover, probiotic immunomodulation may occur through inhibiting the pro- duction of pro-inflammatory cytokines such as IL-12, TNF-α, and INF-α or increase the expression of anti-inflammatory mediators such as IL-10 and transforming growth factor beta (β-TFGF). Thus, by alleviating gut inflammation and suppressing dysregulated immune functions, probiotic supplementation may be effective for improving both gut microbial and behavioral problems in ASD. Lastly, probiotics could act via the vagus nerve-mediated GBA to influence neurotransmission and mood states [86]. They can exert central actions by influencing several neuroactive metabolites such as gamma-aminobutyric acid (GABA) and serotonin which are also associated with neuropsychiatric disorders [87, 88]. Consequently, administration of probiotics regulates the behavior in a way that significantly reduces anxiety, depression, and stress and promotes positive emotional changes. Therefore, probiotic administration might be a useful new therapeutic option to restore normal GM, reduce inflammation, restore epithelial barrier function, and possibly improve some behavioral symptoms associated with ASD.

A role for probiotics has been suggested for children with ASD as well, as preliminary findings from experimental animals studies provide some evidence that administration of selected probiotics may be effective in reducing neurologic signs and symptoms associated with gut dysbiosis. In a summary of studies to date, several probiotic strains, i.e., *Bifidobacterium* sp. (*B. longum*, *B. breve*, *B. infantis*, and *B. bifidum*), *Lactobacillus* sp. (*L. acidophilus*, *L. helveticus*, *L. rhamnosus*, *L. plantarum*, *L. sporogenes*, *L. bulgaricus*, *L. delbrueckii*, *L. salivarius*, *L. casei*, and *L. paracasei*), *Streptococcus* sp. (*S. thermophilus* and *S. salivarius*), and *Bacteroides fragilis*, are presumed to be effective in ameliorating CNS functions related to mental disorders, as shown through several animal models [18, 89–92]. In a mouse model of ASD induced by maternal immune activation, oral administration of human commensal

B. fragilis corrected intestinal permeability, restored GM, improved GI physiology, and abolished ASD-like behavioral disruptions [18]. Interestingly, the probiotic also corrected altered expression of the tight junction proteins in the colon and restored the increase in the pro-inflammatory cytokine IL-6. Moreover, *B. fragilis* treatment mitigated elevations in several maternal immune activation-induced serum metabolites associated with ASD. One metabolite of particular interest was 4-ethyl- phenylsulfate (4EPS) which is chemically similar to p-cresol and is thought to be a possible putative urinary biomarker for ASD [59]. Hence, the administration of *B. fragilis* in a mouse model for ASD was able to reverse autistic symptoms and meta- bolic derangement. Similarly, reconstituting GM with a human breast milk and gut commensal *Lactobacillus reuteri* completely corrected social deficits and reversed aberrant neurotransmission in maternal high-fat diet offspring [89]. This is intriguing in light of reports that risperidone, an FDA-approved treatment for ASD, does not correct social abnormalities [93]. Treatment with *L. reuteri* resulted also in reduction of stress-induced corticosterone, and stimulation of the production of oxytocin, a key regulator of social behaviors, involved in the mesolimbic dopamine reward system, which is thought to be dysregulated in ASD [94]. On another hand, a recent study performed on hamsters in which autistic-like behaviors were induced by propionic acid and clindamycin administration studied the therapeutic effect of a 3-week oral treatment with a mixture of bifidobacteria and lactobacilli strains (Protexin®) [91]. Protexin® administration was effective in rebalancing GM, ameliorating oxidative stress, and counteracting behavioral deficits. Finally, in maternal separation-induced early-life stress mice model, oral administration of *L. plantarum* as a psychobiotic strain significantly reduced anxiety while improving locomotor activities and exploratory behavior [92]. According to various animal studies, *L. plantarum* can modulate the levels of neurotransmitters in the brain by influencing gene expression in the CNS and increasing dopamine level in the prefrontal cortex [92]. Summing up, by conducting behavioral tests in animal models, research provided convincing evidence for the efficacy of psychobiotic strains in improving ASD-like behaviors. However, the experimental evidence for the positive behavioral changes observed in animal models after administration of probiotics raises the interesting question whether the same result also applies to humans.

To date only few studies explored the effects of probiotics on ASD clinical features in humans. **Table 1** synthesizes available evidence on the efficacy and safety of probiotic supplementation as an adjunctive treatment for ASD [49, 51, 73, 95–106]. In these studies, probiotic interventions varied across all of the tri- als. Concentrations of the probiotic strains administered ranged from 107 to 1010 colony-forming units (CFU)/dose, and their usage by recipients differed. Strains were administered alone or as mixed strains with or without other additives (immu- nomodulators [99] and ATBs [73]). In summary, despite the variability in species, strains, dosages, and duration utilized, all studies pointed toward a similar trend

of improvement in both caregiver-reported ASD and GI symptoms after probiotic therapies. Accordingly, a recent survey found that almost 20% of physicians treating ASD patients encourage probiotic use, and almost 60% accept their use [107]. In a recent study, the positive effect of probiotic treatment was represented not only by the ability to consistently normalize the *Bacteroidetes/Firmicutes* ratio and restore the amounts of *Desulfovibrio* and *Bifidobacterium* in children with ASD [51] but also by the tendency to reduce intestinal inflammation and permeability. Similarly, another study associated changes in the GM composition to improvements in GI symptoms and functioning [102]. After probiotic supplementation in the latter study, children with ASD experienced a significant increase in *Bifidobacterium* and *Lactobacillus* in their fecal stool samples with a simultaneous reduction of *Clostridium* species. Furthermore, in a third study, probiotic feeding of children

Authors	Study design	Diagnosis	Probiotic therapy
Blades M., 2000, UK [95]	Case report of a 6 y with ASD	Not reported	Strain and dosage not provided Administration for 2 mo
Sandler et al., 2000, USA [73]	Open-label with self- control study 11 ASD subjects (10 M, 3.5–7 y)	CARS	*L. acidophilus*, *L. bulgaricus*, and *B. bifidum*, 40 × 10^9 CFU/mL QD for 4 wk
Parracho et al., 2010, UK [96]	Randomized, double- blind, PBO-controlled, cross overdesigned feeding study 17 ASD subjects (14 M, 4–16 y)	Not reported	*L. plantarum* WCFS1 4.5 × 1010 CFU QD for 3 wk
Ray et al., 2010,	Open-label with self-	ATEC	*L. rhamnosus*, *B. bifidum*, *L. acidophilus*,

USA [97]	control study 10 ASD subjects (Gender not reported, 4–15 y)		*B. infantis*, *B. longum*, *S. thermophilus*, *L. plantarum*, *L. salivarius*, *L. reuteri*, *L. casei*, *L. bulgaricus*, *L. acidophilus* DDS-1, and *L. sporogenes* Dosage not provided BID for 21 d
Adams et al., 2011 USA [49]	Retrospective case cohort study 57 ASD subjects: 19 PS vs. 38 no PS (50 M, 3–9 y)	ATEC	Strain and dosage not provided QD
Kałzuna- Czaplinska et al., 2012, Poland [98]	Open-label with self- control study 22 ASD subjects (20 M, 4–10 y)	DSM-IV	*L. acidophilus* (strain Rosell-11) 5 × 109 CFU/g BID for 2 mo
West et al., 2013, USA [99]	Open-label with self- control study 33 ASD subjects (Gender not reported, 3–16 y)	Not reported	*L. acidophilus*, *L. casei*, *L. delbrueckii*, *B. longum*, and *B. bifidum* 1 × 1010 CFU/capsule One capsule TID for 6 mo
Partty et al., 2015, Finland [100]	Randomized, double- blind, PBO-controlled prospective follow-up study 75 ASD subjects 40 PS (24 M, 13 y) vs. 35 PBO (16 M, 13 y)	ICD-10	*L. rhamnosus* GG 1 × 1010 CFU QD for 6 mo
Tomova et al., 2015, Slovakia [51]	Case control cohort study 10 ASD subjects (9 M, 2–9 y) 9 non-ASD siblings (7 M, 5–17 y) 10 non-ASD controls (10 M; 2–11 y)	ICD-10	3 strains of *Lactobacillus* (60%, one is *L. casei*), 2 strains of *Bifidobacterium* (25%, one is *B. longum*), and 1 strain of *Streptococcus* (15%, exact strain information not provided) Dosage not provided One capsule TID for 4 mo

Authors	Study design	Diagnosis	Probiotic therapy
Grossi et al., 2016, Italy [101]	Case report of a 12 y boy with ASD	DSM-V + ADOS-2	*B. breve, B. longum, B. infantis, L. acidophilus, L. plantarum, L. paracasei, L. bulgaricus, L. delbrueckii, S. thermophilus*, and *S. salivarius* 9 × 1010 CFU/g *Bifidobacteria* 8 × 1010 CFU/g *Lactobacilli*, and 20 × 1010 CFU/g *Streptococci* QD for 4 months
Shaaban et al., 2017, Egypt [102]	Prospective, open-label cohort study 30 ASD subjects (19 M, 5–9 y) 30 age/sex-matched non-ASD siblings	DSM-V + ADOS + ADI-R	L. acidophilus, L. rhamnosus, and B. longum 10×10^7 CFU/g 5 g QD for 3 mo
Liu et al., 2019, Taiwan [103]	A double-blind, randomized, PBOcontrolled, parallel feeding study 71 ASD subjects: 36 PS (36 M, 7–15 y) vs. 35 PBO (35 M, 7–15 y)	DSM-V + ADI-R	L. plantarum PS128 3×10^{10} CFU/capsule 1 capsule QD for 4 wk
Kobliner et al., 2019, USA [104]	Case report of a 16 y child with ASD	Not reported	*S. boulardii* *3 × 10^9 CFU/capsule* *Week 1: 6 capsules QD. Weekly increases reaching a final dose of 12 capsules BID. After 3 mo, weaning down to 3 capsules BID*
Sanctuary et al., 2019, USA [105]	Double-blind, crossover, randomized clinical trial 8 ASD subjects (7 M, 4–11 y)	ADOS	B. longum subsp. infantis 2×10^{10} CFU QD for 5 wk

y, years; M, male; PBO, placebo; PS, probiotic supplementation; ADI-R, Autism Diagnostic Interview-Revised; ADOS, Autism Diagnostic Observation Schedule; ATEC, Autism Treatment Evaluation Checklist; CARS, Childhood Autism Rating Scale; DSM-IV, Diagnostic and Statistical Manual-4th Edition; ICD-10, International Classification of Diseases-10th Edition; d, days; wk, weeks; mo, months; CFU, colony-forming unit; g, gram; QD, once a day; BID, two times a day; TID, three times a day; QID, four times a day.

Table 1. *Main clinical trials investigating the effectiveness of probiotic supplementation in autism spectrum disorders.*

with ASD significantly increased levels of the beneficial bacteria, specifically, the amount of lactobacilli and enterococci, and reduced their fecal *Clostridium* counts [96]. Additionally, in the same study, the efficacy of probiotic supplementation seemed to be age-dependent, with better effects noticed on younger children, underscoring the importance of early interventions.

Although all aforementioned studies showed changes in GM after the implementation diet with probiotics, only some of them analyzed the correlations with GI symptoms and reported improvement of GI function indices [96, 98, 101, 102, 105]. In fact, a survey conducted on caregivers showed that daily administration of a specific five-strain probiotic decreased the severity of comorbid GI problems [99]. Concurrent with increases in the proportion of normal stools, there was an increased appetite and willingness to consume novel foods that may be due to reduced abdominal pain or improved stool evacuation [105]. Besides, in one case study report, a multi-strain mixture of 10 probiotics not only relieved the GI disturbances but also improved ASD core symptoms [101]. However, benefits reversed after the term of the therapy. In addition to the alteration in GI dysfunction and GM, more than half of the studies also included assessment of change in behavior after probiotic therapy. Most of these studies recorded a reduction in the severity of ASD symptoms, although not all reached significance. Noted positive effects on mood and general behavior were (i) statistically significant differences in behavioral scores for disruptive/ antisocial behavior, anxiety problems, communication disturbances, and self- absorbed behavior compared to baseline [96, 104]; (ii) decrease in the severity of speech/language/communication performance [97, 101]; (iii) progress in sociability, sensory and cognitive awareness, physical health, and behavior [99, 102]; and

(iv) changes in adaptive, repetitive, and aberrant behaviors including irritability, lethargy, stereotypy, and hyperactivity [105]. Conversely, in other investigations, no statistically significant differences in behavioral scores were detected between probiotics and placebo control groups [73, 103]. Lastly, probiotic treatment reversed the metabolic disruptions and improved behavioral performance. In one of the studies, probiotics significantly lowered the concentration of myeloperoxidase (a marker of inflammation and oxidation) in ASD patients compared to other autistic individuals not taking probiotics [106]. Metabolic changes were also observed in another former study in which probiotics substantially diminished fecal levels of propionic acid in ASD individuals [49]. In a third study, probiotic supplementation significantly reduced metabolic products of the pathogenic *Candida* yeast species [98], which are remarkably elevated in the urines of individuals with ASD [108]. This biological improvement also corresponded to behavioral outcomes, where children showed significant ameliorations in their ability to concentrate and fulfill orders. In 2015, an intriguing randomized trial demonstrated that early postnatal probiotic administration has a preventive effect on ASD and may reduce the risk of developing further neuropsychiatric disorder [100]. Results showed that from children randomly assigned to probiotic or placebo groups during the first 6 months of life, 17% in the placebo group had a diagnosis of ASD at the age of 13, while none of the children in the probiotic group did.

At last, though probiotics are considered a relatively risk-free option for individuals with ASD, the current literature cannot confidently state their safety as there is a paucity of systematic reporting of adverse events. However, among the studies that monitored side effects, the reported ones (bloating, skin rash, worsening constipation or diarrhea, and weight loss) appear to be mild, transient, and infrequent [96, 99, 102, 105]. Conclusively, due to the large heterogeneity between trials, studies provide only suggestive but not conclusive evidence regarding the efficacy of probiotics on GI and behavioral symptoms among ASD patients. Thus, future probiotic research holds hopes for discovering the optimal species, strains, strength, and length of probiotic therapy for the particular comorbidity profile of different individuals with ASD.

Prebiotics

The International Scientific Association for Probiotics and Prebiotics (ISAPP) defines prebiotics as substrates "selectively utilized by host microorganisms, confer- ring a health benefit" [75]. Such benefits are not limited to gut homeostasis but can extend elsewhere in the organism, leading to improvements of the immune, meta- bolic, endocrine, or nervous functions. Fructans, comprising fructooligosaccharides (FOS) and inulin, and galactans (galactooligosaccharides (GOS)) are the most rec- ognized prebiotics. Differently from most dietary fibers, which promote growth of a wide variety of microorganisms, prebiotics display a selective effect, being a sub- strate for beneficial strains only, while excluding metabolism by pathogenic bacteria [6]. Thus, the main reason for a potential influence of prebiotics on the treatment of ASD concerns the selective enrichment of *Lactobacillus* and/or *Bifidobacterium* spp. Besides, various other mechanisms have been identified through which prebiotics can act, including generation of SCFA that have an influence on gut energy metabo- lism, barrier function, water fluxes, motility [109], elongation of microvilli, increase in mucus layer thickness, and consequent protection of gut epithelium. However, as with probiotics, prebiotic studies regarding their impact in neurological problems are few and not conclusive. In rats, oral administration of GOS elevates BDNF levels [110], normalizes LPS-induced anxiety, and modulates cortical IL-1β levels [111], thus confirming the potential role for prebiotics in ASD where anxiety and neuroin- flammation are prominent clinical features. Lastly, only two clinical trials examined the use of prebiotics in children with ASD [105, 112]. In the first study, GOS alone did not have a significant effect on GI symptoms, while when combined with bovine colostrum, GI symptoms improved as well as irritability scores and stereotypy [105]. Similarly, in the second trial, GOS intervention did not show a significant impact when provided alone. Nevertheless, when associating GOS treatment with an exclusion diet, notable increases occurred in beneficial bacteria supporting improvement in antisocial behavior [112]. GOS also affected SCFA production decreasing propionic acid as a result of normalized GM composition. Thus, it is sensible to postulate that combined intervention therapies might have a better impact on ASD individuals than single dietary approach and prebiotics could be a useful option for ASD chil- dren treatment in early life.

Fecal microbiota transplantation

The process of FMT consists in the delivery of feces from a healthy donor to a patient with gut dysbiosis, with the aim of replacing an impaired microbiota with a healthy one. Unlike probiotics, where only a restricted number of bacterial species are supplemented, FMT allows the transplantation of thousands of different components of the healthy GM. In fact, FMT is one of the most effective techniques recently considered in treating recurrent ATB-refractory *Clostridium difficile* infection [113] and has shown varying levels of success in patients with other GI diseases such as irritable bower syndrome [114], and non-GI diseases like autoimmune disorders, obesity, and insulin sensitivity [115]. Given the growing evidence for a role of GM disruption and GI symptoms in ASD, clinical trials are under way using FMT to treat children with ASD. In a recent open-label study, ASD children with chronic GI problems underwent a modified FMT protocol, termed microbiota transfer therapy (MTT), consisting of a 2-week vancomycin treatment followed by bowel cleansing, and administration of a high dose of standardized human gut microbiota [24]. After this 10-week treatment, MTT induced an 80%

improvement of GI function and a slow but steady improvement in behavioral ASD scoring, both of them maintained at least 8 weeks after treatment stopped. In line with previous reports, few adverse events were described, and the most commonly reported include mild diarrhea, abdominal tenderness, flatulence, and nausea thus confirming treatment safety and tolerability in autistic patients.

Coincident with these clinical improvements, FMT confirmed its ability in modifying GM composition by significantly increasing bacterial diversity and the relative abundance of *Bifidobacterium* and *Prevotella* [24]. Two years after this original clinical trial was completed, re-evaluation of the participants showed that GI benefits were mostly maintained and ASD symptoms were reported to have improved significantly since the end of treatment [116]. Changes in the GM also persisted for 2 years. According to these encouraging observations, intensive FMT intervention can be considered an effective and well-tolerated promising approach in treating children with ASD who have GI problems. However, the safety of FMT should be considered as it carries many risks including aspiration and transmission of opportunistic pathogens to recipients. Therefore, major efforts at refinement are necessary before FMT can be adopted widely.

Dietary interventions

A number of nutrition intervention strategies have been explored to treat behavioral symptoms and comorbid GI distress [117], but evidence is still relatively weak and sometimes inconsistent, as in the case with gluten-free and/or casein-free (GF/CF) diet and ketogenic diet (KD).

Various observational studies reported alleviation of GI problems and/or significant behavioral improvements in ASD children following an extended GF/CF diet [118, 119]. However, elimination diets for ASD patients should not be recommended as standard treatment and only be considered after reaching a diagnosis of an adverse food reaction since dietary restrictions are likely to limit variety of food intake and provoke nutritional deficiencies in developing children [120]. Moreover, GF/CF diet can also exacerbate the already disrupted gut microbial composition in ASD by reducing beneficial and increasing opportunistic bacteria [121].

Improvements in neurobehavioral symptoms have also been reported in ASD as a result of following a KD. KD is a high-fat and low-carbohydrate diet that, due to its beneficial effects on GM composition, has been suggested as a treatment for ASD. In humans, administration of KD to individuals with ASD results in increased sociability, improved communication, and decreased repetitive comportments [122]. In animal models, both improvement of behavioral symptoms and com- positional remodeling of GM after KD have been described [123, 124]. However, KD causes an "antimicrobial"-like effect by significantly decreasing the total gut microbial abundance. Moreover, it results in an increase in clostridia species and may lead to adverse effects such as dehydration [125]. So, side effects and increase in harmful bacteria in the GM associated with limited number of positive results constitute insufficient evidence for the practicability of KD as a treatment for ASD.

Conclusion and perspectives

In the last few years, the importance of GM in the maintenance of physiologi- cal state into the

CNS is supported by several studies that have shown qualitative and quantitative alterations of the intestinal flora in a number of neuropsychiatric diseases. Within neurobehavioral disorders, it seems that at least a subset of the cases comprising ASD are connected to, and perhaps dependent on, the health and well-being of the GM. In recent years the increased prevalence of ASD, along with the evidence of a significant link between ASD and GI disturbances, raised a special interest in exploring the reciprocal influences between GM and brain under the so-called GBA. Alterations of GM composition in children with ASD presented in the literature mainly consist in reduced levels of *Bifidobacterium* and increased levels of *Clostridium* spp. and *Desulfovibrio*. However, the available data do not allow to define a characteristic and unique profile of ASD. If dysbiosis is confirmed to be a precipitating factor in ASD, then several potential therapeutic approaches ranging from ATBs, probiotics, prebiotics, up to FMT and other nutritional strategies may be useful adjuvant treatments in these patients. Future research, together with the application of state-of-the-art "omics" methods, could address possible unequivocal microbial biomarkers for ASD. Further, as dysbiosis contributes to a significant subset of ASD, specific identification of ASD endophenotypes will allow patient stratification and personalized interventions. Addressing microbial processes could be the aim of the next pharmacological therapy of ASD that will potentially help to alleviate the burden of this disorder for the millions of people affected worldwide.

Abbreviations

4EPS	4-ethylphenylsulfate
ASD	autism spectrum disorders
ATB	antibiotic
BDNF	brain-derived neurotrophic factor
CFU	colony-forming unit
CNS	central nervous system
section	cesarean section
FDA	Food and Drug Administration
FMT	fecal microbiota transplantation
FOS	fructooligosaccharides
GABA	gamma-aminobutyric acid
GBA	gut-brain axis
GF/CF	gluten-free and/or casein-free
GI	gastrointestinal
GM	gut microbiota

GOS	galactooligosaccharides
IL	interleukin
INF	interferon
ISAPP	International Scientific Association for Probiotics and Prebiotics
KD	ketogenic diet
LPS	lipopolysaccharide
MTT	microbiota transfer therapy
PANDAS	pediatric autoimmune neuropsychiatric disorder associated with streptococcal infections
PANS	pediatric acute-onset neuropsychiatric syndrome
PT	preterm
SCFA	short-chain fatty acids
TNF	tumor necrosis factor
TFGF	transforming growth factor

Author details

Rouba El Khatib[1], Dolla Karam-Sarkis[1*], Anne-Judith Waligora-Dupriet[2] and Marie-José Butel[2*]

Microbiology Laboratory, Faculty of Pharmacy, Saint-Joseph University, Beirut, Lebanon

UMR-S U 1139, Pathophysiology and Pharmacotoxicology of the Human Placenta, Pre- and Postnatal Microbiota, INSERM, Hospital University Department Risk in Pregnancy, Faculty of Pharmacy, Paris University, France

*Address all correspondence to: marie-jose.butel@parisdescartes.fr and dolla.sarkis@usj.edu.lb

References

[1] Bik EM. The hoops, hopes, and hypes of human microbiome research. The Yale Journal of Biology and Medicine. 2016;**89**:363-373

[2] Bäckhed F, Roswall J, Peng Y, Feng Q , Jia H, Kovatcheva-Datchary P, et al. Dynamics and stabilization of the human gut microbiome during the first year of life. Cell Host & Microbe. 2015;**17**:690-703

[3] Dinan TG, Cryan JF. The microbiome-gut-brain Axis in health and disease. Gastroenterology Clinics of North America. 2017;**46**:77-89

[4] Carabotti M, Scirocco A, Maselli MA, Severi C. The gut-brain axis: Interactions between enteric microbiota, central and enteric nervous systems. Annals of Gastroenterology. 2015;**28**:203-209

[5] Kim Y-K, Shin C. The microbiota- gut-brain axis in neuropsychiatric disorders: Pathophysiological mechanisms and novel treatments. Current Neuropharmacology. 2018;**16**:559-573

[6] Campion D, Ponzo P, Alessandria C, Saracco GM, Balzola F. The role of microbiota in autism spectrum disorders. Minerva Gastroenterologica e Dietologica. 2018;**64**:333-350

[7] Desbonnet L, Clarke G, Shanahan F, Dinan TG, Cryan JF. Microbiota is essential for social development in the mouse. Molecular Psychiatry. 2014;**19**:146-148

[8] Sharon G, Sampson TR, Geschwind DH, Mazmanian SK. The central nervous system and the gut microbiome. Cell. 2016;**167**:915-932

[9] Collado MC, Rautava S, Aakko J, Isolauri E, Salminen S. Human gut colonisation may be initiated in utero by distinct microbial communities in the placenta and amniotic fluid. Scientific Reports. 2016;**6**:1-13

[10] Willyard BC. Could baby's first bacteria take root before birth. Nature. 2018;**553**:264-266

[11] Milani C, Duranti S, Bottacini F, Casey E, Turroni F, Mahony J, et al. The first microbial colonizers of the human gut: Composition, activities, and health implications of the infant gut microbiota. Microbiology and Molecular Biology Reviews. 2017;**81**:1-67

[12] Yatsunenko T, Rey FE, Manary MJ, Trehan I, Dominguez-Bello MG, Contreras M, et al. Human gut microbiome viewed across age and geography. Nature. 2012;**486**:222-227

[13] Diaz HR. Fetal, neonatal, and infant microbiome: Perturbations and subsequent effects on brain development and behavior. Seminars in Fetal & Neonatal Medicine. 2016;**21**:410-417

[14] Liu J, Zhang M, Kong X-J. Gut microbiome and autism: Recent advances and future perspectives. North American Journal of Medical Sciences. 2016;**9**:104-115

[15] American Psychiatric Association. Diagnostic and Statistical Manual of Mental Disorders (DSM-5®). Washington, DC: American Psychiatric Publishing; 2013

[16] Modabbernia A, Velthorst E, Reichenberg A. Environmental risk factors for autism: An evidence-based review of systematic reviews and meta- analyses. Molecular Autism. 2017;**8**:1-16

[17] McElhanon BO, McCracken C, Karpen S, Sharp WG. Gastrointestinal symptoms in autism spectrum disorder: A meta-analysis. Pediatrics. 2014;**133**:872-883

[18] Hsiao EY, McBride SW, Hsien S, Sharon G, Hyde ER, McCue T, et al. Microbiota modulate behavioral and physiological abnormalities associated with neurodevelopmental disorders. Cell. 2013;**155**:1451-1463

[19] Borre Y, Golubeva A, Crispie F, Scott K, Hyland N, Stanton C, et al. Mode of delivery at birth and behavioral outcomes: Rewiring of the brain- gut-microbiome axis? In: Society for Neuroscience. Washington, USA: Poster presentation; 2014:15-19

[20] Curran EA, O'Neill SM, Cryan JF, Kenny LC, Dinan TG, Khashan AS, et al. Research review: Birth by caesarean section and development of autism spectrum disorder and attention- deficit/hyperactivity disorder: A systematic review and meta-analysis. The Journal of Child Psychology and Psychiatry and Allied Disciplines. 2015;**56**:500-508

[21] Kuzniewicz MW, Wi S, Qian Y, Walsh EM, Armstrong MA, Croen LA. Prevalence and neonatal factors associated with autism spectrum disorders in preterm infants. The Journal of Pediatrics. 2014;**164**:20-25

[22] Grissom NM, Reyes TM. Gestational overgrowth and undergrowth affect neurodevelopment: Similarities and differences from behavior to epigenetics. International Journal of Developmental Neuroscience. 2013;**31**:406-414

[23] Dinan TG, Cryan JF. The impact of gut microbiota on brain and behaviour: Implications for psychiatry. Current Opinion in Clinical Nutrition and Metabolic Care. 2015;**18**:552-558

[24] Kang DW, Adams JB, Gregory AC, Borody T, Chittick L, Fasano A, et al. Microbiota transfer therapy alters gut ecosystem and improves gastrointestinal and autism symptoms: An open-label study. Microbiome. 2017;**5**:1-16

[25] Kumbhare SV, Kumar H, Chowdhury SP, Dhotre DP, Endo A, Mättö J, et al. A cross-sectional comparative study of gut bacterial community of Indian and Finnish children. Scientific Reports. 2017;**7**:1-13

[26] Palleja A, Mikkelsen KH, Forslund SK, Kashani A, Allin KH, Nielsen T, et al. Recovery of gut microbiota of healthy adults following antibiotic exposure. Nature Microbiology. 2018;**3**:1255-1265

[27] Horvath K, Papadimitriou JC, Rabsztyn A, Drachenberg C, Tildon TJ. Gastrointestinal abnormalities in children with autistic disorder. Journal of Pediatrics. 1999;**135**:559-563

[28] Atladottir HO, Henriksen TB, Schendel DE, Parner ET. Autism after infection, febrile episodes, and antibiotic use during pregnancy: An exploratory study. Pediatrics. 2012;**130**:e1447-e1454

[29] Degroote S, Hunting DJ, Baccarelli AA, Takser L. Maternal gut and fetal brain connection: Increased anxiety and reduced social interactions in Wistar rat offspring following peri-conceptional antibiotic exposure. Progress in Neuro-Psychopharmacology & Biological Psychiatry. 2016;**71**:76-82

[30] de Theije CGM, Wopereis H, Ramadan M, van Eijndthoven T, Lambert J, Knol J, et al. Altered gut microbiota and activity in a murine model of autism spectrum disorders. Brain, Behavior, and Immunity. 2014;**37**:197-206

[31] Estes ML, McAllister AK. Maternal immune activation: Implications for neuropsychiatric disorders. Science. 2016;**353**:772-777

[32] O'Mahony SM, Marchesi JR, Scully P, Codling C, Ceolho AM, Quigley EMM, et al. Early life stress alters behavior, immunity, and microbiota in rats: Implications for irritable bowel syndrome and psychiatric illnesses. Biological Psychiatry. 2009;**65**:263-267

[33] Galley JD, Nelson MC, Yu Z, Dowd SE, Walter J, Kumar PS, et al. Exposure to a social stressor disrupts the community structure of the colonic mucosa-associated microbiota. BMC Microbiology. 2014;**14**:1-13

[34] Zhai Q , Cen S, Jiang J, Zhao J, Zhang H, Chen W. Disturbance of trace element and gut microbiota profiles as indicators of autism spectrum disorder: A pilot study of Chinese children. Environmental Research. 2019;**171**:501-509

[35] Jacobson JL, Muckle G, Ayotte P, Dewailly É, Jacobson SW. Relation of prenatal methylmercury exposure from environmental sources to childhood IQ. Environmental Health Perspectives. 2015;**123**:827-833

[36] Finegold SM, Dowd SE, Gontcharova V, Liu C, Henley KE, Wolcott RD, et al. Pyrosequencing study of fecal microflora of autistic and control children. Anaerobe. 2010;**16**:444-453

[37] Martirosian G, Ekiel A, Aptekorz M, Wiechula B, Kazek B, Jankowska-Steifer E. Fecal lactoferrin and Clostridium spp. in stools of autistic children. Anaerobe. 2011;**17**:43-45

[38] Williams BL, Hornig M, Buie T, Bauman ML, Cho Paik M, Wick I, et al. Impaired carbohydrate digestion and transport and mucosal dysbiosis in the intestines of children with autism and gastrointestinal disturbances. PLoS One. 2011;**6**:e24585

[39] Kang DW, Park JG, Ilhan ZE, Wallstrom G, LaBaer J, Adams JB, et al. Reduced incidence of prevotella and other fermenters in intestinal microflora of autistic children. PLoS One. 2013;**8**:68322

[40] de Angelis M, Francavilla R, Piccolo M, De Giacomo A, Gobbetti M. Autism spectrum disorders and intestinal microbiota. Gut Microbes. 2015;**6**:207-213

[41] Strati F, Cavalieri D, Albanese D, De Felice C, Donati C, Hayek J, et al. New evidences on the altered gut microbiota in autism spectrum disorders. Microbiome. 2017;**5**:1-11

[42] Zhang M, Ma W, Zhang J, He Y, Wang J. Analysis of gut microbiota profiles and microbe-disease associations in children with autism spectrum disorders in China. Scientific Reports. 2018;**8**:13981

[43] Liu S, Li E, Sun Z, Fu D, Duan G, Jiang M, et al. Altered gut microbiota and short chain fatty acids in Chinese children with autism spectrum disorder. Scientific Reports. 2019;**9**:1-9

[44] Borre YE, O'Keeffe GW, Clarke G, Stanton C, Dinan TG, Cryan JF. Microbiota and neurodevelopmental windows: Implications for brain disorders. Trends in Molecular Medicine. 2014;**20**:509-518

[45] Kushak RI, Buie TM, Murray KF, Newburg DS, Chen C, Nestoridi E. Evaluation of intestinal function in children with autism and gastrointestinal symptoms. Journal of Pediatric Gastroenterology and Nutrition. 2016;**62**:687-691

[46] Finegold SM, Molitoris D, Song Y, Liu C, Vaisanen M, Bolte E, et al. Gastrointestinal microflora studies in late-onset autism. Clinical Infectious Diseases. 2002;**35**(Suppl 1):s6-s16

[47] Argou-Cardozo I, Zeidán-Chuliá F. Clostridium bacteria and autism spectrum conditions: A systematic review and hypothetical contribution of environmental glyphosate levels. Medical Science. 2018;**6**:29

[48] Wang L, Christophersen CT, Sorich MJ, Gerber JP, Angley MT, Conlon MA. Increased abundance of Sutterella spp. and Ruminococcus torques in feces of children with autism spectrum disorder. Molecular Autism. 2013;**4**:1

[49] Adams JB, Johansen LJ, Powell LD, Quig D, Rubin RA. Gastrointestinal flora and gastrointestinal status in children with autism—Comparisons to typical children and correlation with autism severity. BMC Gastroenterology. 2011;**11**:22

[50] Williams BL, Hornig M, Parekh T. Application of novel PCR-based methods for detection, quantitation, and phylogenetic characterization of. MBio. 2012;**3**:1-11

[51] Tomova A, HusarovaV, Lakatosova S, Bakos J, Vlkova B, Babinska K, et al. Gastrointestinal microbiota in children with autism in Slovakia. Physiology & Behavior. 2015;**138**:179-187

[52] Finegold SM. Desulfovibrio species are potentially important in regressive autism. Medical Hypotheses. 2011;**77**:270-274

[53] Gondalia SV, Palombo EA, Knowles SR, Cox SB, Meyer D, Austin DW. Molecular characterisation of gastrointestinal microbiota of children with autism (with and without gastrointestinal dysfunction) and their neurotypical siblings. Autism Research. 2012;**5**:419-427

[54] Son JS, Zheng LJ, Rowehl LM, Tian X, Zhang Y, Zhu W, et al. Comparison of fecal microbiota in children with autism spectrum disorders and neurotypical siblings in the simons simplex collection. PLoS One. 2015;**10**:1-19

[55] Kantarcioglu AS, Kiraz N, Aydin A. Microbiota-gut-brain axis: Yeast species isolated from stool samples of children with suspected or diagnosed autism spectrum disorders and in vitro susceptibility against nystatin and fluconazole. Mycopathologia. 2016;**181**:1-7

[56] Iovene MR, Bombace F, Maresca R, Sapone A, Iardino P, Picardi A, et al. Intestinal dysbiosis and yeast isolation in stool of subjects with autism spectrum disorders. Mycopathologia. 2017;**182**:349-363

[57] Wang L, Christophersen CT, Sorich MJ, Gerber JP, Angley MT, Conlon MA. Elevated fecal short chain fatty acid and ammonia concentrations in children with autism spectrum disorder. Digestive Diseases and Sciences. 2012;**57**:2096-2102

[58] Andriamihaja M, Lan A, Beaumont M, Audebert M, Wong X, Yamada K, et al. The deleterious metabolic and genotoxic effects of the bacterial metabolite p-cresol on colonic epithelial cells. Free Radical Biology & Medicine. 2015;**85**:219-227

[59] Persico AM, Napolioni V. Urinary p-cresol in autism spectrum disorder. Neurotoxicology and Teratology. 2012;**36**:82-90

[60] Marteau P. Butyrate-producing bacteria as pharmabiotics for inflammatory bowel disease. Gut. 2012;**62**:1673

[61] El-Ansary AK, Ben Bacha AG, Al-Ayahdi LY. Plasma fatty acids as diagnostic markers in autistic patients from Saudi Arabia. Lipids in Health and Disease. 2011;**10**:62

[62] Macfabe DF, Cain N, Boon F, Ossenkopp K, Cain D. Effects of the enteric bacterial metabolic product propionic acid on object-directed behaviour, social behaviour, cognition, and neuro-inflammation in adolescent rats: Relevance to autism spectrum disorder. Behavioural Brain Research. 2011;**217**:47-54

[63] Kelly JR, Kennedy PJ, Cryan JF, Dinan TG, Clarke G, Hyland NP. Breaking down the barriers: The gut microbiome, intestinal permeability and stress-related psychiatric disorders. Frontiers in Cellular Neuroscience. 2015;**9**:392

[64] De Magistris L, Familiari V, Pascotto A, Sapone A, Frolli A, Iardino P, et al. Alterations of the intestinal barrier in patients with autism spectrum disorders and in their first-degree relatives. Journal of Pediatric Gastroenterology and Nutrition. 2010;**51**:418-424

[65] Haba R, Shintani N, Onaka Y, Wang H, Takenaga R, Hayata A, et al. Lipopolysaccharide affects exploratory behaviors toward novel objects by impairing cognition and/or motivation in mice: Possible role of activation of the central amygdala. Behavioural Brain Research. 2012;**228**:423-431

[66] de La Serre C, de Lartigue G, Raybould H. Chronic exposure to low dose bacterial lipopolysaccharide inhibits leptin signaling in vagal afferent neurons. Physiology & Behavior. 2015;**139**:188-194

[67] Grigoleit JS, Kullmann JS, Wolf OT, Hammes F, Wegner A, Jablonowski S, et al. Dose-dependent effects of endotoxin on neurobehavioral functions in humans. PLoS One. 2011;**6**:1-10

[68] Emanuele E, Orsi P, Boso M, Broglia D, Brondino N, Barale F, et al. Low-grade endotoxemia in patients with severe autism. Neuroscience Letters. 2010;**471**:162-165

[69] Neul JL, Sahin M. Therapeutic advances in autism and other neurodevelopmental disorders. Neurotherapeutics. 2015;**12**:519-520

[70] Santocchi E, Guiducci L, Fulceri F, Billeci L, Buzzigoli E, Apicella F, et al. Gut to brain interaction in Autism Spectrum Disorders: A randomized controlled trial on the role of probiotics on clinical, biochemical and neurophysiological parameters. BMC Psychiatry. 2016;**16**:1-16

[71] Liu J, Wan G, Huang M, Agyapong G, Zou T, Zhang X, et al. Probiotic therapy for treating behavioral and gastrointestinal symptoms in autism Spectrum disorder: A systematic review of clinical trials. Current Medical Science. 2019;**39**:173-184

[72] Williams KA, Swedo SE. Post- infectious autoimmune disorders: Sydenham' s chorea, PANDAS and beyond. Brain Research. 1617;**2019**:144-154

[73] Sandler RH, Finegold SM, Bolte ER, Buchanan CP, Maxwell AP, Vaisanen M-L, et al. Short-term benefit from oral vancomycin treatment of regressive- onset autism. Journal of Child Neurology. 2000;**15**:429-435

[74] Frye RE, Slattery J, Macfabe DF, Allen-vercoe E, Parker W, Rodakis J, et al. Approaches to studying and manipulating the enteric microbiome to improve autism symptoms. Microbial Ecology in Health and Disease. 2015;**26**:1-14

[75] Gibson GR, Hutkins R, Sanders ME, Prescott SL, Reimer RA, Salminen SJ, et al. The International Scientific Association for Probiotics and Prebiotics (ISAPP) consensus statement on the definition and scope of prebiotics. Nature Reviews. Gastroenterology & Hepatology. 2017;**14**:491-502

[76] Srinivasjois R, Rao S, Patole S. Probiotic supplementation in children with autism spectrum disorder. Archives of Disease in Childhood. 2015;**100**:505-506

[77] Ng QX, Peters C, Yih C, Ho X, Lim DY, Yeo W. A meta-analysis of the use of probiotics to alleviate depressive symptoms. Journal of Affective Disorders. 2018;**228**:13-19

[78] Messaoudi M, Lalonde R, Violle N, Javelot H, Desor D, Nejdi A, et al. Assessment of psychotropic-like properties of a probiotic formulation (Lactobacillus helveticus R0052 and Bifidobacterium longum R0175) in rats and human subjects. The British Journal of Nutrition. 2011;**105**:755-764

[79] Dinan TG, Stanton C, Cryan JF. Psychobiotics: A novel class of psychotropic. Biological Psychiatry. 2013;**74**:720-726

[80] Sanders ME. Probiotics and microbiota composition. BMC Medicine. 2016;**14**:1-3

[81] Macfabe DF. Enteric short-chain fatty acids: Microbial messengers of metabolism, mitochondria, and mind: Implications in autism spectrum disorders. Microbial Ecology in Health and Disease. 2015;**26**:1-14

[82] Doenyas C. Gut microbiota, inflammation, and probiotics on neural development in autism spectrum disorder. Neuroscience. 2018;**374**:271-286

[83] Critchfield JW, Van Hemert S, Ash M, Mulder L, Ashwood P. The potential role of probiotics in the management of childhood autism spectrum disorders. Gastroenterology Research and Practice. 2011;**2011**:1-8

[84] Bercik P, Denou E, Collins J, Jackson W, Lu JUN, Jury J, et al. The intestinal microbiota affect central levels of brain-derived neurotropic factor and behavior in mice. Gastroenterology. 2011;**141**:599-609

[85] Preidis GA, Versalovic J. Targeting the human microbiome with antibiotics, probiotics, and prebiotics: Gastroenterology enters the metagenomics era. Gastroenterology. 2009;**136**:2015-2031

[86] Liu X, Cao S, Zhang X. Modulation of gut microbiota-brain axis by probiotics, prebiotics, and diet. Journal of Agricultural and Food Chemistry. 2015;**63**:7885-7895

[87] Bravo JA, Forsythe P, Chew MV, Escaravage E, Savignac HM, Dinan TG, et al. Ingestion of Lactobacillus strain regulates emotional behavior and central GABA receptor expression in a mouse via the vagus nerve. PNAS. 2011;**108**:16050-16055

[88] Israelyan N, Gross K. Serotonin as a link between the gut-brain-microbiome axis in autism spectrum disorders. Pharmacological Research. 2018;**132**:1-6

[89] Buffington SA, Di Prisco GV, Auchtung TA, Ajami NJ, Pestrosino JF, Costa-Mattioli M. Microbial reconstitution reverses maternal diet- induced social and synaptic deficits in offspring. Cell. 2016;**165**:1762-1775

[90] Wang X, Yang J, Zhang H, Yu J, Yao Z. Oral probiotic administration during pregnancy prevents autism- related behaviors in offspring induced by maternal immune activation via anti- inflammation in mice. Autism Research. 2019;**12**:576-588

[91] El-ansary A, Bacha A, Ben BG, Al-Orf N, Shafi Bhat R, Moubayed N, et al. Probiotic treatment reduces the autistic-like excitation/inhibition imbalance in juvenile hamsters induced by orally administered propionic acid and clindamycin. Metabolic Brain Disease. 2018;**33**:1155-1164

[92] Liu Y, Liu W, Wu C, Juan Y, Wu Y, Tsai H, et al. Psychotropic effects of Lactobacillus plantarum PS128 in early life-stressed and naïve adult mice. Brain Research. 1631;**2016**:1-12

[93] Accordino RE, Kidd C, Politte LC, Henry CA, Mcdougle C. Psychopharmacological interventions in autism spectrum disorder. Expert Opinion on Pharmacotherapy. 2016;**17**:937-952

[94] Donaldson ZR, Young LJ. Oxytocin, vasopressin, and the neurogenetics of sociality. Science. 2008;**322**:900-904

[95] Blades M. Autism: An interesting dietary case history. Nutrition & Food Science. 2000;**30**:137-139

[96] Parracho HMR, Gibson GR, Knott F, Bosscher D, Kleerebezem M, Mccartney AL. A double-blind, placebo-controlled, crossover- designed probiotic feeding study in children diagnosed with autistic spectrum disorders. International Journal of Probiotics and Prebiotics. 2010;**5**:69-74

[97] Ray S, Sherlock A, Wilken T, Woods T. Cell wall lysed probiotic tincture decreases immune response to pathogenic enteric bacteria and improves symptoms in autistic and immune compromised children. Explore. 2010;**19**:1-5

[98] Kalzuna-Czaplinska J, Blaszczyk S. The level of arabinitol in autistic children after probiotic therapy. Nutrition. 2012;**28**:124-126

[99] West R, Roberts E, Sichel LS, Sichel J. Improvementsingastrointestinal symptoms among children with autism spectrum disorder receiving the Delpro probiotic and immunomodulator formulation. Journal of Probiotics & Health. 2013;**1**:1-6

[100] Partty A, Kalliomaki M, Wacklin P, Salminen S, Isolauri E. A possible link between early probiotic intervention and the risk of neuropsychiatric disorders later in childhood: A randomized trial. Pediatric Research. 2015;**77**:823-828

[101] Grossi E, Melli S, Dunca D, Terruzzi V. Unexpected improvement in core autism spectrum disorder symptoms after long-term treatment with probiotics. SAGE Open Medical Case Reports. 2016;**4**

[102] Shaaban SY, El Gendy YG, Mehanna NS, El-Senousy WM, El-feki HSA, Saad K, et al. The role of probiotics in children with autism spectrum disorder: A prospective, open- label study. Nutritional Neuroscience. 2017;**21**:676-681

[103] Liu Y-W, Liong M-T, Chung Y-C, Huang H-Y, Peng W-S, Cheng Y-F, et al. Effects of Lactobacillus plantarum PS128 on children with autism spectrum disorder in Taiwan: A randomized, double-blind. Placebo-Controlled Trial. Nutrients. 2019;**11**:820

[104] Kobliner V, Nutrition H, Mumper E. Reduction in obsessive compulsive disorder and self-injurious behavior with *Saccharomyces boulardii* in a child with autism: A case report. Integrative Medicine (Encinitas, Calif.). 2019;**17**:14-17

[105] Sanctuary MR, Kain JN, Chen SY, Kalanetra K, Lemay G, Rose DR, et al. Pilot study of probiotic/colostrum supplementation on gut function in children with autism and gastrointestinal symptoms. PLoS One. 2019;**14**:1-30

[106] Russo AJ. Decreased plasma myeloperoxidase associated with probiotic therapy in autistic children. Clinical Medicine Insights: Pediatrics. 2015;**9**:13-17

[107] Golnik AE, Ireland ÆM. Complementary alternative medicine for children with autism: A physician survey. Journal of Autism and Developmental Disorders. 2009;**39**:996-1005

[108] Shaw W, Kassen E, Ascp MT, Chaves E. Assessment of antifungal drug therapy in autism by measurement of suspected microbial metabolites in urine with gas chromatography-mass spectrometry. Clinical Practice of Alternative Medicine. 2000;**1**:15-26

[109] Gibson G, Roberfroid M, De Louuain C. Dietary modulation of the human colonic microbiota: Introducing the concept of prebiotics. The Journal of Nutrition. 1995;**125**:1401-1412

[110] Savignac HM, Corona G, Mills H, Chen L, Spencer JPE, Tzortzis G, et al. Prebiotic feeding elevates central brain derived neurotrophic factor , N-methyl- D-aspartate receptor subunits and serine. Neurochemistry International. 2013;**63**:756-764

[111] Savignac HM, Couch Y, Stratford M, Bannerman DM, Tzortzis G, Anthony DC, et al. Prebiotic administration normalizes lipopolysaccharide (LPS)-induced anxiety and cortical 5-HT2A receptor and IL1-b levels in male mice. Brain, Behavior, and Immunity. 2016;**52**:120-131

[112] Grimaldi R, Gibson GR, Vulevic J, Giallourou N, Castro-mejía JL, Hansen LH, et al. A prebiotic intervention study in children with autism spectrum disorders. Microbiome. 2018;**6**:1-13

[113] Bagdasarian N, Rao K, Malani P. Diagnosis and treatment of *Clostridium difficile* in adults: A systematic review. Journal of the American Medical Association. 2015;**313**:398-408

[114] Cui B, Feng Q , Wang H, Wang M, Peng Z, Li P, et al. Fecal microbiota transplantation through mid-gut for refractory Crohn' s disease: Safety, feasibility, and efficacy trial results. Journal of Gastroenterology and Hepatology. 2015;**30**:51-58

[115] Vrieze A, ELSVAN N, Holleman F, Salojärvi J, Kootte RS, Bloks VW, et al. Transfer of intestinal microbiota from lean donors increases insulin. Gastroenterology. 2012;**143**:913-916.e7

[116] Kang D, Adams JB, Co DM, Pollard EL, Maldonado J, Mcdonough- means S, et al. Long-term benefit of microbiota transfer therapy on autism symptoms and gut microbiota. Scientific Reports. 2019;**9**:5821

[117] Doenyas C. Dietary interventions for autism spectrum disorder: New perspectives from the gut- brain axis. Physiology & Behavior. 2019;**194**:577-582

[118] Whitely P, Rodgers J, Savery D, Shattock P. A gluten-free diet as an intervention for autism and associated spectrum disorders: Preliminary findings. Autism. 1999;**3**:45-65

[119] Adams SJ, Burton N, Cutress A, Adamson A, McColl E, Hare A, et al. Development of double blind gluten and casein free test foods for use in an autism dietary trial. Journal of Human Nutrition and Dietetics. 2008;**21**:374

[120] Hediger ML, England LJ, Molloy CA, Yu KF, Manning- Courtney P, Mills JL. Reduced bone cortical thickness in boys with autism or autism spectrum reduced bone cortical thickness in boys with autism or autism spectrum disorder. Journal of Autism and Developmental Disorders. 2008;**38**:848-856

[121] Autism Speaks. Autism and Health: A Special Report by Autism Speaks. Advances in Understanding and Treating the Health Conditions that Frequently Accompany Autism. Autism Speaks; 2017, New York, USA.

[122] Castro K, Slongo L, Baronio D, Gottfried C, Schweigert I, Riesgo S. Research in autism spectrum disorders effect of a ketogenic diet on autism spectrum disorder: A systematic review. Research in Autism Spectrum Disorder. 2015;**20**:31-38

[123] Ruskin DN, Svedova J, Cote JL, Sandau U, Rho JM, Kawamura M, et al. Ketogenic diet improves core symptoms of autism in BTBR mice. PLoS One. 2013;**8**:4-9

[124] Newell C, Bomhof MR, Reimer RA, Hittel DS, Rho JM, Shearer J. Ketogenic diet modifies the gut microbiota in a murine model of autism spectrum disorder. Molecular Autism. 2016;**7**:1-6

[125] Ballaban-gil K, Dell O, Pappo SM, Moshc S, Shinnar S. Complications of the ketogenic diet. Epilepsia. 1998;**39**:744-748

6

Role of the Microbiome as the First Metal Detoxification Mechanism

Rebeca Monroy-Torres, Marco Antonio Hernández-Luna, Xochitl Sofía Ramírez-Gómez and Sergio López-Briones

Abstract

Exposure to environmental toxins in water, soil and air are increasing with health effects, mainly in older ages and physiological states (childhood and preg- nancy). The role of the microbiota has been widely studied with effects on the maintenance of health but this is only possible with a diet that promotes it. The traditional Mexican diet is rich in fiber, which has prebiotic effects and has found a higher excretion of arsenic and fluoride in adolescents who maintained a diet high in fiber derived from traditional foods. After several descriptive studies in the state of Guanajuato, since 2004, first with arsenic in drinking water in population of several communities, in 2015, it is achieved through an intervention study with a supplementation of several vitamins and minerals in population adolescent, a greater urinary arsenic and fluoride excretion, as well as a greater consumption of traditional foods such as beans, bananas, orange and quelites. Food is key to main- tain a function of the microbiota, so its review and study should be encouraged.

Keywords: microbiome, traditional Mexican diet, soluble fiber, arsenic, fluoride

Introduction

Exposure to toxins in the environment, soil and air increase the effects on the health of the population in all ages and physiological states of higher priority such as childhood and pregnancy [1]. The role of the microbiota has been extensively studied with effects on health maintenance, but this is only possible with a diet that promotes its development, growth and maintenance of key bacteria [2]. An adequate diet (complete, varied, balanced, sufficient and safe) generates intestinal health by preserving the microbiota [3]. Among the key nutrients, it is known that soluble fibers function as a prebiotic for bacteria. The fructoligosaccharides represents one kind of these soluble fibers and are found in food such as banana, bean, onion, garlic, etc., that are part of the traditional Mexican diet [4]. After several descriptive studies in the state of Guanajuato, since 2004, first with arsenic in drinking water in population of several communities, in 2015, it is achieved through an intervention study with a supplementation of several vitamins and minerals in population adolescent, a greater urinary arsenic and fluoride excretion, as well as a greater consumption of traditional foods such as beans, bananas, orange and quelites [5–7]. The traditional Mexican diet composed for quelites, legumes (beans, lentils), a wide variety of fruits and vegetables are rich in flavonoids, vitamins (A, B, C, D, E) have been associated prebiotic effect and a greater excretion of metals. Monroy-Torres, et al found more excretion for fluoride and arsenic in adolescents who maintained these foods in their diet during the study [7]. Knowing the food transition and its impact on the changes in the microbiota is part of a research subline of Environmental Nutrition and Food Security [8].

The contamination of the main environmental resources, necessary to preserve the life of any human being are water, soil and air, which have a deterioration in quantity and quality integrating the presence of various pollutants with their corresponding health risks. Current evidence has integrated and recognized the main toxins present in water, soil, air and therefore in food, with important implications for the health of people at different stages of life, are lead, cadmium, mercury and arsenic [9, 10].

Lead, for many years, is part of the compounds of gasoline and to date is still used in pottery despite the existence of a rule that prohibits it for Mexico and in many countries. Low doses begin to generate problems especially in childhood and when there is anemia or poor nutrition, its toxicity is exacerbated. Sweets have been a source of lead exposure and also some lead-based paints. Pregnant women and children are more susceptible to lead exposure due to their high bioavailability at the gastrointestinal level and its permeability of the blood-brain barrier [11].

Cadmium compounds not only are present in batteries and cigarette smoke, but also food could represent a source of exposure to this metal. Cadmium is used in the manufacture of rechargeable batteries (composition based on nickel and cadmium) whose main problem is that everything containing cadmium is reusable or recyclable being household waste one of the most frequent routes of exposure, so that its exposure has increased and ascending during the twentieth century. One concern is the low dose of this metal to generate kidney damage, bone damage and fractures, although this also varies according to the population group [10, 11].

With respect to mercury, this metal mainly causes deterioration and neurologi- cal damage and the main source of exposure of people is the consumption of fish contaminated with methylmercury, as well as dental amalgams that in rural com- munities remain the main option, given their durability. Due to the exposure of mercury orally through the consumption of fish, food alerts have been generated to avoid or minimize its exposure in pregnant women mainly, since the nervous system develops throughout fetal life, so it is vital to alert the population and avoid the consumption of large fish such as tuna, shark and swordfish [9, 12].

Arsenic is a metalloid present at different concentration in water tables. The toxicity of arsenic depends of its valence states and its organic or inorganic form. Organic As is considered less toxic than inorganic, as it is easier to excrete, this can be found mainly in shellfish and some cereals. Inorganic As is most toxic in its trivalent form that can chemically combine with sulfhydryl groups, these functional groups form intermolecular and intramolecular bridges in proteins and their structure and biological function depend accordingly [1, 12].

El To address in this chapter, with utmost importance, we consider a summary of these four toxic heavy metals since little is known about their interactions and role of the microbiota, in the chemical reactions of oxide-reduction or those that apply for their detoxification of the organism. The health effects will depend on the toxic route of entry, dose and exposure time, as well as the nutritional status of each person. One of the main causes of contamination of food by these metals is that the land and water in which they are contaminated with these environmental toxins and this coupled with the fact that they can also be contaminated in processes of their subsequent industrialization and apart from this. The process generates foods with poor null nutrient content (fiber, vitamins and minerals) and also have more additives and preservatives [9, 10]. Reason for achieving a correct diet that is complete, varied, balanced, sufficient, adequate and safe, has become a challenge, especially safety, defined "that the usual consumption of a diet or food does not involve health risks because it is free of pathogenic microorganisms, toxins and contaminants and is consumed in moderation" [3], whose definition we can affirm is difficult to achieve in the toxicological aspect [1].

Recently, the study of intestinal microbiota has increased significantly, because important protective and metabolic functions have been associated. On this regards, intestinal microbiota competes for nutrients, space and receptors with pathogenic microorganisms, as well as stimulating the production of antimicrobial peptides and immunoglobulins. The microbiota converts many complex substances such as starches, cellulose, pectin and gums into metabolites that are easily absorbed by the host, it also ferments another non-digestible residues from the diet, synthesizes some vitamins and is also involved in the absorption of ions [13, 14].

There are several probiotic bacteria that have been administered gastrointesti- nally to sequester toxins present in food, such as a 250 g yogurt with *1010 CFU* of *Lactobacillus rhamnosus* supplemented with probiotics to reduce exposure to metals in pregnant women and children with elevated blood lead levels in a study con- ducted in Tanzania, changes in blood metal levels were evaluated and children's gut microbiomes were analyzed, finding a protective effect against additional increases in mercury (3.2 nmol/L; P 0.035) and arsenic (2.3 nmol/L; P 0.011), only in women, but not in children [1, 2]. One of the reasons for these differences can be explained by the physiological immaturity of children as well as the stage of growth that are most vulnerable them; In the case of women where benefit was observed, it is explained why it is known that women have an efficiency in the excretion of metals, unlike men [1, 2, 15]. On the other hand, as regards Cadmium, it has been observed in women in the menopause stage, differences in metal metabolism and excretion as well as greater absorption of cadmium orally in women with anemia and osteoporis, causing kidney damage, osteoporosis and osteomalacia [15]. As women, in the menopause period should be more monitored being that is when there is a deterio- ration in nutritional status such as anemia exacerbating increased exposure to lead, for example.

Respect of lifestyle change, nutrition plays an important role in maintain the microbiota; poor quality diets are rich in refined grains and added sugars, salt, unhealthy fats and foods of animal origin; and low in whole grains, fruits, veg- etables, legumes, fish and nuts are the main causes of poor microbiome. Processed red meats are associated with an increase in cardiovascular diseases and strokes.

However, for our population, the relationship between these variables was not conclusive [16].

Gut microbiota is composed of bacteria, fungi, archaea, viruses and protozoa.

Some of the functions of gut microbiota are metabolism and the development of the nervous system and immune system [17]. Furthermore, alterations in microbiota, a process known as dysbiosis, have been associated with the development of obesity and diseases such as diabetes, inflammatory bowel disease and even neoplasms such as colon cancer. Some factors that cause intestinal dysbiosis are antibiotics, alcohol consumption, infection of pathogenic microorganisms and diet [18], this last one can favor the proliferation of specific phyla of bacteria in the gut [19]. Likewise, gut dysbiosis has been associated to exposure to heavy metals as cadmium and mercury [20, 21], halogenated compounds such as fluoride and metalloids such as arsenic, these last two contaminants of water for human consumption [22]. On the other hand, the role of the microbiota in the detoxification of xenobiotics [23], arsenic [24] and heavy metals [25], has been little studied. In the case of heavy metals, it has been observed that probiotics, especially lactic acid bacteria (LAB), could contribute to their elimination, because these kinds of bacteria have a high affinity for heavy metals [26]. Therefore, a diet that favors the proliferation of LAB could be used as the natural detoxification of the organism in populations exposed to environmental pollutants. In this chapter, we review the role of diet in the gut microbiota and its possible use for the natural detoxification of the organism.

Toxicokinetics and heavy metal toxicodynamics

Of the most important aspects to consider in heavy metal poisoning is toxicoki- netics and toxicodynamics. In this regard, the term toxicokinetics refers to the way through which toxins from the source of exposure enter the body, are distributed, biotransformed and eliminated [27]. On the other hand, toxicodynamics studies the mechanisms at the cellular or molecular level by which toxins cause damage [27].

These branches of toxicology are important to know because through toxicokinetics it is possible to prevent the metal from entering the bloodstream, its distribution can be modified and therapeutic measures can be taken to eliminate and reduce the toxic effects caused in the organism [28].

In this context, there are different sources of exposure to heavy metals, these sources can be natural such as water for human consumption or food of animal and plant origin mainly contaminated with arsenic, mercury or lead [29]. On the other hand, there are other sources of exposure such as the use of clay utensils or glazed earthenware that are used in Mexico to prepare food, industrial activities for the manufacture of accumulators, paints, cosmetics, medicines, thermometer manufacturing, etc., which also contribute to heavy metals being available in the environment so that they can enter the body of those exposed to them [27, 30, 31].

Of these sources of exposure, the most worrisome are natural sources such as water and food contaminated with these metals, because the population is unaware that their natural resources are contaminated and use them daily so they have a high risk of these metals that can damage your health, whether in the short, medium and/ or long term [32]. In the case of the sources generated by man as a work practice, it is easier to prevent exposure by taking various hygiene and safety measures in the work environment.

From the sources of exposure, heavy metals enter the body through different routes such as oral, topical and inhalation [27]. However, the main route of entry is the oral and sometimes in the same individual heavy metals can enter through different routes; such as mercury, the most common case is when a glass thermometer containing this metal is broken and the user tries to pick it up with their hands, it will be absorbed through the skin, in addition to room temperature the mercury evaporates in its elementary form so that in this same individual mercury vapors enter by inhalation. But if, in addition, he did not wash his hands after collecting the mercury and immediately consumes food, he can contaminate them by taking them by hand and the mercury will enter orally, although by this route practically the elemental mercury is absorbed very little [33]. In this example, it is clear how the metal can enter via different ways, which increases the amount of metal that enters the body. However, in general, for all metals it is important to consider jointly and not in isolation: the source of exposure, the amount of toxic present, the route of entry of the toxic and the time of exposure to the toxic, since these aspects influence the magnitude of the damage that metals can cause in the organism [27, 28].

From these routes of entry, the four toxicokinetic processes of absorption, distribution, biotransformation and elimination are initiated. The absorption process refers to the passage of the toxic through the route of entry into the blood, for this to happen, it is important to consider whether it is exposed to organic and inorganic compounds or to the elemental form of the metal that is contaminating either water or the food. In the case of organic compounds, they enter the bloodstream more easily because they are fat soluble and diffuse biological membranes faster and therefore their absorption rate may be higher than that of inorganic species. The species in elementary state evaporate so that when they are inhaled, they diffuse through the alveoli [34]. As for the distribution of heavy metals, it is observed that the organic forms are rapidly distributed to the central nervous system since they can easily cross the blood-brain barrier due to their liposolubility, they are also distributed to adipose tissue and other organs with higher fat content such as liver, heart and kidneys [27]. In contrast, inorganic compounds do not pass blood-brain barrier [28]. In general, metals pass placenta, and are attached to the disulfide bonds present in the keratin of the skin, hair and nails. On the other hand, lead is redistributed and deposited in bone and teeth [27].

In the biotransformation process of some heavy metals such as mercury and arsenic mainly, methylation reactions occur [27]. These occur in the liver and kidney, but various biotransformation reactions in the intestine can also be carried out through the microbiota. In this context, there is evidence at the preclinical level in cell cultures and in animal models that the microbiota performs various chemical reactions that can modify the toxicity of heavy metals; since heavy metals when distributed in the body can be concentrated in the bile and subsequently enter the enterohepatic circuit, or during their journey through the gastrointestinal system they may be susceptible to being biotransformed by the microbiota enzymes to less or more compounds toxic to the individual and even to the microbiota itself [35, 36]. This is a new issue that is changing the approach to toxicokinetics and toxicodynamics of heavy metals and other environmental toxins. For example, it has been described in murine models that the intestinal microbiota specifically the presence of bacteria such as *Faecalibacterium* can protect against acute arsenic toxicity [37]. At the preclinical level, there is a very interesting study on the intes- tinal microbiome of conventionally raised mice (with normal microbiomes) and of mice with mammalian microbiomes altered with antibiotics and both groups exposed to sodium arsenite. In this regard, the authors found high levels of As in urine of mice with altered mammalian microbiomes, but the levels of As in the total feces were lower in this group, compared with conventionally raised mice. They also observed that the interruption of the intestinal microbiome with antibiotics significantly modified the biotransformation of arsenic and the urinary ratio of *monomethylarsonic acid/dimethylarsinic acid* increased. Regarding the expression

of carbon metabolism genes (folr2, bhmt and mthfr), they observed a downward regulation, and the levels of S-adenosyl methionine (SAM) in the liver of mice with mammalian microbiome altered with antibiotics and treated with arsenic they also decreased significantly. Finally, they concluded that altering the microbiome with antibiotics also increases the toxic effects of arsenic in mice [38]. In another study, it was observed that dietary supplementation with a galactooligosaccharide produced an increase in fecal excretion of lead, a decrease in plasma and tissue concentra- tion of the metal in mice. This effect was not observed when the microbiota of the mice was modified with antibiotics [39]. On the other hand, the probiotic strain *Lactobacillus plantarum* CCFM8661, prevented the absorption of lead in mice by intestinal sequestration, also significantly induced bile acid synthesis, improved bile flow and bile glutathione excretion, and increased bile acid excretion in feces of mice, the outflow of bile lead and improved fecal excretion of lead. Previous antibiotic treatment eliminated the effects induced by *L. plantarum* CCFM8661 on enterohepatic circulation of bile acids and lead [40].

In another study in rodents with a probiotic strain (*Lactobacillus plantarum CCFM8610*) administered orally, significantly improved hepatic synthesis of bile acids, bile glutathione release and fecal excretion of bile acids. The biliary and fecal excretion of cadmium increased markedly after the administration of *L. plantarum CCFM8610*, which resulted in a marked reduction in Cd levels in tissues. These effects were related to the intestinal microbiota, since prior antibiotic treatment reversed the effects induced by *L. plantarum CCFM8610* on the metabolism of bile acids and cadmium [41].

Finally, heavy metals and some demethylated metabolites are eliminated primarily by urine, but they can also be eliminated by bile, feces, breast milk, skin, nails and hair [27, 28]. Once the processes of absorption, distribution, metabolism and elimination of heavy metals are analyzed, it is very important to consider the pharmacokinetic changes that can occur in the most vulnerable populations who suffer from toxic effects such as children and pregnant women (**Table 1**) [27].

Regarding the toxicodynamics of heavy metals, various toxicity mechanisms have been described, however the most important are:

a. Production of free radicals with oxidative stress generation.

b. Enzymatic inhibition in a reversible and irreversible way.

c. DNA alteration probably due to the products generated in the methylation reactions.

d. Apoptosis.

e. Decoupling oxidative phosphorylation.

These cell targets are primarily responsible for alterations and damage in various organs and systems [42]. Currently, the treatment for heavy metal poisoning is based on the administration of chelating drugs, which form a coordination link with heavy metal, favoring its elimination mainly by urine [27, 43]. However, when analyzing toxicodynamics, an important aspect to consider during treatment is to block the free radicals that are formed to prevent damage caused by the oxidation

	Lead	**Mercury**	**Arsenic**	**Reference**
Children	Oral absorption of lead is more increased compared with an adult. Deficiencies of calcium and iron	Crosses blood-brain barrier.	Crosses blood-brain barrier.	[30, 31]

	increase oral absorption of lead. Children accumulate more lead in bone during the growing. A greater amount of lead is absorbed by inhalation (increased ventilatory capacity). Crosses blood-brain barrier.			
Pregnant woman	In pregnancy, lead is redistributed of bone into the bloodstream by increased calcium requirement. A greater amount of lead is absorbed by inhalation (increased ventilatory capacity). Crosses placenta.	Crosses placenta.	Crosses placenta.	[30, 31]

Table 1. *Most important aspects to consider in the toxicokinetics of vulnerable populations.*

of lipids and proteins of the membranes and thus avoid oxidative cell stress, so that phytochemical compounds present in various foods can be effective in preventing oxidative cell damage. It has been described in the literature that heavy metals readily bind to groups -SH-, -SS-, -NH2-, -OH and -COO-, these groups are present in some endogenous antioxidants such as glutathione and exogenous acids such as acid ascorbic, so many antioxidant compounds such as flavonoids present in food can be an alternative to block the oxidative effect of heavy metals and thereby minimize or prevent oxidative damage in the population that has exposure to heavy metals [28, 44].

Microbiota and its composition

In recent years, the study of intestinal microbiota has increased significantly, because important protective and metabolic functions have been associated.

In this regards, intestinal microbiota compete for nutrients, space and receptors with pathogenic microorganisms, as well as stimulating the production of antimi- crobial peptides and IgA antibodies; whereas in metabolic functions, the microbiota converts many complex substances such as starches, cellulose, pectins and gums into metabolites that are easily absorbed and used by the host, it also ferments another non-digestible residues from the diet, synthesizes some vitamins and is also involved in the absorption of ions (such as calcium, magnesium and iron) and energy recovery [13, 14].

The intestinal microbiota comprised about 35,000 species of microbes and includes at least six bacteria phyla, among them *Firmicutes, Bacteroidetes, Fusobacteria, Proteobacteria, Actinomycetes and Verrucomicrobia*; the predominant bacteria populations constituting about 90% of the total being Firmicutes and Bacteroidetes [45]. In gastrointestinal tract, there is a wide variety of bacteria that survive and have adapted to different pH conditions. In the esophagus, pH <4 is inhabited by bacterial strains from phyla *Bacteroides, Gemella, Megasphaera, Pseudomonas, Prevotella, Rothia sps., Streptococcus and Veillonella.* The phylas such as *Streptococcus, Lactobacillus, Prevotella, Enterococcus* and *Helicobacter* reside mainly in stomach (pH = 2). In small intestine (pH = 5–7), the phylas such as *Bacteroides, Clostridium, Streptococcus, Lactobacillus, g-Proteobacteria* and *Enterococcus* are found mainly (**Figure 1**). In colon (pH = 5–5.7), *Bacteroides, Clostridium, Prevotella, Porphyromonas, Eubacterium, Ruminococcus, Streptococcus, Enterobacterium, Enterococcus, Lactobacillus, Peptostreptococcus* and *Fusobacteria* are the resident phylas and finally in cecum (pH = 5–7) the phylas such as *Lachnospira, Roseburia, Butyrivibrio, Ruminococcus, Fecalibacterium* and *Fusobacteria* are found [46]. The intestinal microbiota has both a symbiotic and mutualism relationship, and also it has an important influence on the health and physiology of host. In this way, intes- tinal dysbiosis has been associated with a large array of human diseases such as, irritable bowel syndrome, inflammatory bowel disease, metabolic diseases (obesity and diabetes), as well as allergic and neurological diseases [47–49].

There are several factors affecting variations on intestinal microbiota, including age and diet [46, 50, 51]. It has been suggested that intestinal microbiota in healthy individuals is relatively stable when the diet remains without major changes, but with a subtle change in diet, the intestinal bacterial composition may also change and produces important alterations on the protective barrier function of the intestine. It is well known that not only the intestinal microbiota supports the protective functions of host. In addition, there is a complex cellular barrier where a variety of cells with different functions are housed, such as enterocytes, goblet cells, enteroen- docrine cells, Paneth cells and intraepithelial lymphocytes; together with the mucus

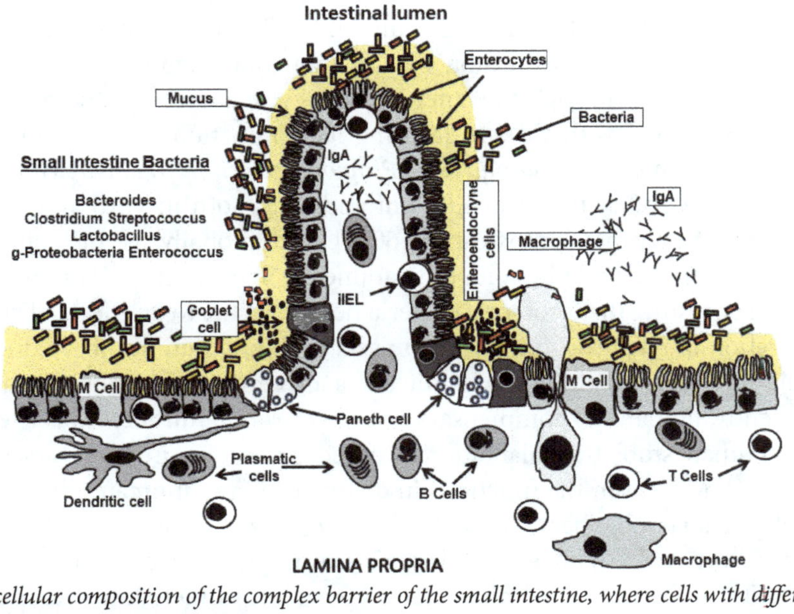

Figure 1. *The cellular composition of the complex barrier of the small intestine, where cells with different functions, such as enterocytes, goblet cells, enteroendocrine cells, Paneth cells and intraepithelial lymphocytes are the main residents. In addition, there is a layer of mucus in the intestinal lumen. Moreover, the lamina propria contains mainly T and B lymphocytes (secretory IgA plasmatic cells), macrophages and dendritic cells. It is possible to find the following bacterial phylus in the small intestine microbiota: Bacteroides, Clostridium, Streptococcus, Lactobacillus, g-Proteobacteria and Enterococcus that adapt to survive at pH = 5–7. These bacteria are in constant contact with all host cells and they have several functions related to health and illness of host.*

and lamina propria (which consist of lymphocytes, macrophages and dendritic cells), the main function is to limit the translocation of bacteria or other antigens and, therefore, proinflammatory processes are avoided and controlled (**Figure 1**). However, some intestinal bacteria from microbiota could produce toxic metabolites that may damage the epithelial barrier, increasing intestinal permeability and allow the entry of bacterial products into the circulatory system [52].

On the other hand, the role of the microbiota in detoxification of xenobiot- ics and contaminants (pesticides and heavy metals) has been poorly studied.

Nevertheless, it has been found that intestinal microbiota contributes to metabolism of xenobiotics [23] and heavy metals such as cadmium [21], mercury [20] and metalloids such as arsenic [24]. In this regards, in rats [53] and mice [54], several bacteria strains have been proposed and used as detoxifying probiotics, including mainly lactic acid bacteria. However, to date, there are no studies about the use of lactic acid bacteria as probiotics in humans. Recently, in *Drosophila melanogaster*, it has been found that use of *L. rhamnosus* has reduced both absorption and toxicity of organophosphate pesticides [55]. Also, lactic acid bacteria [56] and *Pediococcus* [57] have been used to reduce the toxic effects of aflatoxins and mycotoxin present in food, respectively. Likewise, in vitro experiments have shown the ability of lactic acid bacteria to bind and neutralize acrylamide molecules, a carcinogenic component present in some foods [58].

Since 1977, the interaction of the intestinal microbiota with the elimination of heavy metals was demonstrated. Here, elimination of mercury in feces of germ-free mice was lower, while the retention of mercury was slightly higher in organs from germ-free mice than in control mice [59]. In addition, it has been shown that oral administration of probiotics like *Lactobacillus plantarum* inhibit heavy metal cadmium (Cd) absorption by protecting the intestinal barrier upon acute and chronic Cd intoxication in mice [25]. Recently, it has been proposed that the possible mechanism of detoxification of Cd upon oral administration of *Lactobacillus plantarum* as a probiotic is through the enterohepatic cycle, increasing the metabolism of bile salts, which are conjugated with Cd and favoring its elimination in feces [40]. Similarly, the probiotic strains of *Lactobacillus rhamno- sus*, *Propionibacterium freudenreichii* and *Shermanii js.* showed an effective ability to bind cadmium and lead, both in vitro and in experimental mouse model of oral heavy metal intoxication [60, 61]. Additionally, use of probiotics improved the nutritional, biochemical and physiological parameters in experimental rat model intoxicated with chromium, suggesting that probiotic bacteria neutralize the toxic effects of chromium [62]. Also, administration of *Lactobacillus plantarum* as a probiotic significantly reduced both renal and hepatic injury induced by aluminum in mice that are chronically exposed to this toxic heavy metal [63]. Recently, human gut commensal *Faecalibacterium prausnitzii* was used as probiotic in mice. Here human stool transplantation restored protection in acute arsenic toxicity in mouse models [24]. Moreover, the functions in detoxification of inorganic arsenic of arsDABC genes from obligate anaerobic bacteria *Bacteroides vulgatus*, a common resident of the human microbiota have been well characterized [64]. Interestingly, in a clinical pilot study, yogurt supplemented with probiotic was administered to pregnant women and school children living in contaminated areas of Tanzania, resulting in a significant reduction on blood levels of heavy metals such as mercury and arsenic [65].

Finally, the ability of the microbiota to metabolize drugs has been referred as Pharmacomicrobiomics [66]. Intestinal microbiota generates phase I and phase II reactions to metabolize drugs, similar to the one generated by host cells. These include hydrolysis, dealkylation, glucuronidation and others reactions [67]. Thus, it has been strongly suggested that bacteria from the intestinal microbiota produce enzymes that metabolize drugs and contribute to determine the pharmacological properties of several drugs. Some examples of drugs that are metabolized by intestinal microbiota include aminosalicylates and anthranoid laxatives; digoxin; irinotecan and non-steroidal anti-inflammatory drugs, such as rutin, diosmin and baicalin, as well as l-dopa and simvastatin, used for the treatment of Parkinson's disease [68–70].

Evidence in the state of Guanajuato, Mexico

Guanajuato is a Mexican state located in the middle area of the country and according to the 2015 intercensal National Population Census, Guanajuato has 5,853,677 inhabitants, representing 4.9% of the total population of the country. Guanajuato is divided into 6 socioeconomic and geographic regions with 46 municipalities [71]. Since 2005, the risks of arsenic exposure in drinking water have been studied in some rural populations belonging to the State of Guanajuato [72]. In a comparative study in children of 10 years of age on average, arsenic levels were found in hair above the norm (from <0.006 to 1.3 mg/kg) and for the control group <0.006 mg/kg of arsenic [72]; however, it was found that children with non-standard levels of arsenic in hair, referred to drinking only carafe water and not tap water, which was explained because the water used to prepare soups, broths or beans, was used tap water [73]. On the other hand, only 26% had an adequate energy intake and the consumption of protein, folic acid, zinc and fiber was low according to RDA and children used to have mainly soft drinks, fried foods and pro- cessed foods [74]. Another study was applied in 352 households (heads of family) and measuring experiences related to access to water [sufficiency, safety (safety), acceptability, availability and accessibility], 33.4% of households reported concern about not having access to water and 74.8% had no access. About 70.8% had to buy water to drink and 5.7% became ill and related it to water consumption. About 65.6% of households presented food insecurity. The correlation was significant for the level of schooling of female heads of household, the number of households with children aged 1 and 12 with the use of tap water for drinking, preparing powdered milk for children and for food at home and fresh water [70, 75], which reflects the problem of exposure to various pollutants.

Main health problems in the state of Guanajuato

Studies conducted regarding lifestyles and eating habits have found a deteriora- tion in it, where the child and adolescent population are the first years of key life of growth and development, have a low consumption in fruit, vegetables and legumes (beans and lentils), which reflects risks of the absence of having prebiotics that allow maintaining intestinal health (microbiota) [71, 72]. Studies have shown a decrease in breastfeeding that so far during 2019 are changing these figures in Guanajuato but before that, the time of breastfeeding identified was 4.3 months in a range of not breastfeeding up to 7.3 months maximum [71, 72]. It is known that breastfeeding is the best food that promotes intestinal health and thus the microbi- ota. Guanajuato, as well as several regions of Mexico and in the world, has high rates of prematurity, teenage pregnancy and diabetes, which has a strong component with lifestyles and if we combine environmental pollution problems, we will face an obvious risk. For the first known biotransformation mechanism is the microbiota, but studies are still scarce [72, 73].

Regarding the stage of pregnancy, a study found reports of 26% of the mothers who reported a consumption of soda at least twice a week and preferred to buy already processed foods than prepared at home. The consumption of fruit, vegetables, legumes, cereals and tubers were low. This reflects problems of access to food and with its food insecurity, in addition to worrying about not having enough access to water, and in the last 3 months, 50% of households experienced water shortages. Most households used tap water to prepare milk for the children, as well as for personal hygiene. About 25% of the interviews reported that water availability has decreased in their homes, while the cost of water has increased [74].

Dietary and food intervention in the region

Derived from studies with arsenic and knowing its metabolism in addition to deterioration in lifestyles [74], the effect of vitamin and mineral supplementation on nutritional status and urinary excretion of arsenic in a group of 45 exposed adolescents was measured to this metal through drinking water [7], supplementa- tion was provided daily for 4 weeks for subsequent weekly assessment of nutritional status and arsenic levels in urine and drinking water. It was observed that the basal nutritional intake was low for protein, fiber, folic acid, vitamin B2, B6, B12, E, C, selenium and iron, as well as the increase of 1 g/dL of hemoglobin in all participants, at the end of the intervention in addition to an increase in fat-free mass and decrease in the percentage of body fat; the average arsenic consumption of drinking water in participants was 96.2 ± 7.5 μg/L with a urinary excretion of arsenic in the first week of intervention [35.91 μg/g Cr (95% CI = 23.2–74.8 μg/g Cr)] resulting higher, which was statistically significant compared with baseline urinary arsenic levels [43.2 μg/g Cr (95% CI = 30.8–117.6 μg/g Cr)] ($p < 0.05$).

From several species of quelites, which are endemic plants of Mexico, its vast nutritional composition is known. A recent study, evaluated three types of quelites (purslane-*Portulaca oleracea*-;quintonil-*Amaranthus spp.- and quelite ash or green quelite-Chenopodium album-*) from Mexico, determined their nutritional composi- tion, antioxidant capacity and physicochemical analysis finding considerable levels of carotenoids (2.85 mg/g DW) and the highest antioxidant activity. The nutritional content of iron, calcium and magnesium were higher for *quintonyl*, while for phosphorus, potassium and zinc was higher in quelite ash [75]. The ash and purs- lane quelite showed similar antioxidant activity but a higher level in the phenolic and flavonoid content, while the quintonil showed the presence of chlorophyll and protein concentration higher than those found in the quelite ash.

Conclusions

The chapter closes in a compilation of studies that reflect that despite the changes in lifestyle with a greater deterioration in healthy habits and therefore the consumption of an adequate diet (low in fruit and vegetables) and with it a low contribution in nutrients that promote the growth of the microbiota as well as in fiber which is known to have a probiotic effect, the consumption of fruit such as banana, apple and orange that is seasonal and low-cost fruit is preserved; of vegetables such as onion used as a condiment or in the preparation of most Mexican foods such as broths and sauces, as well as the consumption of beans (legume group), which are rich in antioxidants, soluble fiber rich in fructooligosaccharides.

Despite the little evidence, it is clear that the maintenance and promotion of the microbiota is to have a nutritious or adequate, varied, balanced, sufficient diet; as well as considering the alternatives of consuming fermented and probiotic-added dairy products, locally produced nutrients that are known to counteract non-toxic exposures such as those already reviewed, as is the case of prolonged exposure to arsenic, through food and drinking water, as it increases the risk of cancer, diabetes and high blood pressure, among other diseases. The typical Mexican diet is high in soluble fiber, key nutrients such as antioxidants, proteins are beans, tortilla plantain, orange, chili, quelites, showing that they co-help with a greater arsenic excretion through the urinary tract.

It is suggested to continue expanding the evidence of the combination of foods from a diet and lifestyle in general, mainly in women during the menopause period and pregnancy, due to a greater susceptibility to the development of anemia and resorption. Bone with exposure to metals such as lead or cadmium, since arsenic and fluoride are not the only contaminants that have been described in Guanajuato, the presence of lead, chromium and mercury, among the main ones, is known, and it is necessary Scale these experiences to combine nutritional treatment and drug therapy with chelators.

Conflict of interest

The authors declare none conflict interest.

Author details

Rebeca Monroy-Torres[1*], Marco Antonio Hernández-Luna[1], Xochitl Sofía Ramírez-Gómez[2] and Sergio López-Briones[1]

1. Cuerpo Academico de Medicina Traslacional, University of Guanajuato, Campus Leon, Guanajuato, Mexico

2. University of Guanajuato, Campus Celaya Salvatierra, Celaya, Mexico

*Address all correspondence to: rmonroy79@gmail.com

References

[1] Bisanz JE, Enos MK, Mwanga JR, Changalucha J, Burton JP, Gloor GB, et al. Randomized open-label pilot study of the influence of probiotics and the gut microbiome on toxic metal levels in Tanzanian pregnant women and school children. MBio. 2014;**5**(5):e01580-e01514. DOI: 10.1128/ mBio.01580-14

[2] Ibrahim F, Halttunen T, Tahvonen R, Salminen S. Probiotic bacteria as potential detoxification tools: Assessing their heavy metal binding isotherms. Canadian Journal of Microbiology. 2006;**52**(9):877-885. DOI: 10.1139/w06-043

[3] Servicios básicos de salud. Promoción y educación para la salud en materia alimentaria. Criterios para brindar orientación. Available from: https://www.dof.gob.mx/nota_detalle. php?codigo=5285372&fecha=22/01/2013

[4] Escudero Álvarez E, Sánchez PG. La fibra dietética. Nutrición Hospitalaria. 2006;**21**:12

[5] Monroy-Torres R. Dietary exposure to arsenic as an anthropogenic factor: Beyond the recommended diet. In: Environmental Deterioration and Human Health. Netherlands: Editorial Springer Netherlands; 2014

[6] Monroy-Torres R. Food and water security as determinants of the mitigation of health risks due to exposure to arsenic in water. In: Arsenic- Analytical and Toxicological Studies. Rijeka: INTECH; 2018. DOI: 10.5772/intechopen.76977

[7] Monroy Torres R, Espinosa Perez A, Ramirez Gomez X, Carrizales Yanez L, Linares Segovia B, Mejia Saavedra J. Effect of a four-week vitamin and mineral supplementation on the nutritional status and urinary excretion of arsenic in adolescents. Nutrición Hospitalaria. 2018;**35**(4):894-902. DOI: 10.20960/ nh.1600

[8] Monroy-Torres R. Guía para el Desarrollo de proyectos de investigación del área de la salud en una era sostenible. editor Ciudad de México: Pearson; 2018

[9] Jarup L. Hazards of heavy metal contamination. British Medical Bulletin. 2003;**68**:167-182. DOI: 10.1093/bmb/ ldg032

[10] Nordberg G, Jin T, Wu X, Lu J, Chen L, Liang Y, et al. Kidney dysfunction and cadmium exposure— Factors influencing dose-response relationships. Journal of Trace Elements in Medicine and Biology. 2012;**26**(2-3):197-200. DOI: 10.1016/j. jtemb.2012.03.007

[11] Wallace TC, Guarner F, Madsen K, Cabana MD, Gibson G, Hentges E, et al. Human gut microbiota and its relationship to health and disease. Nutrition Reviews. 2011;**69**(7):392-403. DOI: 10.1111/j.1753-4887.2011.00402.x

[12] Nava-Ruíz C, Méndez-Armenta M. Efectos neurotóxicos de metales pesados (cadmio, plomo, arsénico y talio). Archivos de Neurociencias. 2011;**16**(3):140-147

[13] Guarner F, Malagelada JR. Gut flora in health and disease. Lancet. 2003;**361**(9356):512-519. DOI: 10.1007/s00394-018-1703-4

[14] Vahter M, Berglund M, Akesson A. Toxic metals and the menopause. The Journal of the British Menopause Society. 2004;**10**(2):60-64. DOI: 10.1016/S0140-6736(03)12489-0

[15] Collin LJ, Judd S, Safford M, Vaccarino V, Welsh JA. Association of sugary beverage consumption with mortality risk in US adults: A secondary analysis of data from the REGARDS study. JAMA Network Open. 2019;**2**(5):e193121. DOI: 10.1258/136218004774202364

[16] Feng Q , Chen WD, Wang YD. Gut microbiota: An integral moderator in health and disease. Frontiers in Microbiology. 2018;**9**:151. DOI: 10.1001/ jamanetworkopen.2019.3121

[17] Dudek-Wicher RK, Junka A, Bartoszewicz M. The influence of antibiotics and dietary components on gut microbiota. Przegląd Gastroenterologiczny. 2018;**13**(2):85-92. DOI: 10.5114/pg.2018.76005

[18] Graf D, Di Cagno R, Fak F, Flint HJ, Nyman M, Saarela M, et al. Contribution of diet to the composition of the human gut microbiota. Microbial Ecology in Health and Disease. 2015;**26**:26164. DOI: 10.3402/mehd.v26.26164. eCollection 2015

[19] Rothenberg SE, Keiser S, Ajami NJ, Wong MC, Gesell J, Petrosino JF, et al. The role of gut microbiota in fetal methylmercury exposure: Insights from a pilot study. Toxicology Letters. 2016;**242**:60-67. DOI: 10.1016/j.toxlet.2015.11.022

[20] Zhai Q , Liu Y, Wang C, Zhao J, Zhang H, Tian F, et al. Increased cadmium excretion due to oral administration of *Lactobacillus plantarum* strains by regulating enterohepatic circulation in mice. Journal of Agricultural and Food Chemistry. 2019;**67**(14):3956-3965. DOI: 10.1021/acs.jafc.9b01004

[21] Limon-Pacheco JH, Jimenez- Cordova MI, Cardenas-Gonzalez M, Sanchez Retana IM, Gonsebatt ME, Del Razo LM. Potential Co-exposure to arsenic and fluoride and biomonitoring equivalents for Mexican children. Annals of Global Health. 2018;**84**(2):257-273. DOI: 10.29024/aogh.913

[22] Das A, Srinivasan M, Ghosh TS, Mande SS. Xenobiotic metabolism and gut microbiomes. PLoS One. 2016;**11**(10):e0163099. DOI: 10.1371/ journal.pone.0163099

[23] Coryell M, McAlpine M, Pinkham NV, McDermott TR, Walk ST. The gut microbiome is required for full protection against acute arsenic toxicity in mouse models. Nature Communications. 2018;**9**(1):5424

[24] Zhai Q , Tian F, Zhao J, Zhang H, Narbad A, Chen W. Oral administration of probiotics inhibits absorption of the heavy metal cadmium by protecting the intestinal barrier. Applied and Environmental Microbiology. 2016;**82**(14):4429-4440

[25] Kinoshita H. Biosorption of heavy metals by lactic acid bacteria for detoxification. Methods in Molecular Biology. 1887;**2019**:145-157

[26] Brunton LLCB, Knollman B. Goodman & Gilman's 1: The Pharmacological Basis of Therapeutics. McGraw-Hill; 2012

[27] Katzung BGMS, Trevor AJ. Basic & Clinical Pharmacology. McGraw-Hill; 2013

[28] Jose A, Ray JG. Toxic heavy metals in human blood in relation to certain food and environmental samples in Kerala, South India. Environmental Science and Pollution Research International. 2018;**25**(8):7946-7953

[29] Tamayo-Ortiz M, Navia-Antezana J. Reduced lead exposure following a sensitization program in rural family homes producing traditional Mexican ceramics. Annals of Global Health. 2018;**84**(2):285-291. DOI: 10.1007/ s11356-017-1112-x

[30] Borowska S, Brzoska MM. Metals in cosmetics: Implications for human health. Journal of Applied Toxicology. 2015;**35**(6):551-572. DOI: 10.29024/ aogh.916

[31] Rahman Z, Singh VP. The relative impact of toxic heavy metals (THMs) (arsenic (As), cadmium (Cd), chromium (Cr)(VI), mercury (Hg), and lead (Pb)) on the total environment: An overview. Environmental Monitoring and Assessment. 2019;**191**(7):419. DOI: 10.1002/ jat.3129

[32] Cortes J, Peralta J, Diaz-Navarro R. Acute respiratory syndrome following accidental inhalation of mercury vapor. Clinical Case Reports. 2018;**6**(8):1535-1537. DOI: 10.1007/ s10661-019-7528-7

[33] Al Osman M, Yang F, Massey IY. Exposure routes and health effects of heavy metals on children. Biometals. 2019;**32**(4):563-573. DOI: 10.1002/ ccr3.1656

[34] Martin EM, Fry RC. Environmental influences on the epigenome: Exposure- associated DNA methylation in human populations. Annual Review of Public Health. 2018;**39**:309-333. DOI: 10.1007/ s10534-019-00193-5

[35] Koontz JM, Dancy BCR, Horton CL, Stallings JD, DiVito VT, Lewis JA. The role of the human microbiome in chemical toxicity. International Journal of Toxicology. 2019;**38**(4):251-264. DOI: 10.1146/ annurev-publhealth-040617-014629

[36] Chi L, Xue J, Tu P, Lai Y, Ru H, Lu K. Gut microbiome disruption altered the biotransformation and liver toxicity of arsenic in mice. Archives of Toxicology. 2019;**93**(1):25-35. DOI: 10.1177/1091581819849833

[37] Zhai Q , Liu Y, Wang C, Qu D, Zhao J, Zhang H, et al. *Lactobacillus plantarum* CCFM8661 modulates bile acid enterohepatic circulation and increases lead excretion in mice. Food & Function. 2019;**10**(3):1455-1464. DOI: 10.1038/s41467-018-07803-9

[38] Reyes-Becerril M, Angulo C, Sanchez V, Cuesta A, Cruz A. Methylmercury, cadmium and arsenic(III)-induced toxicity, oxidative stress and apoptosis in Pacific red snapper leukocytes. Aquatic Toxicology. 2019;**213**:105223. DOI: 10.1007/ s00204-018-2332-7

[39] Kim JJ, Kim YS, Kumar V. Heavy metal toxicity: An update of chelating therapeutic strategies. Journal of Trace Elements in Medicine and Biology. 2019;**54**:226-231. DOI: 10.1039/ c9fo00587k

[40] Ahuie Kouakou G, Gagnon H, Lacasse V, Wagner JR, Naylor S, Klarskov K. Dehydroascorbic acid S-thiolation of peptides and proteins: Role of homocysteine and glutathione. Free Radical Biology & Medicine. 2019;**141**:233-243. DOI: 10.1039/ c8fo02554a

[41] Frank DN, St Amand AL, Feldman RA, Boedeker EC, Harpaz N, Pace NR. Molecular-phylogenetic characterization of microbial community imbalances in human inflammatory bowel diseases. Proceedings of the National Academy of Sciences of the United States of America. 2007;**104**(34):13780-13785. DOI: 10.1021/acs. jafc.9b01004

[42] Jandhyala SM, Talukdar R, Subramanyam C, Vuyyuru H, Sasikala M, Nageshwar Reddy D. Role of the normal gut microbiota. World Journal of Gastroenterology. 2015;**21**(29):8787-8803. DOI: 10.1016/j. aquatox.2019.105223

[43] Vallianou N, Stratigou T, Christodoulatos GS, Dalamaga M. Understanding the role of the gut microbiome and microbial metabolites in obesity and obesity-associated metabolic disorders: Current evidence and perspectives. Current Obesity Reports. 2019;**8**(3):317-332

[44] Knox NC, Forbes JD, Peterson CL, Van Domselaar G, Bernstein CN. The gut microbiome in inflammatory bowel disease: Lessons learned from other immune-mediated inflammatory diseases. The American Journal of Gastroenterology. 2019;**114**(7):1051-1070. DOI: 10.1016/j. freeradbiomed.2019.06.022

[45] Chernikova D, Yuan I, Shaker M. Prevention of allergy with diverse and healthy microbiota: An update. Current Opinion in Pediatrics. 2019;**31**(3):418-425

[46] Yatsunenko T, Rey FE, Manary MJ, Trehan I, Dominguez-Bello MG, Contreras M, et al. Human gut microbiome viewed across age and geography. Nature. 2012;**486**(7402):222-227

[47] David LA, Maurice CF, Carmody RN, Gootenberg DB, Button JE, Wolfe BE, et al. Diet rapidly and reproducibly alters the human gut microbiome. Nature. 2014;**505**(7484):559-563

[48] Hornig M. The role of microbes and autoimmunity in the pathogenesis of neuropsychiatric illness. Current Opinion in Rheumatology. 2013;**25**(4):488-795

[49] Curro D. The role of gut microbiota in the modulation of drug action: A focus on some clinically significant issues. Expert Review of Clinical Pharmacology. 2018;**11**(2):171-183

[50] Dridi B, Fardeau ML, Ollivier B, Raoult D, Drancourt M. *Methanomassiliicoccus luminyensis* gen. Nov., sp. nov., a methanogenic archaeon isolated from human faeces. International Journal of Systematic and Evolutionary Microbiology. 2012;**62**(Pt 8):1902-1907

[51] Hernandez-Mendoza A, Gonzalez-Cordova AF, Vallejo-Cordoba B, Garcia HS. Effect of oral supplementation of lactobacillus reuteri in reduction of intestinal absorption of aflatoxin B(1) in rats. Journal of Basic Microbiology. 2011;**51**(3):263-268

[52] Jebali R, Abbes S, Salah-Abbes JB, Younes RB, Haous Z, Oueslati R. Ability of *Lactobacillus plantarum* MON03 to mitigate aflatoxins (B1 and M1) immunotoxicities in mice. Journal of Immunotoxicology. 2015;**12**(3):290-299

[53] Daisley BA, Trinder M, McDowell TW, Collins SL, Sumarah MW, Reid G. Microbiota-mediated modulation of organophosphate insecticide toxicity by species-dependent interactions with lactobacilli in a *Drosophila melanogaster* insect model. Applied and Environmental Microbiology. 2018;**84**(9):e02820-e02817

[54] Ahlberg SH, Joutsjoki V, Korhonen HJ. Potential of lactic acid bacteria in aflatoxin risk mitigation. International Journal of Food Microbiology. 2015;**207**:87-102

[55] Martinez MP, Gonzalez Pereyra ML, Pena GA, Poloni V, Fernandez Juri G, Cavaglieri LR. Pediococcus acidolactici and *Pediococcus pentosaceus* isolated from a rainbow trout ecosystem have probiotic and ABF1 adsorbing/ degrading abilities in vitro. Food Additives & Contaminants. Part A, Chemistry, Analysis, Control, Exposure & Risk Assessment. 2017;**34**(12):2118-2130

[56] Zhang D, Liu W, Li L, Zhao HY, Sun HY, Meng MH, et al. Key role of peptidoglycan on acrylamide binding by lactic acid bacteria. Food Science and Biotechnology. 2017;**26**(1):271-277

[57] Nakamura I, Hosokawa K, Tamura H, Miura T. Reduced mercury excretion with feces in germfree mice after oral administration of methyl mercury chloride. Bulletin of Environmental Contamination and Toxicology. 1977;**17**(5):528-533

[58] Bhakta JN, Ohnishi K, Munekage Y, Iwasaki K, Wei MQ. Characterization of lactic acid bacteria-based probiotics as potential heavy metal sorbents. Journal of Applied Microbiology. 2012;**112**(6):1193-1206

[59] Younan S, Sakita GZ, Coluna JGY, Rufino MN, Keller R, Bremer-Neto H. Probiotic mitigates the toxic effects of potassium dichromate in a preclinical study: A randomized controlled trial. Journal of the Science of Food and Agriculture. 2019;**99**(1):183-190

[60] Yu L, Zhai Q , Yin R, Li P, Tian F, Liu X, et al. *Lactobacillus plantarum* CCFM639 alleviate trace element imbalance-related oxidative stress in liver and kidney of chronic aluminum exposure mice. Biological Trace Element Research. 2017;**176**(2):342-349

[61] Li J, Mandal G, Rosen BP. Expression of arsenic resistance genes in the obligate anaerobe *Bacteroides vulgatus* ATCC 8482, a gut microbiome bacterium. Anaerobe. 2016;**39**:117-123

[62] Saad R, Rizkallah MR, Aziz RK. Gut pharmacomicrobiomics: The tip of an iceberg of complex interactions between drugs and gut-associated microbes. Gut Pathogens. 2012;**4**(1):16

[63] Sousa T, Paterson R, Moore V, Carlsson A, Abrahamsson B, Basit AW. The gastrointestinal microbiota as a site for the biotransformation of drugs. International Journal of Pharmaceutics. 2008;**363**(1-2):1-25

[64] Lu L, Wu Y, Zuo L, Luo X, Large PJ. Intestinal microbiome and digoxin inactivation: Meal plan for digoxin users? World Journal of Microbiology and Biotechnology. 2014;**30**(3):791-799

[65] Klaassen CD, Cui JY. Review: Mechanisms of how the intestinal microbiota alters the effects of drugs and bile acids. Drug Metabolism and Disposition. 2015;**43**(10):1505-1521

[66] Información por entidad: Guanajuato. Available from: http:// www.cuentame.inegi.org.mx/ monografias/ informacion/gto/

[67] Monroy-Torres R, Macias AE, Gallaga-Solorzano JC, Santiago-Garcia EJ, Hernandez I. Arsenic in Mexican children exposed to contaminated well water. Ecology of Food and Nutrition. 2009;**48**(1):59-75

[68] Monroy Torres R, Ramírez-Gómez XS, Naves Sanchez J, Macias Hernández AE. Accesibilidad a agua potable para el consumo y preparación de alimentos en una comunidad expuesta a agua contaminada con arsénico. Revista Médica de la Universidad Veracruzana. 2009;**9**(1):4

[69] Monroy Torres R, Arellano-Salgado L, Macías Hernández AE, Claudio L. Food intake and nutritional status of children with high levels of arsenic in hair: Cases study of a historical cohort. Immunology, Endocrine & Metabolic Agents in Medicinal Chemistry. 2017;**17**(2): 127-134. DOI: 10.2174/1871522218666180 130154655

[70] Monroy-Torres R, Espinoza-Pérez A. Factores que intensifican el riesgo toxicológico en comunidades expuestas al arsénico en agua. CienciaUAT. 2018;**12**:148-157. DOI: 10.29059/ cienciauat.v12i2.803

[71] Monroy-Torres R, López López M, Naves Sánchez J. Feeding practices, nutrition and socioeconomic situation in homes with premature children in Guanajuato (Mexico). Anales de Pediatría. 2013;**78**:21-26. DOI: 10.1016/j.anpedi.2012.05.001

[72] Monroy-Torres R, Naves-Sanchez J, Ortega-Garcia JA. Breastfeeding and metabolic indicators in Mexican premature newborns. Revista de Investigación Clínica. 2012;**64**(6 Pt 1):521-528. DOI: 10.1080/03670240802575519

[73] Ortega-Garcia JA, Tellerias L, Ferris-Tortajada J, Boldo E, Campillo-Lopez F, van den Hazel P, et al. Threats, challenges and opportunities for paediatric environmental health in Europe, Latin America and the Caribbean. Anales de Pediatría. 2019;**90**(2):124 e121. DOI: 10.1016/j.anpedi.2018.11.015

[74] Monroy-Torres R, Pérez E, Antonio J, Pérez González RM. Evaluación de las prácticas de alimentación y nutrición en una población expuesta a arsénico: Una propuesta para integrar indicadores de exposición nutricional. Nutrición Clínica y Dietética Hospitalaria. 2016;**2**:140-149. DOI: 10.12873/362monroytorres

[75] Santiago Saenz Y, Hernández-Fuentes AD, Monroy Torres R, Cariño Cortés R, Jiménez Alvarado R. Physicochemical, nutritional and antioxidant characterization of three vegetables (*Amaranthus hybridus* L., *Chenopodium berlandieri* L., *Portulaca oleracea* L.) as potential sources of phytochemicals and bioactive compounds. Journal of Food Measurement and Characterization. 2018;**12**(4):2855. DOI: 10.1007/ s11694-018-9900-7

7

Prebiotics and Probiotics - Potential Benefits in Human Nutrition and Health

Maria Inês Sucupira Maciel and Michelle Maria Barreto de Souza

Abstract

The growing interest of using probiotic bacteria into foods has lately increased due to its beneficial effects on intestinal microbiota. This fact has motivated researchers and the food industry to develop new functional products, such as probiotics. The probiotic fruit juices can be a good alternative for new foodstuff production mainly for people with galactosemia and lactose intolerance. Among the microencapsulation techniques of probiotics in juices, spray drying and freeze drying are two of the most used with 220 and 228 publications found in Science Direct, 8 and 0 Web Science, and 5 and 0 in SCOPUS, respectively. Several studies have reported the addition of probiotics in different fruit juices to produce functional beverages; however, there are no reports on the use of three or more probiotic microorganisms in mixed beverage. Our research group has been directing studies in this area in order to provide results of scientific interest and to food industry.

Keywords: new product, juice fruit, spray dryer

Introduction

Consumers' growing interest in using food to improve their health and wellness has motivated researchers and the food industry to develop new functional products, such as probiotics [1]. Interest in healthy, nutritious, high-end food products has increased worldwide as probiotic-supplemented beverages beneficially affect one or more body functions and have the potential to promote health. Consumers are more aware of the relationship between good nutrition and health, so there is a growing demand for foods that, in addition to nourishing, provide health benefits.

Traditionally, probiotics are incorporated into dairy products. However, non- dairy food matrices have been studied as potential carriers for these microorganisms because of the increasing number of individuals who are affected by lactose intol- erance, milk protein allergy, galactosemia, and hypercholesterolemia. Seventy-five percent of the world's population is suffering from lactose intolerance [2, 3].

Fruit juices, desserts, and cereal-based products featuring probiotics may be other suitable media for delivering probiotics. Most of the probiotic food products is categorized as functional foods. These products include fruit drinks. According to Corbo [4], functional beverages are nonalcoholic beverages containing nontraditional ingredients that offer health benefits. Within this context, fruit juices are indeed promising probiotic carriers due to their essential nutrient content along with their appeal to a niche of consumers who already care about healthier habits [5]. The great advantage of fruit juices as a probiotic food is that it is regularly consumed by high portion of the population, which would allow continuity of the beneficial effect from the probiotic microorganisms carried by this food.

However, the viability of the microorganisms is defined during processing, being a necessary application of methods that can maintain or improve their viability and functionality. With this, new technologies have been proposed, among them, the microencapsulation stands out as a promising technique. The microen- capsulation can be defined as a process in which small solid particles, liquid drop- lets, or gases are evolved by a coating layer, or incorporated into a homogeneous or heterogeneous matrix, yielding small capsules with useful properties [6–9].

The microencapsulation of food ingredients in coating materials can be achieved by various methods. Some of them are spray drying, extrusion, freeze drying, fluidized bed, coacervation, and cocrystallization. Among these methods, atomization is the most used in the manufacture of foodstuffs [10, 11].

The spray drying consists of transforming a product in fluid state into solid state in the form of powder, a continuous operation, through a relatively short time [12–14], being also the most used method to encapsulate probiotic bacteria [15].

Nondairy food matrices as a carrier of microencapsulated probiotic microorganisms microencapsulated

The word probiotic comes from the Greek word "προ-βίος" that means "for life"; thus, probiotics are live microorganisms (mainly bacteria but also yeasts) that confer a beneficial effect on the host if administered in proper amounts [16].

The most common probiotic microorganisms used commercially in food are bacteria from the genera *Lactobacillus* and *Bifidobacterium* [17]. *Lactobacillus* species used as probiotics in food include *L. acidophilus, L. crispatus, L. amylovorus, L. gallinarum, L. gasseri, L. johnsonii, L. helveticus, L. delbrueckii* subsp. *bulgaricus, L. salivarius* subsp. *salivarius, L. casei, L. paracasei* subsp. *paracasei, L. paracasei subsp. tolerans, L. plantarum, L. rhamnosus, L. fermentum,* and *L. reuteri* [18].

Probiotics are widely used in commercial functional products of animal origin, mainly fermented milk, such as yogurt and cheeses; however, the use of milk-based products may be limited by allergies, cholesterol diseases, dyslipidemia, and vegetarianism, and therefore, several raw materials have been extensively investigated to determine if they are suitable substrates to produce novel nondairy functional foods [16, 19].

Nondairy probiotic products are found to a lesser extent in market and are usually restricted to traditional products based on cereals or soy [19]. The number of articles on the use of nondairy matrices for the transport of probiotic microorganisms increases each year (Table 1), which demonstrates the increasing interest they have received in the health sciences literature. From a scientific point of view, it is unquestionable that probiotics constitute an important field of investigation and study, with the digestive tract, more specifically the intestinal microbiota, as the main point [20, 21].

Fruit and vegetable juices can be considered as a new category of food matrices for probiotic bacteria [37] with developments reported in literature [38, 39].

Particularly, fruit juices have been reported as a novel and appropriate medium for probiotic for their content of essential nutrients. Moreover, they are usually referred as healthy foods, designed for young and old people [40].

Probiotic	Food matrices	Food product	References
L. rhamnosus	Apple	Apple cubes	[22]
L. casei NRRL B-442	Cashew	Cashew juice	[23]
L. casei	Melon	Melon juice	[24]
L. casei NRRL B-442	Cashew	Dehydrated cashew juice	[25]
L. plantarum 33	Olive	Olive paste	[26]
L. casei 01	Lychee	Dehydrated lychee juice	[27]
L. casei NRRL B-442	Orange	Dehydrated orange juice	[20]
L. rhamnosus HN001, L. acidophilus LA-5 e L. plantarum	Pineapple, banana, guava, apple, papaya, and mango	Fruit salad	[28]
Bifidobacterium animalis spp. lactis	Jussara	Dehydrated jussara juice	[29]
L. casei NRRL B-442	Orange	Dehydrated orange juice	[30]
L. paracasei	Orange	Orange juice	[31]
L. plantarum	Apple	Apple cubes	[32]
L. salivarius spp. salivarius CECT 4063	Mandarin	Mandarin juice	[33]
L. plantarum LS5	Lemon	Lemon juice	[34]
L. plantarum DW12	Coconut water	Fermented drink of coconut	[35]
L. casei NRRL B-442	Apple	Dehydrated apple cubes	[36]
L. plantarum MTCC2621	Lychee	Dehydrated lychee juice	[21]

Table 1. *Studies on the use of nondairy matrices for the transport of probiotic microorganisms.*

Fruit juices are indeed promising probiotic carriers due to their essential nutrient content along with their appeal to a niche of consumers who already care about healthier habits. Fruits are healthy and refreshing and have good taste and flavor profile and can be suitable for probiotics [41, 42].

The observations that plant compounds, like complex carbohydrates and phenolics, may act synergistically with probiotics [43, 44] in formulations for gut health were important for nondairy probiotic product developers. Every food category from cereals to soy, to fruits and vegetables, has been the subject of research for new product development.

The use of probiotics represents a promising, rapidly growing area for the development of functional foods. Probiotic crops are successfully applied to differ- ent food matrices [45]. However, the development of nondairy products represents a challenge for the industry, as each food matrix has unique characteristics, and it is necessary to optimize and standardize each type of product.

Probiotic foods should be safe and contain sufficient probiotic microorganisms during the shelf life of the product. Therefore, selected probiotic strains should be suitable for large-scale industrial production, with the ability to survive and retain their functionalities during food processing and storage [17]. Several strains of *Lb. plantarum*, *Lb. acidophilus*, and *Lb. casei* can grow in fruit matrices due to their tolerance to acidic environments [46].

During food storage different factors may affect the viability of probiotic bacteria, such as probiotic strains used, pH, the presence of hydrogen peroxide and dissolved oxygen, buffering form, storage temperature, the nature of the ingredients added, and food matrices [23, 47, 48].

In order to exert beneficial effects on health, the number of viable cells of probiotic microorganisms should be located above 106 CFU.g-1 in the product for consumption, available over the entire shelf life. Therefore, the preservation of probiotic cultures in products during storage is of extreme importance [49].

In this context, microencapsulation of probiotic cells has been widely studied as a technique to improve the stability of these microorganisms by protecting them from unfavorable environments [17]. Microencapsulation also has a potential effect on reducing post-fermentation acidification and possible negative sensory effects of probiotic food products [50].

Among the microencapsulation technologies, spray drying and freeze drying are the most commonly used. However, spray drying is the most effective for large- scale industrial production because it is a continuous, rapid process and has relatively low cost and high reproducibility [51, 52]. It is suggested as a technique that improves the survival of probiotics during food processing and storage, as well as confers protection of probiotics against subsequent exposure to the harsh conditions of the gastrointestinal tract, as this process gives a coating to the cells, protecting them from the outside environment [53].

Among the advantages of atomization, a distinguished one is the ability to handle heat-sensitive materials with high surface area/droplets volume ratio, resulting in shorter time of exposition to drying temperature [54]. Besides protecting probiotic cells from adverse conditions, powders obtained through spray dryer have good reconstitution and low water activity and are suited for storage at ambient temperature, what it is desirable, especially in commercial applications, due to higher operational costs associated with cooled storage, transport, and distribution difficulties [55, 56].

The characteristics of the powder produced in driers depend mainly on the operational variables of the drier (air inlet and outlet temperatures), on the product composition, solid concentration, feed flow rate, and also on the type of encapsulating agent used in the formulation [57].

Several studies reported that microencapsulation by spray dryer might provide a more favorable anaerobic environment for sensitive probiotic bacteria, as well as a physical barrier from the harsh acidic conditions of fruit juice [58]. The addition of probiotics in different fruit juices to produce functional beverages microencapsulated by spray drying has also been reported [21, 25, 27, 29].

The most commonly used carrier materials for encapsulation are maltodextrin.

Maltodextrin (MD) is a polysaccharide, which molecular weight and properties depend on the hydrolysis process employed to obtain it from starch. Maltodextrin is classified by its dextrose equivalent (DE) which measures the amount of reducing sugars present in a sugar product, relative to dextrose [59–61].

The wall material is one of the most important parameters in the food microen- capsulation processes. Its chemical composition and structure can affect the quality of the powdered product and criteria, such as solubility, apparent density, absolute density, porosity, particle size distribution, morphology, hygroscopicity, cell viability, water activity, moisture content, and sensory evaluation [61, 62].

However, to date there are no reports on the use of various probiotic microorganisms incorporated into a mixed beverage. In view of this, aiming at complementing nutrients, increasing nutritional characteristics, and developing new flavors, our research group has been directing studies in this area in order to provide results of scientific interest and to food industry.

Microencapsulated probiotic mixed fruit juice

In recent years our research group has been carrying out studies with microen- capsulation of juices from various fruits, using maltodextrin as the main encapsu- lating agent. In a recent study [63], with mixed juice of tropical fruits, the process was optimized in order to obtain products with better sensorial and nutritional characteristics. Based on these results, in order to meet the growing demand for functional foods, probiotic microorganisms were added.

The study related to the process of addition of probiotics to mixed juice in powder is an innovation, consisting of a new food product. Thus, the patent of product and process category was registered, at Brazil's National Institute of Indus- trial Property with Patent nature of Invention, under register number BR 102019 009006 5.

The objective was to develop a novel nondairy probiotic product, composed of mixed juice with three *Lactobacillus* microencapsulated by spray drying using maltodextrin DE 5.

Currently there is a growing market for juices composed of more than one fruit, and this tendency is most observed in products that use tropical fruits. Tropical fruits are widely accepted by consumers and are important sources of antioxidant compounds. For this reason the acerola and siriguela were selected.

Acerola (*Malpighia emarginata* D.C.) is a fruit native to Central America and Northern South America, with some of its largest plant area in Brazil, which has been increasingly produced, because of their high vitamin C contents from 700 to $^{-1}$ 1400 mg/100 g [64–66]. The siriguela (*Spondias purpurea* L.) is a fruit from Anacardiaceae family originally from Central America and widespread in all tropical countries, mainly in the northeast region of Brazil. It is a small yellow fruit, with a pleasant aroma and taste, being a source of carotenoids [67, 68]. Thus, acerola and siriguela juice is an interesting nondairy matrix for a probiotic beverage.

The viability of probiotic bacteria is the most important parameter in the spray drying process using microorganisms due to heat inactivation and exposure to dehydration, and maximum viability is there for the major goal for this type of product [25, 69]. Probiotic mixed powder juice with maltodextrin DE 5 presented viable cell counts above 6.0 log^{-1} CFU.g (**Table 2**), which is the minimum

Analyses	Maltodextrin DE 5
Microbial viability (log CFU.g^{-1})	8.50 ± 0.04
Water activity	0.12 ± 0.00
Moisture content (%)	2.09 ± 0.06
Hygroscopicity (g.100 g^{-1})	24.40 ± 0.70
Apparent density (g/mL)	0.38 ± 0.01
Absolute density (g/mL)	1.97 ± 0.01
Porosity (%)	31.13 ± 0.29
Solubility (%)	92.75 ± 2.03

Table 2. *Physical properties and microbial viability of acerola and siriguela probiotic mixed juice microencapsulated by spray drying using air inlet temperature 140°C, feed flow rate 0.60 L/h, and 10% maltodextrin DE 5.*

recommended level for probiotics in food products necessary to produce therapeutic benefit [25, 70].

Probiotics' survival during spray drying can also be attributed to the strong adherence to the carrier, which protects cells from high acidic and bile conditions. Overall, maltodextrin is confirmed to serve as a good encapsulating matrix as well as a moderate prebiotic for high survival of probiotics. The sugars present in juice may also have contributed to the survival during spray drying since sugars act as thermoprotectants [71].

Moisture and water activity are important indexes for spray-dried powders due to their effects on the shelf life of the product; a large amount of retained water can accelerate degradation reactions and microorganism growth [72]. It is necessary that its equilibrium moisture content is less than 5 g water/100 g of dry solid and its water activity is in the range from 0.1 to 0.4, ensuring greater stability for dry food [73]. The moisture and water activity for probiotic mixed powder juice (Table 2) are in agreement with the study of [73]. Values similar were obtained by [20, 27, 30] studying orange, lychee, and orange juice probiotic microencapsulated by spray drying, respectively.

Hygroscopicity is the ability of a material to attract and hold moisture from the environment. It is generally calculated from the weight gain after storing the food powder in a high humid desiccator (relative humidity more than 60%) for a period (generally 1 week) [74, 75]. The hygroscopicity value (**Table 2**) was similar com- pared to other probiotic microencapsulated formulation [27] and much larger than [21]. The hygroscopicity can be easily controlled by using suitable packaging, maintaining the integrity of the product without significant changes.

Solubility, the ability of a powder to form solution or suspension in water, is defined as the most reliable criterion to evaluate the behavior of powder in an aqueous solution. This parameter shows the ability of the spray-dried powder to form solution or suspension in water [76]. Solubility of powders can be affected by many parameters such as initial composition of the raw material to be microencapsulated by spray-dried, compressed airflow rates, low feed rates, and the carrier agents [77]. In this study, high solubility values were obtained with maltodextrin DE 5 (**Table 2**). The characteristics of solubility normally contradict to apparent density; the powder showing high solubility should have low apparent density [78].

The absolute density of the particle matches the actual density of the solid and does not consider the spaces present between the particles, unlike the apparent density. A similar result with this study (**Table 2**) was reported by [75] using maltodextrin DE 10 for açai pulp drying.

The higher values of porosity indicate the presence of a larger number of spaces between the particles, containing oxygen available for degradation reactions [79]. In this study (**Table 2**), lower value of porosity was obtained. In this way, the malto- dextrin DE 5 is suitable for the spray dryer process. The efficacy of low-DE malto- dextrin as drying carriers is due to the encapsulating property and low moisture diffusivity [80].

Particle size distribution is an important physical property that directly affects the application of microcapsules into food formulations [81]. The microcapsules exhibited lower particle size distribution (7.07 μm ± 0.03). Small particles are preferred in food formulations for ensuring homogeneity and quality, since they have large surface area and enable interaction with microorganisms favoring the microbial reactivation [82] (**Figure 1**).

The morphological characteristics of the particles showed spherical shapes with various sizes that are features of spray-dried powders and surfaces with wrinkled predominance and few particles with smooth surface. The occurrence of roughness

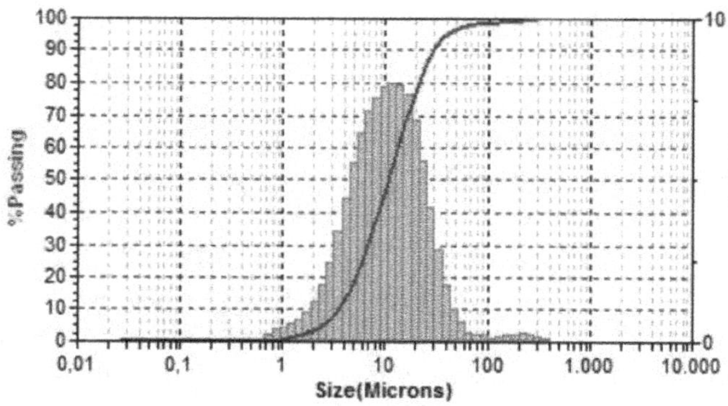

Figure 1. Particle size distribution of acerola and ciriguela probiotic mixed juice microencapsulated by spray drying using air inlet temperature 140 °C, feed flow rate 0.60 L/h and 10% maltodextrin DE 5.

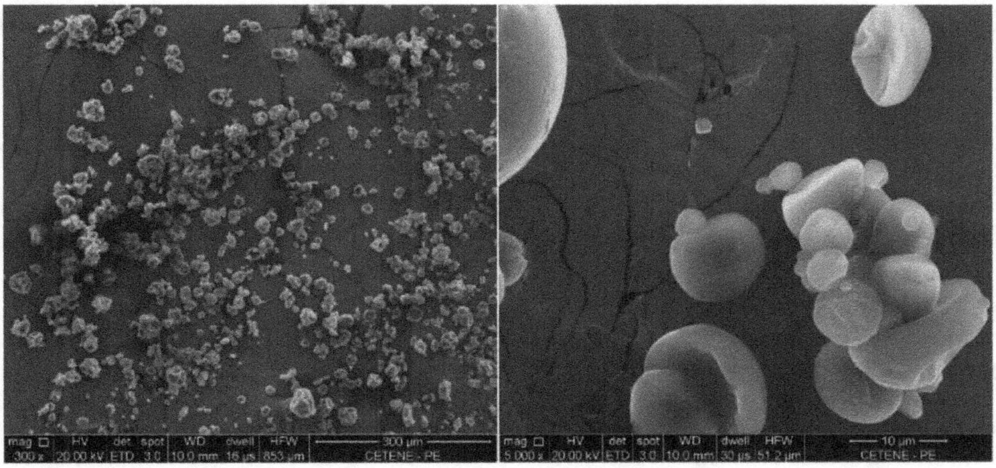

Figure 2. *Scanning electron microscopy (SEM) photographs of acerola and ciriguela probiotic mixed juice microencapsulated by spray drying using air inlet temperature 140 °C, feed flow rate 0.60 L/h and 10% maltodextrin DE 5.*

on the surface of the microparticles may be associated with shrinkage of the wall material during initial stages of the process and temperature used in the drying chamber [83, 84]. With these results, a great potential for the use of such powders in the food industry is observed (**Figure 2**).

Conclusion

The development of new functional food products is very challenging, and it has to complete the consumer's expectations for palatable and healthy products. Probiotic cells can be stabilized with microencapsulation to preserve them from detrimental processing and storage factors such as high acidity and low pH. Therefore, there is a potential market for nondairy probiotics such as vegetable-based products, fruit-based products, cereal- and legumes-based products, confectionary products, and breakfast cereals. The present investigation concludes that it is possible to obtain probiotic foods from several matrices, including fruit juice. The acerola and siriguela mixed juice is a suitable medium for the incorporation of *Lactobacillus* spp. with suitable counts (>6 log CFU.g-1), demonstrating to be an ideal substrate for the culture of probiotics, since it already contains beneficial nutrients such as minerals, vitamins, dietary fibers, and antioxidants. The microencapsulation by spray drying of probiotics in acerola and siriguela mixed juice is a viable alternative, since the powder produced showed favorable physicochemical characteristics, suggesting their production for various applications in the food industry.

Acknowledgements

The authors would like to thank the Coordenação de Aperfeiçoamento de Pessoal de Nível Superior—Brasil (CAPES)—Finance Code 001. Authors are also acknowledged the National Institute of Science and Technology of Tropical Fruits and CNPq for the financial support.

Conflict of interest

The authors declare no conflict of interest.

Author details

Maria Inês Sucupira Maciel[1*] and Michelle Maria Barreto de Souza[2]

1 Graduate Program in Food Science and Technology, Federal Rural University of Pernambuco, Recife, Brazil

2 Technology Center, Federal University of Paraíba, João Pessoa, Brazil

*Address all correspondence to: m.inesdcd@gmail.com

References

[1] Mishra S, Mishra HN. Technological aspects of probiotic functional food development. FEBS Letters. 2013;11: 117-130. DOI: 10.1007/s13749-012- 0055-6

[2] Silanikove N, Leitner G, Merin U. The interrelationships between lactose intolerance and the modern dairy industry: Global perspectives in evolutional and historical backgrounds. FEBS Letters. 2015;7:7312-7331. DOI: 10.3390/nu7095340

[3] Rivera-Espinoza Y, Gallardo-Navarro Non-dairy probiotic products. FEBS Letters. 2010;27:1-11. DOI: 10.1016/j. fm.2008.06.008

[4] Corbo MR, Bevilaqua A, Petruzzelli L, Casanova FP, Sinigaglia M. Functional beverages: The emergenig side of functional foods. FEBS Letters. 2014;13:1192-1206. DOI: 10.1111/1541-4337.12109

[5] Barbosa J, Teixeira P. Development of probiotic fruit juice powders by spray drying: A review. FEBS Letters. 2016;33: 335-358. DOI: 10.1080/87559129.2016. 1175016

[6] Sidira M, Santarmaki V, Kiourtzidis M, Argyri AA, Papadopoulou OS, Chorianopoulos N, et al. Evaluation of immobilized *Lactobacillus plantarum* 2035 on whey protein as adjunct probiotic culture in yoghurt production. FEBS Letters. 2017; 75:137-146. DOI: 10.1016/j. lwt.2016. 08.026

[7] Wang L, Yu X, Xu H, Aguilar ZP, Wei H. Effect of skim milk coated inulin-alginate encapsulation beads on viability and gene expression of *Lactobacillus plantarum* during freeze drying. FEBS Letters. 2016;68:8-13. DOI: 10.1016/j. lwt.2015.12.001

[8] Silva VM, Vieira GS, Hubinger MD. Influence of different combinations of wall materials and homogenisation pressure on the microencapsulation of green coffee oil by spray drying. FEBS Letters. 2014;61:132-143. DOI: 10.1016/ j.foodres.2014.01.052

[9] Gharsallaoui A, Roudaut G, Chambin O, Voilley A, Saurel R. Applications of spray-drying in microencapsulation of food ingredients: An overview. FEBS Letters. 2007;40: 1107-1121. DOI: 10.1016/j.foodres.2007. 07.004

[10] Lisboa MH, Duarte ME, Cavalcanti- Mata ME. Modeling of food drying processes in industrial spray dryers. FEBS Letters. 2018;107:49-60. DOI: 10.1016/j. fbp.2017.09.006

[11] Nazzaro F, Orlando P, Fratianni F, Coppola R. Microencapsulation in food science and biotechnology. FEBS Letters. 2012;23:82-186. DOI: 10.1016/j. copbio.2011.10.001

[12] Martín MJ, Lara-Villoslada F, Ruiz MA, Morales ME. Microencapsulation of bacteria: A review of different technologies and their impact on the probiotic effects. FEBS Letters. 2015; 27:15-25. DOI: 10.1016/j. ifset.2014.09.010

[13] Caliskan G, Dirim SN. The effects of the different drying conditions and the amounts of maltodextrin addition during spray drying of sumac extract. FEBS Letters. 2013;91:539-548. DOI: 10.1016/j. fbp.2013.06.004

[14] Tonon RV, Grosso CRF, Hubinger MD. Influence of emulsion composition and inlet air temperature on the microencapsulation of flaxseed oil by spray drying. FEBS Letters. 2011; 44:282-289. DOI: 10.1016/j.foodres. 2010.10.018

[15] Guerin J, Petit J, Burgain J, Borges F, Bhandari B, Perroud C, et al. *Lactobacillus rhamnosus* GG encapsulation by spray-drying: Milk proteins clotting control to produce innovative matrices. FEBS Letters. 2017; 93:10-19. DOI: 10.1016/j.jfoodeng. 2016.08.008

[16] Perricone M, Bevilacqua A, Corbo MR, Sinigaglia M. Technological characterization and probiotic traits of yeasts isolated from Altamura sourdough to select promising microorganisms as functional starter cultures for cereal-based products. FEBS Letters. 2014;38:26-35. DOI: 10.1016/j. fm.2013.08.006

[17] Tripathi MK, Giri SK. Probiotic functional foods: Survival of probiotics during processing and storage. FEBS Letters. 2014;9:225-241. DOI: 10.1016/j. jff.2014.04.030

[18] Dixit Y, Wagle A, Vakil B. Patents in the field of probiotics, prebiotics, synbiotics: A review. FEBS Letters. 2016;1:1-13. DOI: 10.4172/2476-2059. 1000111

[19] Kandylis P, Pissaridi K, Bekatorou A, Kanellaki M, Koutinas AA. Dairy and non-dairy probiotic beverages. FEBS Letters. 2016;7:58-63. DOI: 10.1016/j. cofs.2015.11.012

[20] Alves NN, Messaoud GB, Desobry S, Costa JMC, Rodrigues S. Effect of drying technique and feed flow rate on bacterial survival and physicochemical properties of a non- dairy fermented probiotic juice powder. FEBS Letters. 2016;189:45-54. DOI: 10.1016/j.jfoodeng.2016.05.023

[21] Kalita D, Saikia S, Gautam G, Mukhopadhyay R, Mahanta CL. Characteristics of synbiotic spray dried powder of litchi juice with *Lactobacillus plantarum* and different carrier materials. FEBS Letters. 2018;87: 351-360. DOI: 10.1016/j.lwt.2017.08.092

[22] Rößle C, Auty MAE, Brunton N, Gormley RT, Butler T. Evaluation of fresh-cut apple slices enriched with probiotic bacteria. FEBS Letters. 2010; 11:203-209. DOI: 10.1016/j.ifset.2009. 08.016

[23] Pereira ALF, Maciel TC, Rodrigues S. Probiotic beverage from cashew apple juice fermented with *Lactobacillus casei*. FEBS Letters. 2011; 44:1276-1283. DOI: 10.1016/j. foodres.2010.11.035

[24] Fonteles TV, Costa MGM, Jesus ALT, Fontes CPML, Fernandes FAN, Rodrigues S. Stability and quality parameters of probiotic cantaloupe melon juice produced with sonicated juice. FEBS Letters. 2013;6: 2860-2869. DOI: 10.1007/s11947-012- 0962-y

[25] Pereira ALF, Almeida FDL, Lima MA, Costa JMC, Rodrigues S. Spray-drying of probiotic cashew apple juice. FEBS Letters. 2014;7:2492-2499. DOI: 10.1007/ s11947-013-1236-z

[26] Alves M, Peres CM, Hernandez- Mendonza A, Bronze MR, Peres C, Malcata FX. Olive paste as vehicle for delivery of potential probiotic *Lactobacillus plantarum* 33. FEBS Letters. 2015;75:61-70. DOI: 10.1016/j. foodres.2015.04.048

[27] Kingwatee N, Apichartsrangkoon A, Chaikham P, Worametrachanon S, Techarung J, Pankasemsuk T. Spray drying *Lactobacillus casei* 01 in lychee juice varied carrier materials. FEBS Letters. 2015;62:847-853. DOI: 10.1016/ j.lwt.2014.12.007

[28] Martins EMF, Ramos AM, Martins ML, Leite Júnior BRC. Fruit salad as a new vehicle for probiotic bacteria. FEBS Letters. 2016;36:540-548. DOI: 10.1590/1678-457X.03316

[29] Paim DRSF, Costa SDO, Walter EHM, Tonon RV. Microencapsulation of probiotic Jussara (*Euterpe edulis* M.) juice by spray drying. FEBS Letters. 2016;74:21-25. DOI: 10.1016/j.lwt.2016.07.022

[30] Alves NN, Sancho SO, Silva ARA, Desobry S, Costa JMC, Rodrigues S. Spouted bed as an efficient processing for probiotic orange juice drying. FEBS Letters. 2017;101:54-60. DOI: 10.1016/j. foodres.2017.08.052

[31] Costa GM, Silva JVC, Mingotti JD, Barão CE, Klososki SJ, Pimentel TC. Effect of ascorbic acid or oligofructose supplementation on *L. paracasei* viability, physicochemical characteristics and acceptance of probiotic orange juice. FEBS Letters. 2017;75:195-201. DOI: 10.1016/j. lwt.2016.08.051

[32] Emser K, Barbosa J, Texeira P, Morais AMMB. *Lactobacillus plantarum* survival during the osmotic dehydration and storage of probiotic cut apple. FEBS Letters. 2017;38:519-528. DOI: 10.1016/j. jff.2017.09.021

[33] Betoret E, Calabuig-Jimenez L, Patrignani F, Lanciotti R, Dalla RM. Effect of high pressure processing and trehalose addition on functional properties of mandarin juice enriched with probiotic microorganisms. FEBS Letters. 2017;85:418-422. DOI: 10.1016/ j.lwt.2016.10.036

[34] Hashemi SMB, Khaneghah AM, Barba FJ, Nemati Z, Shokofti SS, Alizadeh F. Fermented sweet lemon juice (*Citrus limetta*) using *Lactobacillus plantarum* LS5: Chemical composition, antioxidant and antibacterial activities. FEBS Letters. 2017;38:409-414. DOI: 10.1016/j. jff.2017.09.040

[35] Kantachote D, Ratanaburee A, Hayisama-ae W, Sukhoom A, Nunkaew T. The use of potential probiotic *Lactobacillus plantarum* DW12 for producing a novel functional beverage from mature coconut water. FEBS Letters. 2017;32:401-408. DOI: 10.1016/j.jff.2017.03.018

[36] Rodrigues S, Silva LCA, Mulet A, Cárcel JA, Fernandes FAN. Development of dried probiotic apple cubes incorporated with *Lactobacillus casei* NRRL B-442. FEBS Letters. 2018; 41:48-54. DOI: 10.1016/j.jff.2017.12.042

[37] Prado FC, Parada JL, Pandey A, Soccol CR. Trends in non-dairy probiotic beverages. FEBS Letters. 2008; 41:111-123. DOI: 10.1016/j.foodres.2007. 10.010

[38] Nagpal R, Kumar A, Kumar M. Fortification and fermentation of fruit juices with probiotic lactobacilli. FEBS Letters. 2012;62:1573-1578. DOI: 10.1007/s13213-011-0412-5

[39] Nematollahi A, Sohrabvandi S, Mortazavian AM, Jazaeri S. Viability of probiotic bacteria and some chemical and sensory characteristics in cornelian cherry juice during cold storage. FEBS Letters. 2016;21:49-53. DOI: 10.1016/j. ejbt.2016.03.001

[40] Luckow T, Sheehan V, Fitzgerald G, Delahunty C. Exposure, health information and flavour-masking strategies for improving the sensory quality of probiotic juice. FEBS Letters. 2006;47:315-323. DOI: 10.1016/j. appet.2006.04.006

[41] Panghal A, Virkar K, Kumar V, Dhull SB, Gat Y, Chhikara N. Development of probiotic beetroot drink. FEBS Letters. 2017;5:257-262. DOI: 10.12944/ CRNFSJ.5.3.10

[42] Galvano F, Caruso MC, Condelli N, Favati F. Probiotics in Fruits and Vegetables: Technological and Sensory Aspects. 1nd ed. Chichester: Wiley; 2015. pp. 189-206. DOI: 13: 978-1- 4822-0663-0

[43] Puupponen-Pimiä R, Aura AM, Oksman-Caldentey KM, Myllärinen P, Saarela M, Mattila-Sandholm T, et al. Development of functional ingredients for gut health. FEBS Letters. 2002;13: 3-11. DOI: 10.1016/S0924-2244(02) 00020-1

[44] Selma MV, Espín JC, Tomás- Barberán FA. Interaction between phenolics and gut microbiota: Role in human health. FEBS Letters. 2009;57: 6485-6501. DOI: 10.1021/jf902107d

[45] Bark SA. The potential applications of probiotics on dairy and non-dairy foods focusing on viability during storage. FEBS Letters. 2015;4:423-431. DOI: 10.1016/j. bcab.2015.09.010

[46] Peres CM, Peres C, Hernández- Mendoza A, Malcata FX. Review on fermented plant materials as carriers and sources of potentially probiotic lactic acid bacteria-with an emphasis on table olives. FEBS Letters. 2012;26: 31-42. DOI: 10.1016/j.tifs.2012.01.006

[47] Fonteles TV, Costa MG, De Jesus ALT, Rodrigues S. Optimization of the fermentation of cantaloupe juice by *Lactobacillus casei* NRRLB-442. FEBS Letters. 2011;5:2819-2826. DOI: 10.1007/ s11947-011-0600-0

[48] Costa MGM, Fonteles TV, de Jesus ALT, Rodrigues S. Sonicated pineapple juice as substrate for *L. casei* cultivation for probiotic beverage development: Process optimisation and product stability. FEBS Letters 2013; 139: 261–266. DOI: 10.1016/j. foodchem.2013.01.059

[49] Nualkaekul S, Lenton D, Cook MT, Khutoryanskiy V, Charalampopoulos D. Chitosan coated alginate beads for the survival of microencapsulated *Lactobacillus plantarum* in pomegranate juice. FEBS Letters. 2012;90:1281-1287. DOI: 10.1016/j.carbpol.2012.06.073

[50] Sohail A, Turne RMS, Prabawati EK, Coombes AG, Bhandari B. Evaluation of *Lactobacillus rhamnosus* GG and *Lactobacillus acidophilus* NCFM encapsulated using a novel impinging aerosol method in fruit food products. FEBS Letters. 2012;157:162-166. DOI: 10.1016/j.ijfoodmicro.2012.04.025

[51] Burgain J, Gaiani C, Linder M, Scher J. Encapsulation of probiotic living cells: From laboratory scale to industrial applications. FEBS Letters. 2011;104:467-483. DOI: 10.1016/j. jfoodeng.2010.12.031

[52] De Vos P, Faas MM, Spasojevic M, Sikkema J. Encapsulation for preservation of functionality and targeted delivery of bioactive food components. FEBS Letters. 2010;20: 292-302. DOI: 10.1016/j. idairyj.2009.11.008

[53] Kent RM, Doherty SB. Probiotic bacteria in infant formula and follow-up formula: Microencapsulation using milk and pea proteins to improve microbiological quality. FEBS Letters. 2014;64:567-576. DOI: 10.1016/j. foodres.2014.07.029

[54] Roccia P, Martínez ML, Llabot JM, Ribotta PD. Influence of spray drying operating conditions on sunflower oil powder qualities. FEBS Letters. 2014; 254:307-313. DOI: 10.1016/j. powtec.2014.01.044

[55] Ranadheeraa CS, Evans CA, Adams MC, Baines SK. Microencapsulation of *Lactobacillus acidophilus* LA-5, *Bifidobacterium animalis* subsp. lactis BB-12 and Propionibacterium jensenii 702 by spray drying in goat's milk. FEBS Letters. 2015;123:155-159. DOI: 10.1016/j. smallrumres.2014.10.012

[56] Mishra P, Mishra S, Mahanta CL. Effect of maltodextrin concentration and inlet temperature during spray drying on physicochemical and antioxidant properties of amla (Emblica officinalis) juice powder. FEBS Letters. 2014;92:252-258. DOI: 10.1016/j. fbp.2013.08.003

[57] Oliveira GHH, Corrêa PC, Araújo EF, Valente DSM, Botelho FM. Desorption isotherms and thermodynamic properties of sweet corn cultivars (*Zea mays* L.). FEBS Letters. 2010;45:546-554. DOI: 10.1111/ j.1365-2621.2009.02163.x

[58] Ding WK, Shah NP. Survival of free and microencapsulated probiotic bacteria in orange and apple juices. FEBS Letters. 2008;15:219-232. DOI: http://www.ifrj.upm.edu.my/pre viousissues.html

[59] Domian E, Brynda-Kopytowska A, Cenkier J, Swirydow E. Selected properties of microencapsulated oil powders with commercial preparations of maize OSA starch and trehalose. FEBS Letters. 2015;152:72-84. DOI: 10.1016/j.jfoodeng.2014.09.034

[60] Chranioti C, Nikoloudaki A, Tzia C. Saffron and beetroot extracts encapsulated in maltodextrin, gum arabic, modified starch and chitosan: Incorporation in a chewing gum system. FEBS Letters. 2015;127: 252-263. DOI: 10.1016/j. carbpol.2015.03.049

[61] Araujo-díaz SB, Leyva-Porras C, Aguirre-Banuelos P, Álvare Z-Salas C, Saavedra-Leos Z. Evaluation of the physical properties and conservation of the antioxidants content, employing inulin and maltodextrin in the spray drying of blueberry juice. FEBS Letters. 2017;167:317-325. DOI: 10.1016/j. carbpol.2017.03.065

[62] Campelo HP, Carmo LE, Zacarias DR, Yoshida IM, Ferraz PV, Fernandes RAB, et al. Effect of dextrose equivalent on physical and chemical properties of lime essential oil microparticles. FEBS Letters. 2017;102: 105-114. DOI: 10.1016/j.indcrop.2017. 03.021

[63] Ribeiro CMCM, Magliano LCSA, Costa MMA, Bezerra TKA, Silva FLH, Maciel MIS. Optimization of the spray drying process conditions for acerola and seriguela juice mix. FEBS Letters. 2018;38:48-55. DOI: 10.1590/fst.36217

[64] Ribeiro HL, Oliveira AV, Brito ES, Ribeiro RV, Filho SMM, Azeredo MCH. Stabilizing effect of montmorillonite on acerola juice anthocyanins. FEBS Letters. 2018;245:966-973. DOI: 10.1016/j.foodchem.2017.11.076

[65] Moura CFH, Oliveira LS, Souza KO, França LG, Ribeiro LB, Souza PA, et al. Acerola–*Malpighia emarginata*. FEBS Letters. 2018;12:7-14. DOI: 10.1016/ B978-0-12-803138-4.00003-4

[66] Malegori C, Nascimento Marques EJ, de Freitas ST, Pimentel MF, Pasquini C, Casiraghi E. Comparing the analytical performances of micro-NIR and FT-NIR spectrometers in the evaluation of acerola fruit quality, using PLS and SVM regression algorithms. FEBS Letters. 2017;165:112-116. DOI: 10.1016/j.talanta.2016.12.035

[67] Augusto ED, Cristianini M, Ibarz A. Effect of temperature on dynamic and steady-state shear rheological properties of siriguela (*Spondias purpurea* L.) pulp. FEBS Letters. 2012;108:283-289. DOI: 10.1016/j.jfoodeng.2011.08.015

[68] Engels C, Gräter D, Esquivel P, Jiménez VM, Gänzle MG, Schieber A. Characterization of phenolic compounds in jocote (Spondias purpurea L.) peels by ultra-high- performance liquid chromatography/ electrospray ionization mass spectrometry. FEBS Letters. 2012;46: 557-562. DOI: 10.1016/j.foodres.2011. 04.003

[69] Fu N, Chen XD. Towards a maximal cell survival in convective thermal drying processes. FEBS Letters. 2011;44: 1127-1149. DOI: 10.1016/j.foodres.2011. 03.053

[70] FAO/WHO. Guidelines for the Evaluation of Probiotics in Food. Food and Agriculture Organization of the United Nations and World Health Organization Working Group Report. 2002

[71] Carvalho AS, Silva J, Ho P, Teixeira P, Malcata FX, Gibbs P. Effects of addition of sucrose and salt, and of starvation upon thermotolerance and survival during storage of freeze-dried *Lactobacillus delbrueckii* ssp. *bulgaricus*. FEBS Letters. 2003;68: 2538-2541. DOI: 10.1111/ j.1365-2621.2003.tb07057.x

[72] Escalona-García LA, Pedroza-Islas R, Natividad R, Rodríguez-Huezo ME, Carrillo-Navas H, Pérez-Alonso C. Oxidation kinetics and thermodynamic analysis of chia oil microencapsulated in a whey protein concentrate- polysaccharide matrix. FEBS Letters. 2016;175:93-103. DOI: 10.1016/j. jfoodeng.2015.12.009

[73] Tontul I, Topuz A. Influence of emulsion composition and ultrasonication time on flaxseed oil powder properties. FEBS Letters. 2014; 26:454-460. DOI: 10.1016/j. powtec.2014.05.002

[74] Zhang L, Zeng X, Fu N, Tang X, Sun Y, Lin L. Maltodextrin: A consummate carrier for spray-drying of xylooligosaccharides. FEBS Letters. 2018;18:963-996. DOI: 10.1016/j. foodres.2018.01.004

[75] Tonon RV, Baroni AF, Brabet C, Gibert O, Pallet D, Hubinger MD. Water sorption and glass transition temperature of spray dried açai (*Euterpe oleracea* Mart.) juice. FEBS Letters. 2008;94:215-221. DOI: 10.1016/j. jfoodeng.2009.03.009

[76] Bicudo MOP, Jó J, Oliveira GA, Chaimsohn FP, Sierakowski MR, Freitas RA, et al. Microencapsulation of juçara (*Euterpe edulis* M.) pulp by spray drying using different carriers and drying temperatures. FEBS Letters. 2015;33:153-161. DOI: 10.1080/ 07373937.2014.937872

[77] Kha TC, Nguyen MH, Roach PD. Effects of spray drying conditions on the physicochemical and antioxidant properties of the Gac (*Momordica cochinchinensis*) fruit aril powder. FEBS Letters. 2010;98:385-392. DOI: 10.1016/ j.jfoodeng.2010.01.016

[78] Fazaeli M, Emam-Djomeh Z, Ashtari AK, Omid M. Effect of spray drying conditions and feed composition on the physical properties of black mulberry juice powder. FEBS Letters. 2012;90:667-675. DOI: 10.1016/j. fbp.2012.04.006

[79] Santhalakshmy S, Bosco SJD, Francis S, Sabeena M. Effect of inlet temperature on physicochemical properties of spray-dried jamun fruit juice powder. FEBS Letters. 2015;274: 37-43. DOI: 10.1016/j.powtec.2015. 01.016

[80] Rowe RC, Sheskey PJ, Quinn ME, editors. Handbook of Pharmaceutical Excipients. 6nd ed. Chichester: Wiley; 2009. pp. 222-418

[81] Rajam R, Anandharamakrishnan C. Microencapsulation of lactobacillus plantarum (MTCC 5422) with fructooligosaccharide as wall material by spray drying. FEBS Letters. 2015;60: 773-780. DOI: 10.1016/j. lwt.2014.09.062

[82] Arslan S, Erbas M, Tontul I, Topuz A. Microencapsulation of probiotic Saccharomyces cerevisiae var. boulardii with different wall materials by spray drying. FEBS Letters. 2015;63: 685-690. DOI: 10.1016/j.lwt.2015. 03.034

[83] Rajabi H, Ghorbani M, Jafari SM, Mahoonak AS, Rajabzadeh G. Retention of saffron bioactive components by spray drying encapsulation using maltodextrin, gum Arabic and gelatin as wall materials. FEBS Letters. 2015;51: 327-337. DOI: 10.1016/j.foodhyd.2015. 05.033

[84] Botrel DA, de Barros Fernande RV, Borges SV, Yoshida MI. Influence of wall matrix systems on the properties of spray-dried microparticles containing fish oil. FEBS Letters. 2014;62:344-352. DOI: 10.1016/j. foodres.2014.02.003

8

Comparision of Antioxidant Activity of Cow and Goat Milk During Fermentation with *Lactobacillus acidophilus* LA-5

Jessica Dayara Álvarez-Rosales, César Ozuna, Rubén Salcedo-Hernández and Gabriela Rodríguez-Hernández

Abstract

This research aimed to evaluate the effects on the antioxidative capacity of cow and goat milk during fermentation with *Lactobacillus acidophilus* LA-5. The anti- oxidative capacity of milk samples during 28 days of storage was measured using a spectrophotometric decolorization assay by using 2,2-diphenyl-1-picrylhydrazyl (DPPH) radical scavenging activity. Also a principal component analysis (PCA) was used, to study the development of antioxidative activity during storage, and the connection to proteolysis and peptide concentration. The results of this study suggest that milk type (cow or goat) was a significant parameter for the proteolytic and antioxidative activity of fermented milk. Additionally, high degree positive and negative correlations were observed between the variables analyzed (0.511–0.787).

Keywords: probiotics, bioactive peptides, proteolysis, fermented milk, DPPH radical assay

Introduction

Milk and milk products are a source of vitamins, minerals, lipids, and proteins of high biological value [1]. For decades, fermented dairy products are considered beneficial foods for the health of human beings; this is due to a large part of the microorganisms involved in the fermentation and to the products released during this process [2]. The lactic acid bacteria (LAB), such as *Lactobacillus acidophilus,* are important in fermentation processes, which is why they are widely used in the food industry, due to their ability to acidify the food and preserve it from spores as well as to intervene in the texture, taste, and smell of fermented products [3]. The proteolytic activity of LAB in the milk fermentation process produces bioactive peptides that provide additional benefits in consumer health [2], such as those with antioxidant activity [4]. The composition of milk determines its nutritional quality and its properties in the manufacture of food products; goat milk has high nutritional values only surpassing human breast milk. Among the proteins of cow's milk and goat's milk, there are many differences in their composition [5], which is why fermented beverages based on cow's and goat's milk, inoculated with the probiotic *Lactobacillus acidophilus*, will be made in the present study. Additionally, the correlation between proteolytic activities, peptide concentrations, and antioxidant activity were determined.

Background

Milk and its components

Milk is composed of water, carbohydrates, lipids and proteins, as well as enzymes, vitamins, and mineral salts [6]. Fat is the component that varies the most in milk and is the main determinant of its physical and organoleptic properties [7]. Lactose is the major carbohydrate in milk; it is formed by glucose and galactose, two simple sugars that the human body uses directly as an energy source [8]. It partici- pates in the synthesis of cerebrosides and glycoproteins; also it acts to facilitate the absorption of calcium. Lactose and other sugars in milk also favor the growth of probiotics in the intestine [9]. In milk, two main classes of proteins are identified: the caseins, which represent 80% of the total proteins, comprised of several types (αs1, αs2, β, K, and γ), and serum proteins (α-lactalbumin, β lactoglobulin, and small amounts of serum albumin, immunoglobulins, and protease-peptone) [10].

These proteins are separated by the acidification of milk at pH 4.6, the isoelectric point of caseins, which produces its precipitation [6]. The biological value of casein in the diet is due to its content of essential amino acids [7]. The protein concentration, as well as the concentration and amino acid sequence of each of the milk proteins, depends on where the species comes from. Casein micelles are smaller in goat's milk (50 nm) than that in cow's milk (75 nm); these caseins present in goat's milk contain more glycine, less arginine, as well as sulfur-containing amino acids. Another difference between these types of milk is that cow's milk is slightly acidic and goat's milk is almost neutral (pH 6.7) because it has higher protein content and different combinations of phosphates [5]; also in cow's milk the largest fraction of protein is comprised of αS1-casein; however, in the case of goat's milk, the larger fractions include β-casein and αS2-casein [11]. Of the mineral elements, the milk contains sodium, potassium, magnesium, calcium, manganese, iron, cobalt, copper, fluorides, iodides, and phosphorus. Of which calcium, copper, iron, magnesium, manganese, phosphorus, and zinc are in the highest concentration in the membrane of the fat globules. In addition, the milk contains vitamins such as A, D, E, K, B1, B2, B6, B12, and C, carotenes, nicotinamide, biotin, and folic acid [7].

Proteolytic activity

Proteolysis is the process of degradation of proteins by the breakdown of their peptide bonds. From the abasement of milk proteins, peptides and free amino acids are formed, which can later present diverse bioactivities in the organism [12]. The proteolytic system of lactic acid bacteria is basically composed of proteinases that initially cleave the milk protein to peptides, and then these peptides are divided into peptides and smaller amino acids by intracellular peptidases. Subsequently, the amino acids are catabolized, producing a variety of low molecular weight compounds responsible for the formation of odor and taste compounds in fermented milks [13].

The initial step in the degradation process is carried out by the proteinase PrtP bound to the extracellular wall that decompose the proteins into peptides of 5–30 amino acids that are transported to the cells [2]. The action of the proteinases and peptidases provides the cells with peptides and free amino acids, which are transported through the membrane by specific transport systems, where the peptides are hydrolyzed by cytoplasmic peptidases [13].

Bioactive peptides

The biologically active peptides derived from milk are initially in inactive form within the sequence of the precursor molecules but can be released in different ways: by hydrolysis with digestive enzymes such as pepsin, trypsin, chymotrypsin, etc., proteolysis by enzymes derived from proteolytic microorganisms, and by fermentation of milk with proteolytic starter cultures [14]. During the fermentation of milk, LAB are able to produce bioactive peptides by the fermentation process; this is because they contain an active proteolytic system that allows the degradation and release of amino acids from milk proteins [2]. This system consists of a series of different intracellular peptidases, including endopeptidases, aminopeptidases, dipeptidases, and tri-pepti- dases. The production of various bioactive peptides, including antimicrobial peptides, immunomodulators, and antioxidants, has been demonstrated through microbial proteolysis [14]. Bioactive peptides derived from milk are generally composed of 2–20 amino acids and become reactive after the release of the precursor protein. Several lactic acid bacteria such as *Lactococcus lactis* and *Lactobacillus helveticus* have been reported to release bioactive peptides through the fermentation process [15].

Bioactive peptides with antioxidant activity

Oxidizing compounds can cause damage to proteins, lipids, or DNA. These damages are related to the development of various diseases and to aging.

Antioxidant peptides present in dietary proteins can limit oxidative damage, both in food and in the oxidation of body cells when they are ingested in the diet [16]. There are dairy peptides with antioxidant activity, and caseins and whey proteins are considered as precursors of these peptides [4]. Antioxidant peptides derived from milk are formed from 5 to 11 hydrophobic amino acids, including proline, histidine, tyrosine, and tryptophan, in sequence, that are widely distributed among the caseins, which can work by eliminating or preventing the formation of radicals as well as inhibiting enzymatic and nonenzymatic lipid peroxidation [15].

1,1-Diphenyl-2-picrylhydrazyl (DPPH) is known as a stable free radical due to the delocalization of an unpaired electron over the entire molecule. The delocalization of the electron intensifies the violet color of the radical, which it absorbs in methanol at 517 nm. When the DPPH solution reacts with an antioxidant substrate that can donate a hydrogen atom, the violet color fades. The antioxidant activity cannot be measured directly, but it can be determined by the effects of the antioxidant compound in a controlled oxidation process [17].

Lactic acid bacteria

Lactic acid bacteria are microorganisms that have various applications, including the fermentation of foods such as milk, meat, and vegetables. These bacteria, in addition to contributing to the biopreservation of foods, help to improve taste, smell, texture, and nutritional quality [18]. In addition, beneficial effects on health are attributed to them through the direct effects of live microorganisms known as probiotics as well as indirect effects during fermentation, in which these micro- organisms participate in the generation of secondary metabolites such as peptides with biological activities [19].

There are several genera of LAB, which are classified as homofermentative and heterofermentative based on the final product of their fermentation. The homofer- mentative produce lactic acid as a product of the fermentation of glucose. On the other hand, the heterofermentative produce lactic acid in addition to other products such as acetates, ethanol, and carbon dioxide as a product of its fermentation [18]. The LAB are characterized by Gram-positive cocci or bacilli, catalase and oxidase negative, facultative anaerobic, non-sporulated, and non-motile. In addition, they are tolerant acid, being able to grow some at pH values as low as 3.2 and others at values up to 9.6; however, most grow between pH of 4 and 4.5 [19].

Lactobacillus acidophilus

There are many probiotic bacteria that are used for human consumption, although the most used are *Lactobacillus* spp. These lactic acid bacteria have been used for food preservation through fermentation for hundreds of years, in addition to providing flavor and texture, and they increase the nutritional value and are also found in the gastro-intestinal tract of humans [20]. In addition, they are a key factor in the processes of competitive exclusion and immunomodulation carried out by commensal organisms. *Lactobacillus acidophilus* is a Gram-positive, non-spore- forming bacterium, homofermentative anaerobic, and catalase negative, 2–10 μm in diameter, which has an optimal growth temperature of 37°C and is a typical intes- tinal bacterium in humans [21]. This microorganism is not part of the natural flora of milk and acts on it very slowly, which is why it is essential to avoid contamination during the manufacture of a product [8].

Lactobacillus acidophilus uses the glycolysis or Embden-Meyerhof-Parnas route (EMP) to ferment hexoses and produce lactic acid. Lactic acid does not contribute to the aroma because it is odorless, but it helps the sour taste of dairy products [13]. *Lactobacillus acidophilus* LA-1/LA-5 is one of the main species of microorganisms that can potentially be used as probiotic cultures in dairy products. Some examples in the market of fermented milk products that include *Lactobacillus acidophilus* are Bioghurt, Aktifit, Actimel, Bifilac, Kaiku, and Kefir [22].

Fermented beverages

Today, the main function of fermented milk is to prolong shelf life, improve flavor and digestibility, and manufacture a wide range of dairy products. In fermented milk products, probiotic bacteria can act in the treatment of some infectious, atopic, and tumoral diseases, among others [23]. Fermented milks can be classified based on different criteria, among them are its fat content, the concentration of milk, separation of the whey, the use of milk from different species, and the type of fermentation process. Based on the type of fermentation, there are the products with a lactic fermentation such as ymer, langfil, viili, yogurt, and acidified milk.

Among the products in which lactic fermentation is combined with the production of alcohol are kefir and koumiss [8]. Today, the main function of fermented milk is to prolong shelf life, improve flavor and digestibility, and manufacture a wide range of dairy products. In fermented milk products, probiotic bacteria can act in the treatment of some infectious, atopic, and tumoral diseases, among others [23]. Fermented milks can be classified based on different criteria, among them are its fat content, the concentration of milk, separation of the whey, the use of milk from different species, and the type of fermentation process. Based on the type of fermentation, there are products with a lactic fermentation such as ymer, langfil, viili, yogurt, and acidified milk.

Among the products in which lactic fermentation is combined with the production of alcohol are kefir and koumiss [8].

Materials and methods

Treatment of fermented beverages

Cow's milk (10 L) and goat's milk (10 L) were separately subjected to a heat treatment at 95°C for 20 minutes and then cooled down to 37°C. Four treatments were prepared by making three batches of each of them, all incubated at 37°C until a pH of 4.5. Two of the treatments were fermented with *Lactobacillus acidophilus* LA-5 (Chr. Hansen) at 1%, only one of them with goat's milk and the other with cow's milk; the other two treatments that were the controls were not added with probiotics, and their fermentation occurred due to thermodynamic microorganisms (persistent after pasteurization). The beverages were kept refrigerated at 8°C during their shelf life.

Physicochemical analysis

The physicochemical analysis was formed by fat (%), nonfat solids (%), density (kg/m3), lactose (%), protein (%), total solids (%), added water (%), and freezing point (°C), and electrical conductivity (Ms/cm) was performed in triplicate of the raw material (cow's and goat's milk) in the Lactoscan Milk Analyzer (Lactoscan SA, Milkotronic Ltd., Bulgaria).

Determination of titratable acidity

It was determined based on the norm NOM-155-SCFI-2012 [24], taking 10 mL of sample, 20 mL of distilled water, and two drops of phenolphthalein, carrying out the titration with 0.1 N NaOH. The calculation of titratable acidity was carried out using the following equation:

$$\text{Acidity}\,(g/L) = \frac{(V)\,(N)\,(90)}{M} \tag{1}$$

where V = milliliters of 0.1 N NaOH solution, spent in the titration; N = nor- mality of the NaOH solution; M = volume of the sample in mL; 90 = lactic acid equivalent.

Preparation of the filtrates

The samples were treated as described by Donkor [25], in which 5 mL of each was taken and mixed with 10 mL of 0.75% trichloroacetic acid (TCA), passing the mixture through filter paper (Whatman No. 1 of 150 mm), obtaining the filtered beverages (FB), which were frozen (−20°C) until analysis. The filtrates were carried out at 0, 7, 14, 21, and 28 days to determine the proteolytic activity, the total peptide concentration, and the antioxidant activity.

Proteolytic activity

The proteolysis of each of the FB was determined in triplicate based on the reaction of the free primary amines (NH3) with O-phthaldialdehyde (OPA) and b-mercaptoethanol, according to the Church method [26]. The OPA reagent was prepared as follows: 25 mL of 100 mM sodium tetraborate, 2.5 mL of 20% sodium dodecyl sulfate (SDS), 40 mg of OPA in 1 mL of methanol, 100 mL of b-mercapto- ethanol, and capacity to 50 mL with tridestilated water. For the readings, 100 mL of each sample was taken and mixed with 2 mL of the OPA reagent by inversion of the quartz cell, with 2 minutes of incubation at room temperature and inside the equipment to avoid exposure to light; the absorbance in a spectrophotometer (Genesys 10S UV-Visible, Thermo, USA) at a wavelength of 340 nm was read. The degree of proteolysis was determined as the difference between the proteolytic activity in the treatments (beverages fermented with probiotics) and the control samples (fermented beverages without probiotics).

Total peptide concentration

The concentration of the peptides contained in each of the FB was determined in triplicate using the Bradford method [27]. This is based on the reaction of the proteins with the bright blue dye of Comassie G-250, to form a colorful compound that absorbs strongly at 595 nm. For which, a calibration curve was made using eight bovine serum albumin (BSA) standards; at concentrations of 0.1–0.01 mg/mL, the standards were prepared using 0.15 M saline solution. The absorbance reading was performed in the spectrophotometer (Genesys 10S UV-Visible, Thermo, USA), and a calibration curve was made. A linear regression was made from the given curve, obtaining the following equation:

$$Y = 0.3123X - 0.1007, R^2 = 0.9977 \tag{2}$$

Based on the equation, the peptide concentration of each one of the filtrates during its shelf life could be determined from the given absorbance reading.

Antioxidant activity

This activity was evaluated by means of the spectrophotometric technique described by Pritchard [28], which determines antioxidant activities with the DPPH radical (1,1-diphenyl-2-picrylhydrazyl) in the presence of an antioxidant substance (in this case the content of FB), measuring the inactivation potential of said radical in aqueous medium. For this, we started from an initial concentration of free radical at 0.1 mM DPPH in ethanol, respectively, diluting 1500, 1000, and 750 μL plus 500 μL of the FB adjusting to a volume of 2 mL with water HPLC grade, which generated three concentrations of the radical (0.075, 0.05 and 0.0375 mM). Water HPLC grade dissolved in DPPH was used as control, according to the concentration used. Subsequently, the samples were subjected to centrifugation at 9470 g (Spectrafuge 16 M, Labnet, USA) for 2 minutes, and the absorbance at 517 nm was measured in the spectrophotometer (Genesys 10S UV-Visible, Thermo, USA). The percentages of inhibition were calculated by the following equation:

$$\%inibición = \frac{A^{control} - A^{extracto}}{A^{control}} \, x\, 100 \tag{3}$$

Statistical analysis

The analysis will be carried out using the SAS statistical package [29], in which an analysis of variance was carried out with the GLM procedure; considering a block design (three lots), treatments were used as qualifying variables and as variables of response (proteolysis, peptide concentration, and antioxidant activity). Considering the following model:

$$yijkl = \mu + \tau i + Dj + (\tau D)ij + \beta k + \theta(ij) + \varepsilon ijkl \quad (4)$$

where yijkl = response variable measured over time; μ = general average; ti = fixed effect of the i-th treatment; Dj = effect of the j-th day (0, 7, 14, 21, 28); (tD) ij = fixed effect of the interaction between the i-th treatment and the j-th day; βk = random effect of the k-th block; Θ (ij) = random effect of the j-th experimental unit, nested in the i-th treatment; eijkl = random error distributed in normal form with zero mean and variance; and eij = N (0, s2).

A principal component analysis was also performed using the PRINCOMP pro- cedure, and it was determined as response variables (proteolysis, peptide concen- tration, and antioxidant activity) within which its correlations will be determined.

Results and discussion

Treatment of fermented beverages

The time elapsed after the pasteurization of the cow's milk until it reached a pH of 4.5 for the beverages inoculated with *Lactobacillus acidophilus* was 16 hours, while for the controls the necessary time was 27 hours. On the other hand, in goat's milk, the beverages inoculated with *Lactobacillus acidophilus* needed a time of 11 hours, whereas the controls 19 hours. In both types of milk for the controls, a longer time lapse is observed to reach the ideal pH; this because the fermentation of the milk in these treatments was carried out by the thermoduric microorganisms, which tolerate the thermal treatments applied to the milk. In the pasteurization process, it has also been observed that as the incubation temperature of the milk increases, there is evidence of greater microbial development of thermoduric species [30].

Physicochemical analysis

A physicochemical analysis was carried out in triplicate in cow and goat milk, as shown in **Table 1**. Between each parameter analyzed by the type of milk, a significant difference ($p \leq 0.05$) occurred, because milk differs in its composition depending on the species where it comes from. For cow's milk, the average percent- age of total solids that it must contain is 12.7 [31], fat 4.2, protein 3.3, lactose 4.7, and nonfat solids 8.8%, while in goat's milk, its fat content should be 4.5, protein

2.9–4.60, lactose 4.1, nonfat solids 8.9%, and total solids from 11.70 to 15.21%; how- ever, all these values depend on several factors such as the breed of the animal, its age, the period of lactation, and feeding, among others [32]. For cow milk analyzed, the percentage of protein and total solids that was obtained is within the reported parameters, although a smaller amount was registered in the parameter corresponding to fat and a slight increase in the percentage of lactose and nonfat solids. On the other hand, in goat's milk the percentages of total solids and protein are within the established ranges; there was a slight increase in both fat and lactose and a lower percentage in nonfat solids. However the percentage of protein in goat's milk is within the parameters reported for a good quality milk compared to the percentage obtained in cow's milk that present a significant difference ($p \leq 0.05$), surpassing the milk of cow.

Regarding the physical properties, at 20°C the density of the milk is approximately 1030 kg/m3, but this depends on its composition [8]. Cow's milk showed an optimum density, while a lower density was registered in goat's milk (1027.5 kg/m3). Based on the freezing point, this is relatively constant and is between −0.510 and −0.560°C due to the natural fluctuations of the composition of the milk [32], the freezing point recorded in the sample of cow's milk was −0.580°C, so it may be that an balance in

Parameter	Parameter reading	
	Cow milk	Goat milk
Fat	3.66% ± 0.01b	5.57% ± 0.02a
NFS	9.13% ± 0.01a	8.52% ± 0.02b
Lactose	5.01% ± 0.00a	4.68% ± 0.01b
Protein	3.33% ± 0.00a	3.09% ± 0.00b
Total solids	12.79% ± 0.01b	14.09% ± 0.01a
Added water	0% ± 0a	0% ± 0a
Density	1031.39 kg/m3 ± 0.04a	1027.5 kg/m3 ± 0.09b
Freezing point	−0.58°C ± 0.00b	−0.55°C ± 0.00a
Electric conductivity	4.64 mS/cm ± 0.01b	5.49 mS/cm ± 0.02a

NFS: non-fatty solids.
[a, b]Different literals indicate significant differences ($p \leq 0.05$) between parameters by type of milk.

Table 1. *Physicochemical parameters of cow's and goat's milk (raw material).*

the salt-lactose ratio has occurred in the cow's milk. In the goat's milk there was a freezing point of −0.550°C, being within the acceptable range. On the other hand, the electrical conductivity of milk is given by the presence of ions such as chlorides, phosphates, calcium, as well as sodium, and its value is between 4.0 and 6.0 mS/cm for a good quality milk. Mastitis is part of the risk conditions in the process of milk production, but through electrical conductivity it is possible to identify the beginning of this disease, because mastitis causes an increase in the concentration of sodium and chloride in the milk, increasing the conductivity values [33]. The conductivity of cow and goat milk was 4.64 and 5.49 mS/cm, respectively, so the animals from which the milk came were free from mastitis.

Determination of titratable acidity

After the incubation of the beverages, the titratable acidity of each of the treatments was measured, which is shown in **Table 2**. The predominant acid in the fermented beverages is lactic acid, although bacterial fermentation can determine the production of other acids other than lactic acid, such as acetic acid [8]. A significant difference ($p \leq 0.05$) was observed between the two treatments, where the beverages inoculated with *Lactobacillus acidophilus* showed higher values of titratable acidity; this may be due to the fact that the probiotic favored the production of lactic acid. Regarding the type of milk, there was no significant difference ($p \leq 0.05$).

Mexican standard NOM-181-SCFI-2010 [34] for yogurt indicates a minimum acidity of 0.5% lactic acid, equivalent to 5 g/L of lactic acid, while CODEX STAN [35] establishes a minimum acidity of 0.6% lactic acid; therefore, both controls and treatments inoculated with *Lactobacillus acidophilus* in the two types of beverages presented higher values than those established as minimum required acidity.

Proteolytic activity

The production of fermented beverages is a process that involves many physical and chemical changes during its production and shelf life. One of these changes is proteolysis, which consists in the progressive hydrolysis of milk caseins to

Fermented beverage type	Lot number	Titratable acid (g/L)	
		LA-5	Control
Cow milk	1	9.67 ± 0.31a	7.15 ± 1.97b
	2	6.34 ± 0.57a	7.42 ± 0.31b
	3	7.56 ± 0.38a	5.31 ± 0.12b
Goat milk	1	8.05 ± 0.06a	5.94 ± 0.38b
	2	8.01 ± 0.12a	6.70 ± 0.31b
	3	7.38 ± 0.12a	6.79 ± 0.19b

LA-5: fermented beverage with Lactobacillus acidophilus.
[a, b]*Different literals indicate significant differences ($p \leq 0.05$) between treatments.*

Table 2. *Titratable acid values (g/L lactic acid) for each treatment of the fermented beverages.*

smaller polypeptides, peptides, and amino acids by intracellular peptidases [25]. In **Figure 1**, the percentages of proteolytic activity of each type of fermented beverage are shown, which were compared with their respective control, taking it as 0%, to observe the percentage of proteolytic activity obtained in each type of beverage by the effect of the addition of the probiotic *Lactobacillus acidophilus.* The proteolytic activity of fermented beverages based on cow's milk ranged from 17.0 to 49.9% during their shelf life, while beverages based on goat milk ranged from 15.8 to 58.8%. For the two types of fermented beverages, a significant difference ($p \leq 0.05$) occurred during their shelf life, showing a tendency to increase the percentage of proteolysis over time.

Based on the type of beverage, there was also a significant difference ($p \leq 0.05$), where from day 0 to 7 the cow milk-based beverages had the highest percentage of proteolysis, while from 14 to 28 beverages, fermented milk-based goats presented the highest percentages; this may be due to the fact that the casein concentration is higher in goat's milk [10], which generates a greater proteolytic activity in the beverages.

Considering the absorbance at 340 (**Figure 2**), the proteolytic activity was estimated by determining the free amino groups using the OPA method. There was a significant difference ($p \leq 0.05$) between the days of monitoring, the treatments, and the type of fermented beverage. The absorbance of the controls was lower compared to the absorbance of the beverages inoculated with *Lactobacillus acidophilus* throughout their shelf life; these values are influenced by the effect of the probiotic in the milk. However, species and strains of lactic acid bacteria differ in their ability to hydrolyze proteins, due in part to the organization of proteolytic enzymes [8]. It is observed that cow's milk-based beverages have greater absor-bance throughout their shelf life. However, in **Figure 1**, these beverages only show higher proteolytic activity on days 0 and 7; this is because the beverages are based on goat's milk; although they have less absorbency, from day 14 they have greater absorbance than their control, unlike cow milk-based beverages, which is why their percentage of proteolytic activity is higher. On the other hand, for beverages based on cow's milk, their absorbance is in a range of 0.96–1.49 during their shelf life, values higher than those reported by Donkor [25], who estimated values of 0.80–1.03 during the same days of monitoring; these differences may be due to the fact that in their research they also used the probiotics *Lactobacillus delbrueckii* and *Streptococcus thermophilus*. The ability of LAB to grow at high cell densities in milk depends on a proteolytic system that can release essential amino acids from

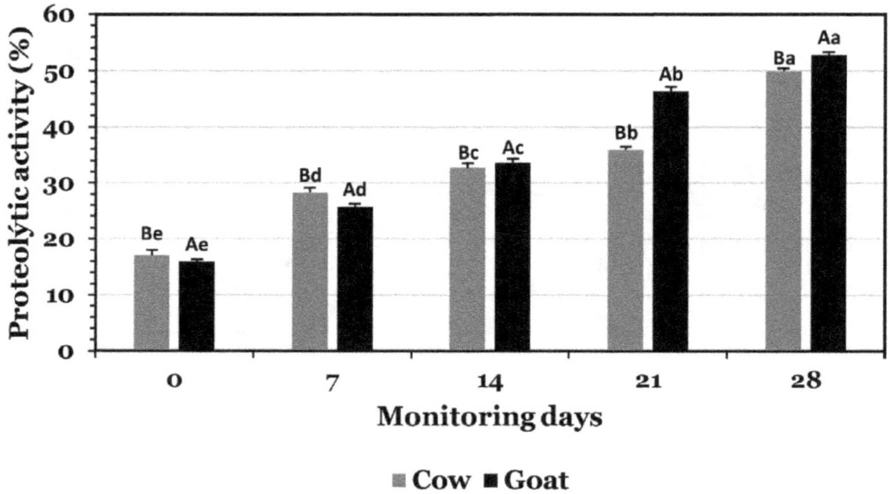

Figure 1. *Percentage of proteolytic activity of each type of beverage fermented during its shelf life compared to its respective control.* [A, B]*Different literals in uppercase indicate significant difference (p ≤0.05) per type of fermented drink.* [a, b, c, d, e]*Different literals in lowercase indicate significant difference (p ≤ 0.05) due to the effect of monitoring day.*

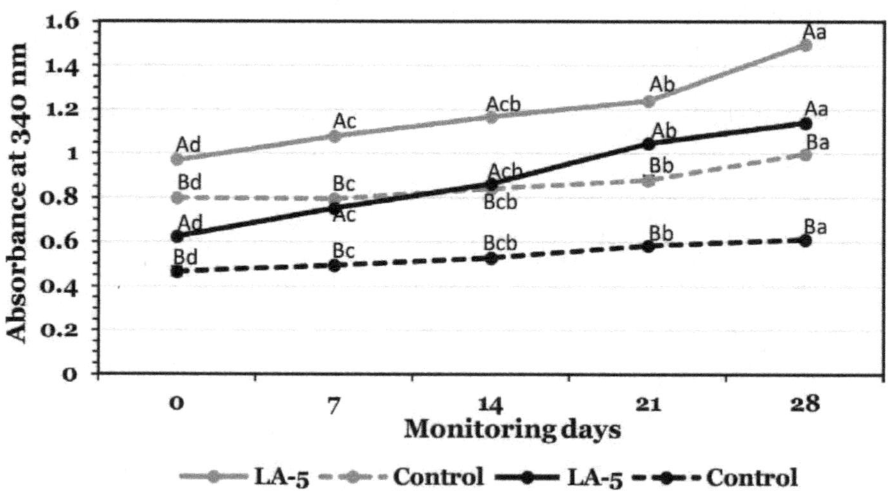

Figure 2. *Proteolytic activity of fermented beverages during their shelf life. Gray color, drinks based on cow's milk; black color, drinks based on goat's milk. LA-5: fermented beverages with Lactobacillus acidophilus. With significant difference (p ≤ 0.05) between milk types. A, BDifferent literals in uppercase indicate significant difference (p ≤ 0.05) between treatments by the type of milk. a, b, c, dDifferent literals in lowercase indicate significant difference (p ≤ 0.05) due to the effect of monitoring day.*

casein-derived peptides; ultimately these amino acids are catabolized producing many low molecular weight compounds such as aldehydes, alcohols, carboxylic acids, esters, and sulfur compounds [13]. That is why as they lived their shelf life, a more intense aroma in the drinks was perceived, due to the compounds that were forming.

Total peptide concentration

To determine the concentration of the peptides contained in each of the filtrates, a standard calibration curve was first performed (**Figure 3**), which is used to determine the protein concentration in unknown samples. The Bradford method [27] is based on the specific binding of the Coomassie G-250 bright blue dye (GBB) to the Arg, Trp, Tyr, His, and Phe residues of the proteins, producing a maximum absorbance at 595 nm, whereas the free dye has an absorbance at 470 nm.

Regarding the total peptide concentration (**Figure 4**), the highest value recorded was 0.105 mg/mL, which corresponds to the zero day control of goat milk-based beverages, and the lowest value was of 0.018 mg/mL corresponding to the LA-5 of day 28 of the drinks based on cow's milk. Based on the monitoring day, there was a significant difference ($p \leq 0.05$), observing that as time went by the peptide concentration was decreased, both in the controls and in the beverages fermented with *Lactobacillus acidophilus.* Considering the type of fermented beverage, beverages based on goat's milk always had the highest peptide concentration; this is due to the fact that goat's milk contains a higher concentration of caseins [10]. Added to this, in the controls there was a higher peptide concentration; this was because beverages fermented with *Lactobacillus acidophilus* had more microorganisms than degraded milk proteins, since LAB proteinases are able to hydrolyze more than 40% of the peptide bonds of the caseins, which generates a large amount of peptides, which can be degraded by peptidases to generate various volatile compounds [2]. This could be observed in the decrease of the peptide concentration along the length of the fermentation time during the shelf life.

Antioxidant activity

The two types of fermented beverages were analyzed for the presence of antioxi- dant activity by determining whether the generated peptides inhibit 1,1-diphenyl- 2-picrylhydrazyl (DPPH), a free radical. The antioxidant activity of the peptides is due to the unique physicochemical properties conferred by their amino acid sequences. The fermented beverages presented antioxidant activity as shown in **Table 3**.

The highest percentage of inhibition occurred in the concentration of 0.05 mM DPPH on day 0 for the drink based on cow's milk fermented with *Lactobacillus acidophilus* with a value of 73.30%, while the lowest percentage of inhibition was obtained by the control of the drink based on cow's milk on day 28 of monitoring at the same concentration of DPPH, with a value of 23.71%. There was a significant difference ($p \leq 0.05$) between the concentrations of the DPPH radical used, where the treatments containing a concentration of 0.075 mM DPPH obtained on average the highest percentages of radical inhibition. Regarding shelf life, there was also significant difference ($p \leq 0.05$), where on day 7 the highest percentages of inhibition were presented, followed by day 0; however, from day 14 the percentages of inhibition were decreasing considerably for all the concentrations of the radical; this is related to the peptide concentration that was obtained, where at a lower concentration of peptides in the product, the percentage of inhibition is also lower.

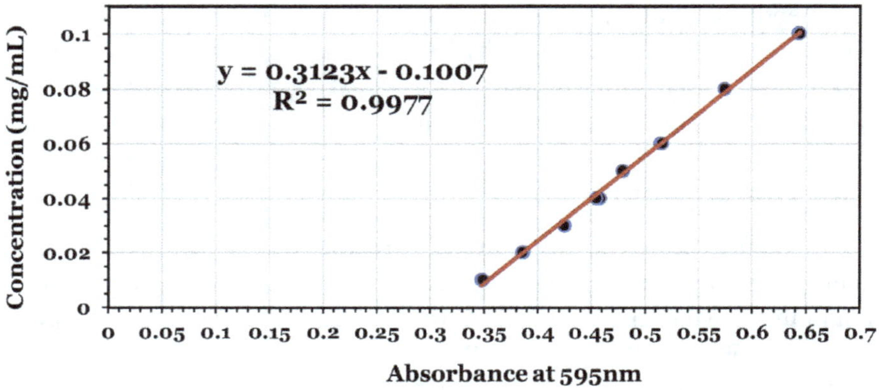

Figure 3. *Calibration curve for the determination of protein concentration (serum bovine albumin).*

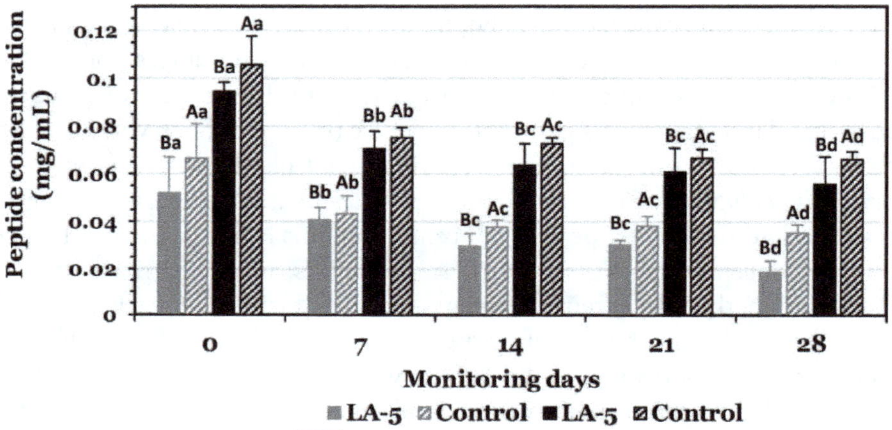

Figure 4. *Total peptide concentration of the fermented beverages during their shelf life. Gray color, drinks based on cow's milk; black color, drinks based on goat's milk. LA-5: fermented beverages with Lactobacillus acidophilus. With significant difference (p ≤ 0.05) between mil types. A, BDifferent literals in uppercase indicate a significant difference (p ≤ 0.05) between treatments. a, b, c, dDifferent literals in lowercase indicate significant difference (p ≤ 0.05) due to the effect of monitoring day.*

Day	Treatment	DPPH0.075mM		DPPH 0.05mM		DPPH 0.037mM	
		Fermented beverage		Fermented beverage		Fermented beverage	
		Cow milk[a]	Goat milk[b]	Cow milk[a]	Goat milk[b]	Cow milk[a]	Goat milk[b]
0	LA-5[a]	69.57 ± 1.07^{Aab}	43.62 ± 1.20^{Aab}	73.30 ± 0.75^{Bab}	38.14 ± 2.01^{Bab}	50.58 ± 1.45^{Bab}	50.58 ± 1.45^{Bab}
	Control[b]	37.80 ± 0.51^{Aab}	43.06 ± 0.69^{Aab}	56.88 ± 3.76^{Bab}	35.65 ± 0.75^{Bab}	39.39 ± 1.21^{Bab}	51.51 ± 0.80^{Bab}
7	LA-5[a]	75.05 ± 0.19^{Aa}	48.21 ± 1.35^{Aa}	69.48 ± 1.25^{Ba}	47.42 ± 0.75^{Ba}	69.69 ± 1.75^{Ba}	45.92 ± 2.45^{Ba}
	Control[b]	73.71 ± 0.69^{Aa}	39.03 ± 1.27^{Aa}	51.24 ± 3.58^{Ba}	43.44 ± 2.35^{Ba}	33.79 ± 3.15^{Ba}	49.41 ± 2.45^{Ba}
14	LA-5[a]	73.26 ± 1.35^{Abc}	40.60 ± 1.00^{Abc}	59.70 ± 0.86^{Bbc}	40.13 ± 2.07^{Bbc}	40.32 ± 0.80^{Bbc}	45.22 ± 1.06^{Bbc}
	Control[b]	72.14 ± 1.20^{Abc}	34.34 ± 2.60^{Abc}	48.59 ± 2.01^{Bbc}	33.00 ± 2.24^{Bbc}	27.73 ± 2.82^{Bbc}	39.16 ± 1.85^{Bbc}
21	LA-5[a]	65.65 ± 0.96^{Acd}	40.38 ± 1.07^{Acd}	54.72 ± 1.79^{Bcd}	37.97 ± 2.24^{Bcd}	37.06 ± 2.52^{Bcd}	39.86 ± 3.04^{Bcd}
	Control[b]	50.22 ± 0.38^{Acd}	34.67 ± 0.84^{Acd}	29.51 ± 3.19^{Bcd}	32.17 ± 1.74^{Bcd}	26.57 ± 3.20^{Bcd}	36.13 ± 1.75^{Bcd}
28	LA-5[a]	55.59 ± 1.17^{Ad}	40.26 ± 0.58^{Ad}	37.97 ± 2.01^{Bd}	34.82 ± 0.99^{Bd}	32.86 ± 1.85^{Bd}	35.89 ± 2.13^{Bd}
	Control[b]	48.21 ± 1.17^{Ad}	29.86 ± 0.67^{Ad}	23.71 ± 1.74^{Bd}	28.02 ± 2.55^{Bd}	25.87 ± 1.21^{Bd}	24.70 ± 1.75^{Bd}

LA-5: drink fermented with Lactobacillus acidophilus.
A,BDifferent literals in uppercase indicate a significant difference (p ≤ 0.05) between concentrations of DPPH.
a, b, c, dDifferent literals in lowercase indicate significant difference (p ≤ 0.05) due to the effect of monitoring day, with a significant difference (p ≤ 0.05) between the type of milk and treatment.

Table 3. *Percentage (%) of inhibition of DPPH radical in three different concentrations.*

The levels of antioxidant activity determined in the present study are higher than those reported by Amirdivani and Salihin [36], who reported values of 28–34% inhibition of the radical during the shelf life of their drinks; these differences may be due to the concentration of DPPH used as well as the LAB used in the process of making the drink, since they used *Streptococcus thermophilus*, *Lactobacillus acidophilus*, *Lactobacillus bulgaricus*, and *Bifidobacterium bifidum* as probiotics.

Likewise, Amirdivani and Salihin [36] also reported the highest percentages of antioxidant activity on day 7 of refrigeration, which can be attributed to metabolically active BAL even at low temperature. In this sense, the consumption of fermented beverages is highly recommended within 7 days after its preparation to

Variables	Proteolytic activity	Total peptide concentration	Antioxidant activity
Proteolytic activity	1		
Total peptide concentration	−0.787	1	
Antioxidant activity	0.511	−0.606	1

Table 4. *Pearson correlation coefficients (R) given by the principal component analysis determined for different variables obtained in fermented beverages.*

benefit from the high content of biopeptides and high antioxidant activities useful for consumer health. Free amino acids are generally not effective as antioxidants in food, so extensive proteolysis of proteins results in a decrease in antioxidant activity [37], which is reflected in the decrease in the percentage of inhibition of the radical when proteolysis increases during shelf life.

Principal component analysis

The antioxidant activity is given for 0.075 μM DPPH. Level of significance of the correlations ($p < 0.01$).

For the variables analyzed (proteolytic activity, total peptide concentration, and antioxidant activity) of the fermented beverages during their shelf life, a correlation coefficient was performed as shown in **Table 4**. The Pearson correlation coefficient is an index that measures the degree of covariation between different linearly related variables, where the correlation between directly proportional variables is positive and inversely proportional negative [38].

In this analysis, it is observed that the correlation between equal variables is positive, because exactly as one variable increases, the other increases [39] because the same data is analyzed in the two axes. The proteolytic activity and the total peptide concentration showed a negative correlation with a value of −0.787, since there is a tendency between the increase in proteolysis and the decrease in the peptide concentration.

Conclusion

The analyzed physicochemical parameters of cow and goat milk showed values of a good quality product. The beverages fermented with *Lactobacillus acidophilus*, as well as the controls, presented an acidity higher than the mini- mum required for commercial yogurts; in addition, no significant variations were observed in the titratable acidity for the two types of milk. Regarding the proteolytic activity, this was significantly augmented during the shelf life of the beverages compared to the antioxidant activity and the peptide concentration, which were decreasing. For the proteolytic activity and the peptide concentration, goat's milk-based beverages had the highest values; however, in the antioxidant activity, cow's milk beverages had the highest percentages of radical inhibition. In the peptide concentration, the controls showed the highest concentration, confirming the effect of the addition of lactobacilli to trans- form the proteins into different compounds, for which a continuation study is suggested where the volatile compounds that are suggested are quantified they have formed.

Acknowledgements

The authors would like to thank Chr. Hansen de México S.A de C.V. for the donation of the culture *Lactobacillus acidophilus* LA-5R employed in this scientific research and Juan Luis Velazquez Hernández for his translation work on this chapter.

Author details

Jessica Dayara Álvarez-Rosales, César Ozuna, Rubén Salcedo-Hernández and Gabriela Rodríguez-Hernández*

Division of Life Science, University of Guanajuato, Guanajuato, México

*Address all correspondence to: gabriela.rodriguez@ugto.mx

References

[1] Jauhiainen T, Korpela R. Milk peptides and blood pressure. The Journal of Nutrition. 2007;**137**:825-829

[2] González LG, Jiménez J, Cruz A, Rodríguez Serrano G, Gómez Ruiz L, García Garibay M. Liberación de péptidos bioactivos por bacterias lácticas en leches fermentadas comerciales. Revista Mexicana de Ingenieria Quimica. 2011;**10**:179-188

[3] Huertas P, Adolfo R. Bacterias Ácido Lácticas: Papel funcional en los alimentos. Vol. 8. Córdoba: Facultad de Ciencias Agropecuarias. Universidad Nacional de Córdoba; 2010. pp. 94-105

[4] Hernández Ledesma B, Amigo L. La leche como fuente de antioxidantes naturales. Alimentación, Nutrición y Salud. 2004;**11**:61-65

[5] Bidot Fernández A. Composición, cualidades y beneficios de la leche de cabra: Revisión bibliográfica. Revista de Producción Animal. 2017;**29**:32-41

[6] Hernández Ledesma B. Caracterización y bioactividad de péptidos obtenidos a partir de proteínas lácteas mediante hidrólisis enzimática y procesos fermentativos [tesis]. Madrid, España: Universidad Complutense de Madrid, Instituto de Fermentaciones Industriales; 2002

[7] Agudelo DA, Bedoya O. Composición nutricional de la leche de Ganado vacuno. Revista Lasallista de Investigación. 2005;**2**(1):38-42

[8] Walstra P, Geurts TJ, Noomen A, Jellema A, van Boekel MAJS. Leche: Composición, estructura y propiedades. In: Ciencia de la leche y tecnología de los productos lácteos. Zaragoza, España: Acribia; 2001. pp. 7-10

[9] Fernández Alonzo R. Hidrólisis enzimática de Seroalbúmina Bovina. Producción de péptidos bioactivos [thesis]. Ovideo, España: Universidad de Oviedo; 2015

[10] Cozma A, Andrei S, Miere D, Filip L, Loghin F. Proteins profile in milk from three species of ruminants. Notulae Scientia Biologicae. 2011;**3**(1):26-29

[11] Chacón Villalobos A. Aspectos nutricionales de la leche de cabra (*Capra hircus*) y sus variaciones en el proceso agroindustrial. Agronomía Mesoamericana. 2005;**16**(2):239-252

[12] Smit G, Smit BA, Engels WJ. Flavour formation by lactic acid bacteria and biochemical flavour profiling of cheese products. FEMS Microbiology Reviews. 2005;**29**:591-610

[13] Gutiérrez Méndez N. Metabolic flavor production pathways of *Lactococcus lactis*. In: Food Science and Food Biotechnology Essentials: A Contemporary Perspective. Chihuahua, México: AMECA; 2012. pp. 87-91

[14] Elena Elena MC. Papel de los péptidos bioactivos de los lácteos de la dieta en la prevención de las infecciones [thesis]. Palma, España: Universidad de las Islas Baleares; 2017

[15] Mohanty DP, Mohapatra S, Misra S, Sahu PS. Milk derived bioactive peptides and their impacton human health-A review. Saudi Journal of Biological Sciences. 2016;**23**:577-583

[16] Vioque J, Millán J. Los péptidos bioactivos en alimentación: Nuevos agentes promotores de salud. Agrocsic. 2005;**26**:103-107

[17] Molyneux P. The use of the stable free radical diphenylpicrylhydrazyl (DPPH) for estimating antioxidant activity. Macrophile Associates. 2003;**26**:211-219

[18] Ramírez JC, Rosas P, Velázquez MY, Ulloa JA, Arce F. Bacterias lácticas: Importancia en alimentos y sus efectos en la salud. Revista Fuente. 2011;**2**(7):1-16

[19] Rosetti L. Bacterias lácteas. Identificación de bacterias acido lácticas (BAL) por biología molecular. Buenos Aires, Argentina: INTI; 2016. p. 44

[20] Guarner F, Ellen Sanders M, Eliakim R, Fedorak R, Gangl A, Garish J. Probióticos y prebióticos. Wisconsin, Estados Unidos: Guías Mundiales de la Organización Mundial de Gastroenterología; 2017

[21] Molina Pulloquinga MP. Efecto probiótico de *Lactobacillus acidophilus* y *Bacillus subtilis* en cuyes (*Cavia porcellus*) de engorde [thesis]. Sangolquí, Ecuador: Escuela Politécnica del Ejercito, Departamento de Ciencias de la Vida; 2008

[22] Tamime AY, Saarela M, Korslund A, Mistry AA, Shah NP. Production and maintenance of viability of probiotic micro-organisms in dairy products. In: Probiotic Dairy Products. Blackwell Publishing; 2005. pp. 39-49

[23] Ebringer L, Ferencik M, Krajcovic J. Beneficial health effects of milk and fermented dairy products—Review. Folia Microbiologica. 2008;**53**(5):378-394

[24] Norma Oficial Mexicana. NOM- 155-SCFI-2012. Leche-Denominaciones, especificaciones fisicoquímicas, información comercial y métodos de prueba. Cuauhtémoc: Diario Oficial de la Federación; 2012

[25] Donkor ON, Henriksson A, Singh TK, Vasiljevic T, Shah NP. ACE- inhibitory activity of probiotic yoghurt. International Dairy Journal. 2007;**17**:1321-1331

[26] Church FC, Swaisgood HE, Porter DH, Catignani GL. Department of food science spectrophotometric assay using o-phthaldialdehyde for determination of proteolysis in milk and isolated milk proteins. Journal of Dairy Science. 1983;**66**:1219-1227

[27] Bradford MM. A rapid and sensitive method for the quantitation of microgram quantities of protein utilizing the principle of protein-dye binding. Analytical Biochemistry. 1976;**72**:248-254

[28] Pritchard SR, Phillips M, Kailasapathy K. Identification of bioactive peptides in commercial Cheddar cheese. Food Research International. 2010;**43**:1545-1548

[29] SAS, Statistical Analysis System. Version 9.1.3 for Windows. Cary, N.C.: SAS Institute Inc.; 2006

[30] Moreno Ramírez E. Evaluación de microorganismos termodúricos y principales fuentes de contaminación en el tambo [thesis]. Montevideo, Uruguay; Universidad de la República, Departamento de Agronomía; 2011

[31] Viera MA. Parámetros de calidad de leche de vacuno en los distritos de Apata, Matahuasi y Concepción en el valle del Mantaro [thesis]. Lima, Perú: Universidad Nacional Agraria La Molina; 2013

[32] Schlimme E, Buchheim W. La leche: Aspectos químicos y físicos de su composición. In: La leche y sus componentes, propiedades químicas y físicas. Zaragoza, España: Acribia; 2002. pp. 1-5

[33] Roca A. Estudio de la conductividad eléctrica de la leche de oveja manchega como método de detección de mamitis [tesis]. Orihuela, España: Universidad Miguel Hernández, Departamento de Tecnología Agroalimentaria; 2017

[34] Norma Oficial Mexicana. NOM- 181-SCFI-2010. Yogurt-Denominación, especificaciones fisicoquímicas y microbiológicas, información comercial y métodos de prueba. Cd. De México: Diario Oficial de la Federación;

[35] FAO (Food and Agriculture Organization). Norma del Codex Para Leches Fermentadas: CODEX STAN 243-2003. Leche y Productos Lácteos. 2nd ed. Roma, Italia: FAO; 2003

[36] Amirdivani S, Salihin A. Changes in yogurt fermentation characteristics, and antioxidant potential and in vitro inhibition of angiotensin-1 converting enzyme upon the inclusión of peppermint, dill and basil. Food Science and Technology. 2011;**44**:1458-1464

[37] Campos SM, Guerrero CL, Ancona BD. Proteínas y péptidos biológicamente activos con potencial nutracéutico. In: Bioactividad de péptidos derivados de proteínas alimentarias. Yucatán, México: OmniaScience; 2013. pp. 11-29

[38] González L, Restrepo F. De Pearson a Sperman. Revista Colombiana de Ciencias Pecuarias. 2007;**20**:183-192

[39] Mondragón Barrera MA. Uso de la correlación de Spearman en un estudio de intervención en fisioterapia. Movimiento Científico. 2014;**8**(1):98-104

9

The Use of Probiotics in Poultry Production for the Control of Bacterial Infections and Aflatoxins

Daniel Hernandez-Patlan, Bruno Solis-Cruz, Billy M. Hargis and Guillermo Tellez

Abstract

In intensive poultry production, a large number of antimicrobials are frequently employed to prevent (prophylactic use) and treat (therapeutic use) diseases, as well as for growth promotion (subtherapeutic use), in order to increase productivity. However, it has been reported that the use of antimicrobials at subtherapeutic doses is closely related to the increase in bacterial resistance and with the treatment failure. In addition to antimicrobial resistance, another problem derived from the use of antimicrobials is the presence of residues in animal products. Therefore, these problems and the ban of antimicrobial as growth promoters have prompted the poultry industry to look for alternatives with similar benefits to antibiotics.

Among these alternatives, probiotics are one of the most widely studied and interesting groups. Hence, in the present chapter, the effect of probiotics and direct-fed microbial against foodborne pathogens and mycotoxins will be summarized.

Keywords: probiotics, direct-fed microbial, foodborne pathogens, antimicrobial resistance, aflatoxins

1. Introduction

Since the discovery and application of penicillin in 1940, antibiotics have played an unprecedented role in the prevention, control, and treatment of infectious diseases in both humans and animals [1]. However, in animal production, they have also been used at subtherapeutic doses [2]. It is estimated that the global consumption of antibiotics in animal production could increase by 67% in the coming years [3] mainly because of the growing global demand for animal protein [2, 4]. Although it has been reported that in developed countries the total consumption of antibiotics has decreased by around 4%, consumption of antibiotics in the USA increased slightly [5]. Furthermore, it has been reported that the amount of antibiotics used in animal production in the USA is 100–1000 times higher than human medicine, being used ~80–90% at subtherapeutic doses, and for prophylactic purposes, while the remaining 10–20% at therapeutic doses [6, 7].

The inclusion of antibiotics at subtherapeutic doses into the feed was general- ized in the early 1950s, both in the EU and the USA since they could be used to prevent diseases and positively influence the promotion of growth and feed efficiency of animals [3, 8, 9].

Nevertheless, in the last decades, these practices have changed considerably due to the concern of the increase of bacteria resistant to antibiotics, since they can be transmitted zoonotically from animals to humans, causing serious problems in public health and even death because of the failure of the antibiotic at therapeutic doses [10]. Furthermore, another problem for human health is the presence of antibiotic residues in animal-derived food, by the use of antibiotics for long periods of time, since it is associated in some cases with allergic reactions, imbalance of the intestinal microbiota, and especially, the development of antibacterial resistance [11].

Consequently, one of the measures taken in the face of the problems of bacterial resistance was the restriction of antibiotics at subtherapeutic doses in the EU in 2006 [12] and the USA in 2017 [13], and although in countries as Mexico they have not been officially banned, the Ministry of Agriculture and Rural Development (SADER), through its decentralized administrative body, the National Health Service, Food Safety and Food Quality (SENASICA), has promoted initiatives to prevent their use since 2012 [14–17]. However, as a consequence of this measure, the incidence of enteric diseases in animals has increased significantly [18], as well as the use of antibiotics, but at therapeutic doses for the purpose of controlling and preventing diseases, which could lead to a worse scenario of bacterial resistance [2, 19–21]. In this context, the European One Health Action Plan against antimicrobial resistance calls for the phasing out of routine prophylactic (Prevent) and metaphy- lactic (Control) antimicrobial use in animal production and investment in the research of new alternatives [22], since they could be regulated in the coming years.

Therefore, the poultry industry has been under pressure to seek and investigate new alternatives to reduce the problems of bacterial resistance, prevent and control diseases, reduce the mortality rate, and finally promote the growth of animals.

Among these alternatives, the most popular are probiotics (yeasts or bacteria) since it has been reported that they can improve the performance [23, 24], as well as prevent and control enteric pathogens in poultry [25–27]. Furthermore, it has been reported that probiotics could be an interesting alternative to prevent and control the toxic effects of aflatoxins. For these reasons, the probiotic market has expanded rapidly and is expected to grow to around 7% in 2020. However, this market is led mainly by Asia and Europe given the growing demand for dietary supplements [18].

Probiotics

Probiotics are defined as "live strains of strictly selected microorganisms which, when administered in adequate amounts, confer a health benefit on the host" [28]. The most common microorganisms used as probiotics in livestock production are lactic acid bacteria (LAB) from the genus *Lactobacillus*, *Pediococcus*, *Lactococcus*, *Enterococcus*, *Streptococcus*, and *Leuconostoc*. Nevertheless, only the genera *Lactobacillus*, *Streptococcus*, *Pediococcus*, *Enterococcus*, and *Weissella* are the most frequently used in poultry production [29]. Although the efficacy of probiotics reducing enteric pathogens is evident, one of the disadvantages is that they require refrigeration or lyophilization to survive for long storage periods or can be encapsulated to increase their stability/viability when included in the feed, which would increase the cost of production at the industrial level, making it unprofitable [30]. Unlike LAB, direct-fed microbials (DFM) as *Bacillus* spores, other types of probiotics, have several potential applications since they can be included as feed additives in poultry diets, due to their remarkable heat stability and long shelf life [31, 32]. Bacteria of the genus *Bacillus* are Gram-positive, frequently found in the soil. However, several studies have shown that *Bacillus* spores can also be present, germinate, and survive in the gastrointestinal tract (GIT) of different animal species [25]. The survival rate and persistence of some *Bacillus* strains in the GIT could be related to their capacity to synthesize biofilms, thus protecting themselves against the different conditions present in the gut [33]. Furthermore, another advantage of *Bacillus* strains is that they are frequently used by biotechnology companies for the production of enzymes and antibiotics. Therefore, these multifunctional microorganisms have different applications, since they are useful inside or outside a host [34, 35].

Mechanisms of action probiotics

Pathogenicbacteria

Although a large number of studies have shown the possible mechanisms by which probiotics have a beneficial action in inhibiting of pathogens, more studies are needed to elucidate them.

The possible modes of action of probiotics for the inhibition of pathogens include two basic mechanisms [29, 36, 37]: competitive exclusion and modulation of the host immune system (**Figure 1**). Competitive exclusion involves mechanisms such as (1) production of inhibitory compounds, that is, hydrogen peroxide, bacteriocins, and defensins [38, 39], (2) prevention of the pathogen adhesion [38],

(3) competition for nutrients [40], and (4) reduction of toxin bioavailability [36]. Meanwhile, in the modulation of the host immune system, both innate and adaptive immune responses are involved [29]. The adaptive immune response depends on B and T lymphocytes to induce an antigen-specific response and produce antibodies [29, 41]. In contrast, physical and chemical barriers (innate immunity), such as intestinal epithelial cells (IEC), are the first line of defense to prevent the spread of pathogens and subsequent infections. Furthermore, IEC are the target cells for probiotics, which can improve the function of the intestinal barrier by stimulating the production of mucus and antimicrobial peptides such as defensins [42, 43].

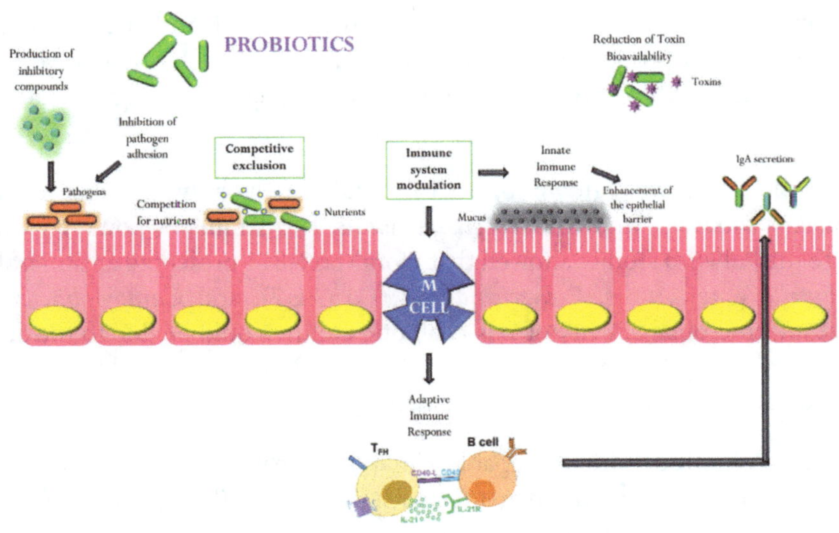

Figure 1. *Mechanism of action of probiotics.*

Aflatoxins

Similar as for pathogenic bacteria, probiotics can (1) compete for space and nutrients with aflatoxigenic mold strains, (2) degrade aflatoxins by the production of enzymes, or (3) avoid the intestinal absorption of AFB1 by its binding to the cell walls of probiotic strains [44].

Probiotic application in poultry industry

Although probiotics are considered potential alternatives to antibiotic use in poultry because they leave no residues in the meats and eggs given their modes of action, the variety of microorganisms in terms of species and even between strains of the same species, as well as their variation in metabolic activity, could affect their effectiveness. Furthermore, other factors that influence the effectiveness of probiotics in poultry are the species of origin, the probiotic preparation method, the survival of colonizing microorganisms in the gastrointestinal tract conditions, the environment where the birds are raised, the application time and administration route of probiotics, the immunologic state, the lineage of poultry, as well as age and concomitant use of antibiotics [45, 46]. Below are some of the applications of probiotics in poultry.

Effects of lactic acid bacteria against pathogens of importance in poultry

Several articles published by our laboratory have shown that the use of probiotics as a replacement of antibiotics in poultry production has had positive effects by reducing the growth of pathogens in *in vitro* models that simulate or not the three main compartments in birds (crop, proventriculus, and intestine) [47, 48], as well as the colonization of pathogens through the gastrointestinal tract in both turkeys and broiler chickens [26, 27, 49–51]. Although the results obtained have been promising, it is a fact that the isolated probiotics were characterized biochemically and by 16S rRNA sequence analyses (Microbial ID Inc., Newark, DE 19713, USA), subsequently, they were evaluated using *in vitro* models to determine their activity against pathogens, and, finally, the candidates were tested in *in vivo* models with the purpose of obtaining a well-characterized functional product.

Extensive research conducted by our laboratory determined the antimicrobial capability of several lactic acid bacteria (LAB) isolates mainly against *Salmonella* in *in vitro* models. However, only 11 were selected to produce a product called FloraMax®-B11 given their effect against *Salmonella*. Subsequently, these LAB were characterized by 16S rRNA sequence analyses (**Table 1**) [52].

However, since these LAB were grown together in a culture, the only LAB that remained viable were *Lactobacillus salivarius* and *Pediococcus parvulus*, two strains of poultry gastrointestinal origin. Despite this, *in vitro* studies showed that FloraMax®-B11 presented antimicrobial activity against *Salmonella* enter- itidis, *Escherichia coli* (O157:H7), and *Campylobacter jejuni* [47] (**Table 2**). The antimicrobial activity of this probiotic culture could be due to the accumulation of primary metabolites such as lactic acid, ethanol, and carbon dioxide and to the production of other antimicrobial compounds such as bacteriocins [53].

Furthermore, the probiotic culture was capable of maintaining its viability under acidic conditions (pH = 3) for 4 h, which agrees with other studies where *Lactobacillus* spp. isolates were resistant to low pH, with high survival rates at

LAB identification	16S rRNA sequence analyses (Microbial ID Inc.)
18	*Pediococcus parvulus*
24	*Weissella confusa*
27	*Weissella confusa*
29	*Pediococcus parvulus*
36	*Lactobacillus salivarius*
37B	*Weissella confusa*
40	*Weissella confusa*
44	*Weissella paramesenteroides*
46	*Lactobacillus salivarius*
48	*Lactobacillus salivarius*
42	*Pediococcus parvulus*

Table 1. *Identifications of FloraMax®-B11 (FM-B11) lactic acid bacteria (LAB).*

	Salmonella enteritidis	Escherichia coli (O157:H7)	Campylobacter jejuni
Lactobacillus salivarius	+	+	+
Pediococcus parvulus	+	+	+
Symbols: +, inhibition.			

Table 2. *In vitro assessment of antimicrobial activity of Lactobacillus salivarius and Pediococcus parvulus present in FloraMax®-B11 against enteropathogenic bacteria.*

pH 3.0 for 1 h [54]. Although probiotic bacteria need to survive passage through the stomach (pH 1.5–2.0) [55], and maintain their viability for 4 h or more [56] before reaching the intestine, the feed passage rate for birds is faster; therefore, bacterial acid tolerance is not as critical in chickens as it is in other animals [57].

Additionally, this probiotic culture grew at low and high temperatures for 4 h of incubation. However, the ability to grow at high temperatures is an impor- tant advantage since the production of lactic acid increases, and, therefore, the bacterial load decreases [58]. The probiotic culture was also able to tolerate high osmotic concentrations of NaCl, but it is extremely important since it has been reported that a high salt concentration could affect the physiology of probiotics, as well as their enzymatic activity, water activity, and metabolism [58]. Finally, this probiotic culture has its ability to tolerate bile salt concentrations of 0.4, 0.5, and 0.6% for 2, 4, and 24 h of incubation. Bile resistance of probiotics is related to their enzyme activity of bile salt hydrolase that helps to hydrolyze conjugated bile, reducing its toxic effect [59, 60].

Furthermore, the effect of this commercial product (FloraMax®-B11) has been evaluated in different models of infection both in broiler chickens and turkeys.

In neonatal broilers, the administration of 1×106 cfu/bird FloraMax®-B11 by oral gavage 1 h after the chicks were challenged with *Salmonella* enteritidis (SE) and *Salmonella* typhimurium (ST) (1×104 cfu/bird) reduced the incidence of SE and ST, as well as the SE counts by >2.9 log, 24 h post-LAB administration [61] (**Table 3**). In contrast, there were no significant differences at 6- and 12-h post-LAB administration, but a slight reduction was observed at 12-h post-LAB

Rep.	Treatment	ST cecal tonsil +/− (%)	SE cecal tonsil +/− (%)	Log SE cecal recovery (all samples)	Log SE cecal recovery (only positive samples)
1	Control	20/25 (80)	22/25 (88)	3.81 ± 0.32	4.33 ± 0.17
	LAB	2/25 (8)*	8/25 (32)*	0.62 ± 0.19*	1.95 ± 0.09*
2	Control	18/25 (72)	25/25 (100)	3.59 ± 0.23	3.59 ± 0.23
	LAB	2/25 (8)*	7/25 (28)*	0.42 ± 0.18*	1.91 ± 0.29*
3	Control	20/25 (80)	25/25 (100)	3.91 ± 0.19	3.91 ± 0.19
	LAB	1/25 (4)*	11/25 (40)*	1.00 ± 0.25*	2.22 ± 0.24*

**A significant ($p \leq 0.05$) difference was observed between control and treated within a single experiment in each column.*

Table 3. *Effect of lactic acid bacteria (LAB) on Salmonella typhimurium (ST) or Salmonella enteritidis (SE) recovered from cecal tonsils or ceca of broiler chicks 24-h post-LAB administration.*

administration. These data suggest that the mechanism to reduce *Salmonella* was initiated within the first 12 h after treatment. Probably the reduction of *Salmonella* is due to the set of mechanisms of action of probiotics: bacterial interactions (competitive exclusion) or stimulation of a host innate immune response. The competi- tive exclusion could have included competition for receptor sites, production of volatile fatty acids that are inhibitors of certain enteric pathogens, production of bacteriocins, or competition with pathogens and native flora for limiting nutrients [62]. Furthermore, since the *Salmonella* recovery was performed in the early stages of infection, the innate immune response could be responsible for the reduction of *Salmonella*.

In our other studies, the administration of FloraMax®-B11 in drinking water (106 cfu/mL) for 3 days post-SE challenge (104 cfu/bird) using two presentations, liquid and lyophilized significantly reduced the incidence of *Salmonella* [63], which agrees with other studies [64]. Furthermore, the administration of FloraMax®-B11 at the same concentration as the previous study after 1-h post-*Salmonella* Heidelberg (SH) challenge practically eliminated the concentration of SH, as well as its incidence, since only one sample was positive. However, in turkey poults under the same experimental conditions (**Table 4**), although similar significant results were observed at day 3 post-FloraMax®-B11 administration, it is clear that poults were more susceptible to SH colonization than chicks [51].

Finally, trying to find FloraMax®-B11 applications in poultry, we opted for spray application since it could be more efficient and has lower cost than its application in

Treatment	24 h		72 h	
	Cecal tonsils1	SH2 (log10 cfu/g of ceca content)	Cecal tonsils1	SH2 (log10 cfu/g of ceca content)
Control SH	20/20 (100)	7.04 ± 0.19a	20/20 (100)	6.05 ± 0.28a
FloraMax®-B11	13/20 (65)*	4.36 ± 0.74b	9/20 (45)*	2.15 ± 0.75b

[a,b]*Different superscripts within columns indicate significant differences ($p < 0.05$).*
[1]*Data expressed as positive/total poults (%).*
[2]*Data expressed as mean ± SE.*
$^*p < 0.001$.

Table 4. *In vivo evaluation of FloraMax-B11 against Salmonella Heidelberg (SH) at 24 and 72 h in poults.*

drinking water since it is important to take into account water quality and medicator/proportioner function [65]. The results obtained were promising since when the probiotic was applied by spray and in drinking water, there was a reduction in the recovery of SE (55 and 50%, respectively; controls 85%) when chicks were held for 8 h prior to SE challenge and placement. In the same way, when probiotic was applied by spray or in drinking water and SE challenge occurred simultaneously, with placement 8 h after treatment, a marked and significant reduction of SE recovery was noted after 5d (10 and 40%, respectively; controls 55%). Furthermore, when the probiotic was sprayed and chickens were SE challenged simultaneously,

with placement 8 h after treatment, a significant reduction of SE recovery was again noted in both the spray and DW application (80% controls, 15% spray, 15% drinking water) (**Table 5**). These results suggest that the spray application of this probiotic can be effective in protecting chicks against *Salmonella* infection. Furthermore, hatchery administration could prove to be a more effective way to administer probiotics because the chicks will be receiving the beneficial bacteria at the earliest possible time, in the absence of *in ovo* administration.

In this regard, an *in ovo* study was performed to know the effectiveness of FloraMax®-B11 [66]. For this, 18-day-old embryos were candled and inoculated with either saline or 104 cfu FloraMax®-B11 via *in ovo* injection into the amnion. On day 21, chicks were pulled from hatchers to measure hatchability. Subsequently, all chickens were then orally gavaged with SE on the day of hatch (~104 cfu/chick) and maintained for 7 days. *Salmonella* recovery was done 24-h post-SE challenge. Body weight (BW) was determined at days 1, 3, and 7. In this experiment, a significant increase in BW was observed. Furthermore, chickens that received the probiotic culture showed a significant reduction in the incidence and counts of SE in cecal tonsils when compared with saline control chickens (**Table 6**).

These results agree with another study where the *in ovo* colonization with a probiotic could become an important method to reduce *Salmonella* and other intestinal bacterial infections in poultry [67]. Regarding the increase of BW in the group treated with the probiotic, this could be due to the significant morphometric changes in the duodenum and ileum observed at day 1 of age.

Treatment regimen		Group	Cecal tonsils
		Exp. 1	Exp. 2
Treat-challenge-place immediately	Control	95% (19/20)	95% (19/20)
	Probiotic (drinking water)	75% (15/20)	25% (5/25)**
	Probiotic spray	90% (18/20)	80% (16/20)
Treat-hold 8 h-challenge-place	Control	85% (17/20)	70% (14/20)
	Probiotic (drinking water)	50% (10/20)*	70% (14/20)
	Probiotic spray	55% (11/20)*	80% (16/20)
Treat-challenge-hold 8 h-place	Control	55% (11/20)	80% (16/20)
	Probiotic (drinking water)	44% (7/20)*	15% (2/20)*
	Probiotic spray	20% (2/20)**	15% (2/20)*

**Indicates significant ($p < 0.05$) differences were observed between control and treated within a single experiment and treatment regime in each column.*
***Significantly ($p < 0.01$) different than all groups within a single experiment and treatment regime in each column.*

Table 5. *Salmonella enteritidis recovery from cecal tonsils of broiler chicks 5-day post-challenge.*

Treatment	Day 1 BW (g)	Day 3 BW (g)	Day 7 BW (g)	SE incidence cecal tonsils 24 h PI	Log SE/g of ceca content 24 h PI
Saline	49.13 ± 0.30a	62.53 ± 0.81b	132.89 ± 3.06b	20/20 (100%)	7.13 ± 1.01a
FloraMax®-B11	49.72 ± 0.36a	65.42 ± 0.77a	144.98 ± 3.02a	9/20 (45%)*	5.45 ± 1.25b

a,bSuperscripts within columns indicate significant differences $p < 0.05$, $n = 12$/group.
**Indicates significant differences $p < 0.001$, $n = 20$/group.*

Table 6. *Evaluation of in ovo administration of FloraMax®-B11 on body weight and Salmonella enteritidis (SE) recovery in broiler chickens.*

The use of direct-fed microbials (DFM) for the control of pathogens in poultry

Although the use of LAB has been promising for the control of pathogens such as *Salmonella* spp., as described above, it is important to mention that one limitation is their sensitivity to pelletizing processes for feed production (heating) [30, 68, 69], environmental factors [70], and the low pH of the stomach and the presence of bile salts in the small intestine [71, 72]. For this reason, some strategies to increase the viability of these bacteria include their microencapsulation in polymer matrices [73, 74], as well as their freezing or lyophilization [75, 76]. However, production costs increase, so it becomes nonviable in animal production. Although LAB are better probiotics than *Bacillus*, the latter is more stable due to their ability to form spores, which are more resistant to severe environmental conditions, feed pelleting process with extreme temperatures, as well as tolerance to extremes of pH, dehydration, high pressures, and chemicals, and therefore, stability to long period storage conditions, making them suitable for commercialization [77, 78] since they could be used as direct-fed microbials (DFM) [68].

Previously in our laboratory, we have screened and identified *Bacillus* spp. isolates as DFM. Some of these demonstrated to be effective as potential DFM candidates by reducing *Salmonella* colonization and having a positive effect on the increase in body weight gained in both chickens and turkeys, as well as tolerance to acidic condition (pH = 2), high osmotic pressure (NaCl at 6.5%), and 0.037% bile salts after 24 h of incubation [79–81].

Several studies have reported that some *Bacillus* species are capable of producing different exogenous enzymes such as protease, lipase, cellulase, xylanase, phytase, and keratinase [82–86], which agrees with one of our studies already published [25].

These enzymes could improve the digestion of nutrients, making them more bioavailable, and also, they help to reduce intestinal viscosity in non-starch polysac- charide diets and decrease the substrates available for the growth of pathogenic bacteria. Considering this information, we performed a study in order to evaluate the effect of three *Bacillus*-DFM candidates with excellent to good relative enzyme activity values (cellulase and xylanase) on digesta viscosity and *Clostridium perfringens* (CP) proliferation in different poultry diets using an *in vitro* digestive model [87]. One of the three *Bacillus* strains was identified as *Bacillus subtilis* and the other two isolates as *Bacillus amyloliquefaciens* by 16S rRNA sequence analysis. Subsequently, *Bacillus* candidate strains were sporulated and mixed in equal amounts during the *Bacillus*-DFM preparation process [88] and incorporated into the experimental diets (108 spores/g). The results of this study demonstrated that *Bacillus* candidate significantly reduced the viscosity of non-corn-based diets.

This could be due to the capability of these *Bacillus* strains to produce cellulase and xylanase, which could help improve the digestibility of cereals with high-soluble non-starch polysaccharides [89]. Furthermore, *Bacillus*-DFM candidate demon- strated effective antimicrobial properties against CP (**Table 7**), given their capability to produce antimicrobial-like compounds and/or compete for nutrients. Likewise, it was shown that the persistence of *Bacillus*-DFM candidate spores changes in each compartment of the *in vitro* digestive model mainly due to the conditions of pH and suggests that their full life cycle is developed in the gastrointestinal tract.

Based on the previous results, the effect of *Bacillus*-DFM candidate spores formed by an isolate of *Bacillus subtilis* and two of *Bacillus amyloliquefaciens* on growth performance, intestinal integrity, necrotic enteritis (NE) lesions, and ileal microbiota in broiler chickens using a previously established NE-challenged model [90] was evaluated [24]. This study consisted of three experimental groups: negative control (NC), positive control (PC), and *Bacillus*-DFM group (DFM). The last two groups were challenged with *Salmonella* typhimurium (ST, day 1), *Eimeria maxima* (EM, day 13), and *Clostridium perfringens* (CP, day 18–19). The overall results of performance showed that chickens supplemented with DFM had a significant body weight (BW) higher than PC. Furthermore, the body weight gain (BWG) and feed conversion ratio (FCR) were 59 g higher and 17 points lower, respectively, in the DFM group than PC (**Table 8**).

This enhancement in the performance of chickens supplemented with *Bacillus*- DFM could be due to better digestibility of nutrients, maintenance of the benefi- cial gut microbiota, and promotion of a healthy intestinal integrity [48, 87, 91].

Furthermore, these results could relate to the low-serum FITC-d concentration, bacterial translocation (BT), ileal lesion (IL), and total intestinal IgA levels in the DFM group compared to the PC group given the low impact of EM and CP challenge since DFM could produce beneficial chemical compounds, has immuno- regulatory capacity, and stimulates the homeostasis of the intestinal microbiota, resulting in a proper intestinal health status [92].

Microbiota analysis confirms that DFM played a vital role in restoring gut dysbiosis. Although only the phylum *Proteobacteria* was significantly lower in DFM group than PC group, it could be explained due to the antimicrobial properties

of DFM against ST [25], a predisposing factor in the NE model. In contrast, the genus *Lactobacillus* was significantly predominant in both NC and DFM groups with respect to PC, but it was higher in the DFM group than NC group (**Figure 2**). It has been reported that DFM is capable of increasing the genus *Lactobacillus*, which plays a crucial role in preventing dysbiosis and maintaining gut integrity (homeostasis) [36, 93].

Diet	Control diet	Bacillus-DFM
Corn-based	6.44 ± 0.19a	6.68 ± 0.08a
Wheat-based	7.12 ± 0.07a	5.20 ± 0.18b
Barley-based	7.50 ± 0.13a	6.86 ± 0.11b
Rye-based	7.15 ± 0.09a	6.68 ± 0.12b
Oat-based	6.96 ± 0.13a	5.76 ± 0.07b

[a,b]*Different superscripts within a row indicate significant differences* $p < 0.05$.
[1]*Inoculum used 105 cfu of CP.*
[2]*Data expressed in log10 cfu/mL.*

Table 7. *Concentration of Clostridium perfringens (CP)1 in different digested diets with or without inclusion of Bacillus-DFM candidate spore2.*

Item	Negative control	Positive control	DFM
BW, g/broiler			
d 0	46.88 ± 0.64^{b}	46.54 ± 0.64^{b}	49.23 ± 0.68^{a}
d 7	127.14 ± 2.90^{a}	115.58 ± 3.27^{b}	123.05 ± 3.80^{ab}
d 14	273.80 ± 11.02^{b}	295.78 ± 12.10^{ab}	318.08 ± 13.57^{a}
d 18	457.79 ± 18.97^{ab}	456.32 ± 19.39^{b}	525.58 ± 17.92^{a}
d 21	603.81 ± 24.32^{a}	445.96 ± 18.50^{c}	507.77 ± 20.60^{b}
BWG, g/broiler			
d 0–7	80.39 ± 3.06^{a}	67.74 ± 3.24^{b}	75.08 ± 3.64^{ab}

d 7–14	147.01 ± 9.51[b]	182.60 ± 9.48[a]	196.22 ± 10.56[a]
d 14–18	183.99 ± 9.85[ab]	160.55 ± 9.02[b]	198.31 ± 9.61[a]
d 14–21	325.78 ± 15.58[a]	152.13 ± 9.67[b]	185.27 ± 10.52[b]
d 0–21	552.72 ± 24.35[a]	399.42 ± 19.79[b]	458.58 ± 20.48[b]
FI, g/broiler			
d 0–21	808.21 ± 29.86[a]	772.34 ± 10.66[a]	805.21 ± 71.07[a]
FCR			
d 0–21	1.46 ± 0.04[b]	1.93 ± 0.10[a]	1.76 ± 0.18[ab]

[1]*Data expressed as mean ± SE from 40 chickens (four replicates with 10 chicks each pen). $p < 0.05$.*
[a–c]*Values within columns with different superscripts differ significantly ($p < 0.05$).*

Table 8. *Evaluation of body weight (BW), body weight gain (BWG), feed intake (FI), and feed conversion ratio (FCR) in chickens supplemented with or without DFM on a necrotic enteritis challenge model1.*

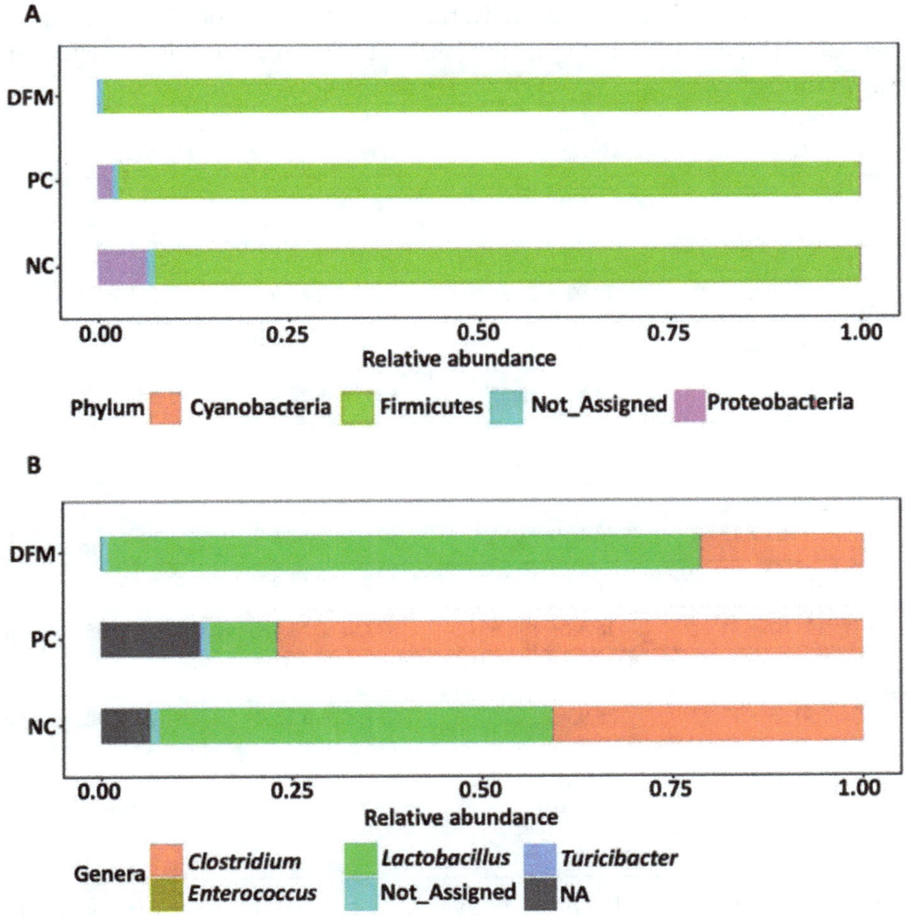

Figure 2. *Relative abundance of different phyla (A), families (B), and genera (C) in different treatment groups (NC, PC, and DFM). NA refers to those reads that were not assigned to the respective taxonomic levels.*

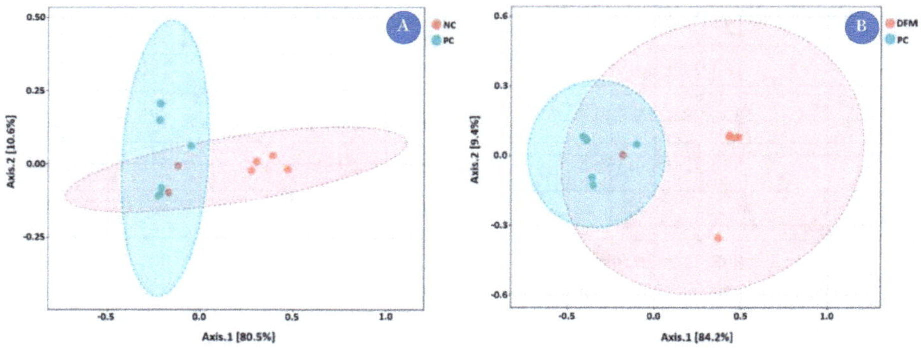

Figure 3. *PCoA plot showing difference in microbial community structure between (A) NC and PC (ANOSIM; R = 0.40 and p < 0.05) and (B) DFM and PC (ANOSIM; R = 0.73 and p < 0.01).*

Furthermore, *Clostridium* was significantly higher in PC group due to the change in the ileum microbiota caused by NE [94], whereas the genera *Lactobacillus* and *Bacillus* were more abundant in the DFM group, suggesting that these genera could alleviate the negative impacts caused by CP [95].

Finally, significant differences in beta diversity were found between NC versus PC and PC versus DFM (**Figure 3**), which agrees with another study where NE causes significant changes in the intestinal microbiota [96]. Interestingly, there was no difference in bacterial community structure between NC and DFM. It confirms again that DFM played a vital role in restoring the gut dysbiosis in this study.

Item	NC	AFB1	DFM	SEM2	p-value
BW, g/broiler					
d 0	46.23 ± 0.68[a]	47.92 ± 0.72[a]	48.12 ± 0.74[a]	0.4174	0.1275
d 7	133.29 ± 4.64[a]	129.92 ± 2.78[a]	137.02 ± 4.19[a]	2.2763	0.4502
d 14	320.92 ± 17.53[a]	272.06 ± 8.54[b]	318.42 ± 14.65[a]	8.4215	0.0263
d 21	640.10 ± 31.51[a]	474.81 ± 15.57[b]	571.60 ± 25.47[a]	16.2361	0.0001
BWG, g/broiler					
d 0–7	87.06 ± 4.24[a]	82.00 ± 2.71[a]	88.90 ± 4.15[a]	2.1705	0.4103
d 7–14	187.63 ± 13.82[a]	142.13 ± 7.06[b]	181.40 ± 11.38[a]	6.7337	0.0097
d 14–21	319.17 ± 16.08[a]	202.75 ± 9.77[c]	253.17 ± 14.89[b]	9.5832	<0.0001
d 0–21	593.87 ± 31.21[a]	426.88 ± 15.66[c]	523.48 ± 25.42[b]	16.2105	0.0001
FI, g/broiler					
d 0–21	750.55 ± 17.23[a]	775.93 ± 3.51[a]	731.97 ± 82.35[a]	25.1292	0.8193
FCR					
d 0–21	1.27 ± 0.06[b]	1.82 ± 0.06[a]	1.40 ± 0.06[b]	0.0875	0.0016

[a-c]Superscripts within rows indicate significant difference at p < 0.05.

Table 9. *Evaluation of body weight (BW), body weight gain (BWG), feed intake (FI), and feed conversion ratio (FCR) in broiler chickens consuming a corn-soybean-based diet contaminated with aflatoxin B1 (2 ppm) supplemented with or without DFM.*

The use of Bacillus-DFM candidate to prevent the toxic effects of aflatoxin B1 (AFB1) in poultry

Aflatoxin B1 (AFB1) is the predominant mycotoxin produced by several species of *Aspergillus* [97]. This mycotoxin has hepatotoxic and hepatocarcinogenic effects [98]. It has been reported that AFB1 has detrimental effects on performance parameters, which can cause serious economic problems in the poultry industry [99]. Therefore, the control of AFB1 is critical for producers. In this sense, the use of probiotics has proven effective in preventing and controlling the toxic effects of AFB1.

An *in vitro* study performed in our laboratory showed that 3 of 69 *Bacillus* spp. candidates were capable of biodegrading AFB1 since they reduced the fluorescence and area of clearance around each colony [100]. However, when these *Bacillus* spp. were tested in broiler chickens, no significant differences in performance parameters were observed when the groups were compared [101].

Despite the previous results, the *Bacillus*-DFM candidate spores formed by the isolate of *Bacillus subtilis* and the two of *Bacillus amyloliquefaciens* were included in the diets containing AFB1 to determine their effect on performance in broiler chickens fed with 2-ppm AFB1-contaminated diet [unpublished work from our laboratory]. The results are promising since the *Bacillus*-DFM improved performance of broilers, and even, there were no significant differences between the negative control (NC) and DFM group. It was due to the capacity of DFM to produce certain essential nutrients, extracellular enzymes, and growth factors to promote host growth [99, 102] (**Table 9**).

Conclusions

As it can be seen, probiotics could be considered a potential alternative to the use of antibiotics in poultry since it has been reported that they can improve the performance, as well as prevent and control enteric pathogens in poultry.

However, their applications depend on the type of microorganism. In this regard, since lactic acid bacteria (LAB) are very sensitive to pelletizing processes for feed production (heating), environmental factors, and the low pH of the stomach, as well as the presence of bile salts in the small intestine, their administration in a single dose could be the most viable application especially to prevent bacterial diseases in both *in ovo* and broiler chickens. In contrast, *Bacillus* spp. direct-fed microbials (DFM) can be a better alternative since they are more stable because they can form spores. Therefore, DFM can be included in the feed, and, in addition, the production costs are lower than the microencapsulation and freezing or lyophilization processes that are used to maintain the viability of LAB. Finally, probiotics as *Bacillus*-DFM have also shown beneficial effects in preventing and controlling toxic effects of AFB1. Although the mechanisms by which the DFM reduce the effect of AFB1 are still known, our laboratory is working to elucidate the mechanism.

Acknowledgements

This research was supported by the Arkansas Biosciences Institute under the project: Development of an avian model for evaluation early enteric microbial colonization on the gastrointestinal tract and immune function. The authors thank the CONACyT for the doctoral scholarship number 270728.

Author details

Daniel Hernandez-Patlan1, Bruno Solis-Cruz[1], Billy M. Hargis[2] and Guillermo Tellez[2*]

1 Laboratorio 5: LEDEFAR, Unidad de Investigacion Multidisciplinaria, Facultad de Estudios Superiores (FES) Cuautitlan, Universidad Nacional Autonoma de Mexico (UNAM), Cuautitlan Izcalli, Estado de Mexico, Mexico

2 Department of Poultry Science, University of Arkansas, Fayetteville, AR, USA

*Address all correspondence to: gtellez@uark.edu

References

[1] Cheng G, Hao H, Xie S, Wang X, Dai M, Huang L, et al. Antibiotic alternatives: The substitution of antibiotics in animal husbandry? Frontiers in Microbiology. 2014;**5**:217

[2] Hao H, Cheng G, Iqbal Z, Ai X, Hussain HI, Huang L, et al. Benefits and risks of antimicrobial use in food-producing animals. Frontiers in Microbiology. 2014;**5**:288

[3] Van Boeckel TP, Brower C, Gilbert M, Grenfell BT, Levin SA, Robinson TP, et al. Global trends in antimicrobial use in food animals. Proceedings of the National Academy of Sciences. 2015;**112**(18):5649-5654

[4] Van Boeckel TP, Glennon EE, Chen D, Gilbert M, Robinson TP, Grenfell BT, et al. Reducing antimicrobial use in food animals. Science. 2017;**357**(6358):1350-1352

[5] Klein EY, Van Boeckel TP, Martinez EM, Pant S, Gandra S, Levin SA, et al. Global increase and geographic convergence in antibiotic consumption between 2000 and 2015. Proceedings of the National Academy of Sciences. 2018;**115**(15):E3463-E3470

[6] Bartlett JG, Gilbert DN, Spellberg B. Seven ways to preserve the miracle of antibiotics. Clinical Infectious Diseases. 2013;**56**(10):1445-1450

[7] Mehndiratta PL, Bhalla P. Use of antibiotics in animal agriculture & emergence of methicillin-resistant *Staphylococcus aureus* (MRSA) clones: Need to assess the impact on public health. The Indian Journal of Medical Research. 2014;**140**(3):339

[8] Huyghebaert G, Ducatelle R, Van IF. An update on alternatives to antimicrobial growth promoters for broilers. The Veterinary Journal. 2011;**187**(2):182-188. Available from: http://www.sciencedirect.com/science/ article/pii/S1090023310000869

[9] Allen HK, Levine UY, Looft T, Bandrick M, Casey TA. Treatment, promotion, commotion: Antibiotic alternatives in food-producing animals. Trends in Microbiology. 2013;**21**(3):114-119

[10] Zaman SB, Hussain MA, Nye R, Mehta V, Mamun KT, Hossain N. A review on antibiotic resistance: Alarm bellsareringing.Cureus.2017;**9**(6):e1403. DOI: 10.7759/ cureus.1403

[11] Muaz K, Riaz M, Akhtar S, Park S, Ismail A. Antibiotic residues in chicken meat: Global prevalence, threats, and decontamination strategies: A review. Journal of Food Protection. 2018;**81**(4):619-627

[12] EPC. Ban on Antibiotics as Growth Promoters in Animal Feed Enters into Effect. European Commission—IP/05/1687 [Internet]. 2005. Available from: http://europa.eu/rapid/ press-release_IP-05-1687_en.htm

[13] Editors A. U.S. Bans Antibiotics Use for Enhancing Growth in Livestock [Internet]. 2017. McGraw-Hill Education. Available from: https://www. accessscience.com:443/content/u-s-bans- antibiotics-use-for-enhancing-growth-in- livestock/BR0125171 OP—AccessScience

[14] DOF. Acuerdo por el que se dan a conocer los Lineamientos para la Operación Orgánica de las actividades agropecuarias. México; 2013

[15] SENASICA. Guía para la presentación e integración de los documentos para la autorización o registro de los productos y aditivos alimenticios para consumo por animales. 2015. Available from: http://analav.com. mx/ wp-content/uploads/2016/08/Guía- regulación-ALIMENTICIOS-oct15.pdf

[16] Zaidi MB, Dreser A, Figueroa IM. A collaborative initiative for the containment of antimicrobial resistance in Mexico. Zoonoses and Public Health. 2015;**62**:52-57

[17] DOF. Acuerdo por el que se declara la obligatoriedad de la Estrategia Nacional de Acción contra la Resistencia a los Antimicrobianos. México; 2018

[18] Park YH, Hamidon F, Rajangan C, Soh KP, Gan CY, Lim TS, et al. Application of probiotics for the production of safe and high-quality poultry meat. Korean Journal for Food Science of Animal Resources. 2016;**36**(5):567

[19] Marshall BM, Levy SB. Food animals and antimicrobials: Impacts on human health. Clinical Microbiology Reviews. 2011;**24**(4):718-733

[20] Borck Høg B, Korsgaard HB, Wolff Sönksen U, Bager F, Bortolaia V, Ellis-Iversen J, et al. DANMAP 2016- Use of Antimicrobial Agents and Occurrence of Antimicrobial Resistance in Bacteria from Food Animals, Food and Humans in Denmark [Internet]. 2016. Available from: http://orbit.dtu. dk/files/140535625/DANMAP_2016_LOW_241017.pdf

[21] Founou LL, Founou RC, Essack SY. Antibiotic resistance in the food chain: A developing country- perspective. Frontiers in Microbiology. 2016;**7**:1881. Available from: http:// www.ncbi.nlm.nih.gov/pmc/articles/PMC5120092/

[22] European Parliament. European Parliament Resolution of 13 September 2018 on a European One Health Action Plan against Antimicrobial Resistance (AMR) (2017/2254(INI)). 2018; Available from: http://www.europarl.europa.eu/doceo/document/ TA-8-2018-0354_EN.pdf

[23] Latorre JD, Hernandez-Velasco X, Vicente JL, Wolfenden R, Hargis BM, Tellez G. Effects of the inclusion of a *Bacillus* direct-fed microbial on performance parameters, bone quality, recovered gut microflora, and intestinal morphology in broilers consuming a grower diet containing corn distillers dried grains with solubles. Poultry Science. 2017;**96**(8):2728-2735

[24] Hernandez-Patlan D, Solis-Cruz B, Pontin KP, Hernandez X, Merino-Guzman R, Adhikari B, et al. Impact of a *Bacillus* direct-fed microbial on growth performance, intestinal barrier integrity, necrotic enteritis lesions and ileal microbiota in broiler chickens using a laboratory challenge model. Frontiers in Veterinary Science. 2019;**6**:108

[25] Latorre JD, Hernandez-Velasco X, Wolfenden RE, Vicente JL, Wolfenden AD, Menconi A, et al. Evaluation and selection of *Bacillus* species based on enzyme production, antimicrobial activity, and biofilm synthesis as direct-fed microbial candidates for poultry. Frontiers in Veterinary Science. 2016;**3**:95

[26] Yang Y, Latorre JD, Khatri B, Kwon YM, Kong BW, Teague KD, et al. Characterization and evaluation of lactic acid bacteria candidates for intestinal epithelial permeability and *Salmonella* typhimurium colonization in neonatal Turkey poults. Poultry Science. 2017;**97**(2):515-521

[27] Prado-Rebolledo OF, Delgado- Machuca J d J, Macedo-Barragan RJ, Garcia-Márquez LJ, Morales-Barrera JE, Latorre JD, et al. Evaluation of a selected lactic acid bacteria-based probiotic on *Salmonella* enterica serovar enteritidis colonization and intestinal permeability in broiler chickens. Avian Pathology. 2017;**46**(1):90-94

[28] Tsuda H, Miyamoto T. Guidelines for the evaluation of probiotics in food. Report of a joint FAO/WHO working group on drafting guidelines for the evaluation of probiotics in food guidelines for the evaluation of probiotics in food. Report of a joint FAO/WHO working group on drafting guidelines for the evaluation of probiotics in food, 2002. Food Science and Technology Research. 2010;**16**(1):87-92

[29] Vieco-Saiz N, Belguesmia Y, Raspoet R, Auclair E, Gancel F, Kempf I, et al. Benefits and inputs from lactic acid bacteria and their bacteriocins as alternatives to antibiotic growth promoters during food-animal production. Frontiers in Microbiology. 2019;**10**:57

[30] Tellez G, Pixley C, Wolfenden RE, Layton SL, Hargis BM. Probiotics/direct fed microbials for *Salmonella* control in poultry. Food Research International. 2012;**45**(2):628-633

[31] Cartman ST, La Ragione RM, Woodward MJ. *Bacillus subtilis* spores germinate in the chicken gastrointestinal tract. Applied and Environmental Microbiology. 2008;**74**(16):5254-5258

[32] Nguyen ATV, Nguyen DV, Tran MT, Nguyen LT, Nguyen AH, Phan T. Isolation and characterization of *Bacillus subtilis* CH 16 strain from chicken gastrointestinal tracts for use as a feed supplement to promote weight gain in broilers. Letters in Applied Microbiology. 2015;**60**(6):580-588

[33] Barbosa TM, Serra CR, La Ragione RM, Woodward MJ, Henriques AO. Screening for *Bacillus* isolates in the broiler gastrointestinal tract. Applied and Environmental Microbiology. 2005;**71**(2):968-978

[34] Priest FG. Extracellular enzyme synthesis in the genus *Bacillus*. Bacteriological Reviews. 1977;**41**(3):711

[35] Azevedo EC, Rios EM, Fukushima K, Campos-TakakiGM.Bacitracinproduction by a new strain of *Bacillus subtilis*. Applied Biochemistry and Biotechnology. 1993;**42**(1):1

[36] Caly DL, D'Inca R, Auclair E, Drider D. Alternatives to antibiotics to prevent necrotic enteritis in broiler chickens: A microbiologist's perspective. Frontiers in Microbiology. 2015;**6**:1336

[37] Alagawany M, El-Hack MEA, Farag MR, Sachan S, Karthik K, Dhama K. The use of probiotics as eco-friendly alternatives for antibiotics in poultry nutrition. Environmental Science and Pollution Research. 2018;**25**(11):10611-10618

[38] Tiwari G, Tiwari R, Pandey S, Pandey P. Promising future of probiotics for human health: Current scenario. Chronicles of Young Scientists. 2012;**3**(1):17

[39] Pan D, Yu Z. Intestinal microbiome of poultry and its interaction with host and diet. Gut Microbes. 2014;**5**(1):108-119

[40] Kabir SML, Rahman MM, Rahman MB, Rahman MM, Ahmed SU. The dynamics of probiotics on growth performance and immune response in broilers. International Journal of Poultry Science. 2004;**3**(5):361-364

[41] Khalighi A, Behdani R, Kouhestani S. Probiotics: A Comprehensive Review of their Classification, Mode of Action and Role in Human Nutrition. IntechOpen; 2016

[42] Schlee M, Harder J, Köten B, Stange EF, Wehkamp J, Fellermann K. Probiotic lactobacilli and VSL# 3 induce enterocyte β-defensin 2. Clinical and Experimental Immunology. 2008;**151**(3):528-535

[43] Bermudez-Brito M, Plaza-Díaz J, Muñoz-Quezada S, Gómez-Llorente C, Gil A. Probiotic mechanisms of action. Annals of Nutrition & Metabolism. 2012;**61**(2):160-174

[44] Solis-Cruz B, Hernandez-Patlan D, Hargis B, Téllez G. Control of Aflatoxicosis in Poultry Using Probiotics and Polymers [Online First]. In: Micotoxins-Impact and management strategies. IntechOpen; 2018. DOI: 10.5772/intechopen.76371. Available from: https://www.intechopen.com/

[45] Otutumi LK, De Moraes Garcia ER, Góis MB, Loddi MM. Variations on the efficacy of probiotics in poultry. In: Rigobelo EC, editor. Probiotic in Animals. Rijeka: InTech; 2012:203-230

[46] Harimurti S, Hadisaputro W. Probiotics in Poultry. In: Beneficial microorganisms in agriculture, aquaculture and other areas. Cham: Springer; 2015:1-19

[47] Menconi A, Kallapura G, Latorre JD, Morgan MJ, Pumford NR, Hargis BM, et al. Identification and characterization of lactic acid bacteria in a commercial probiotic culture. Bioscience of Microbiota, Food and Health. 2014;**33**(1):25-30

[48] Latorre JD, Hernandez-Velasco X, Bielke LR, Vicente JL, Wolfenden R, Menconi A, et al. Evaluation of a *Bacillus* direct-fed microbial candidate on digesta viscosity, bacterial translocation, microbiota composition and bone mineralisation in broiler chickens fed on a rye-based diet. British Poultry Science. 2015;**56**(6):723-732

[49] Higgins JP, Higgins SE, Wolfenden AD, Henderson SN, Torres-Rodriguez A, Vicente JL, et al. Effect of lactic acid bacteria probiotic culture treatment timing on *Salmonella* enteritidis in neonatal broilers. Poultry Science. 2010;**89**(2):243-247

[50] Higgins SE, Wolfenden AD, Tellez G, Hargis BM, Porter TE. Transcriptional profiling of cecal gene expression in probiotic-and *Salmonella*- challenged neonatal chicks. Poultry Science. 2011;**90**(4):901-913

[51] Menconi A, Wolfenden AD, Shivaramaiah S, Terraes JC, Urbano T, Kuttel J, et al. Effect of lactic acid bacteria probiotic culture for the treatment of *Salmonella* enterica serovar Heidelberg in neonatal broiler chickens and Turkey poults. Poultry Science. 2011;**90**(3):561-565

[52] Tellez G, Higgins SE, Donoghue AM, Hargis BM. Digestive physiology and the role of microorganisms. Journal of Applied Poultry Research. 2006;**15**(1):136-144

[53] Rattanachaikunsopon P, Phumkhachorn P. Lactic acid bacteria: Their antimicrobial compounds and their uses in food production. Annals of Biological Research. 2010;**1**(4):218-228

[54] Fontana L, Bermudez-Brito M, Plaza-Diaz J, Munoz-Quezada S, Gil A. Sources, isolation, characterisation and evaluation of probiotics. The British Journal of Nutrition. 2013;**109**(S2):S35-S50

[55] Dunne C, O'Mahony L, Murphy L, Thornton G, Morrissey D, O'Halloran S, et al. In vitro selection criteria for probiotic bacteria of human origin: Correlation with in vivo findings. The American Journal of Clinical Nutrition. 2001;**73**(2):386s-392s

[56] Bakari D, Tatsadjieu NL, Mbawala A, Mbofung CM. Assessment of physiological properties of some lactic acid bacteria isolated from the intestine of chickens use as probiotics and antimicrobial agents against enteropathogenic bacteria. Innovative Romanian Food Biotechnology. 2011;**8**:33

[57] Boonkumklao P, Kongthong P, Assavanig A. Acid and bile tolerance of *Lactobacillus* thermotolerans, a novel species isolated from chicken feces. Kasetsart Journal. 2006;**40**:13-17

[58] Ibourahema C, Dauphin RD, Jacqueline D, Thonart P. Characterization of lactic acid bacteria isolated from poultry farms in Senegal. African Journal of Biotechnology. 2008;**7**(12):2006-2012

[59] Du Toit M, Franz C, Dicks LMT, Schillinger U, Haberer P, Warlies B, et al. Characterisation and selection of probiotic lactobacilli for a preliminary minipig feeding trial and their effect on serum cholesterol levels, faeces pH and faeces moisture content. International Journal of Food Microbiology. 1998;**40**(1-2):93-104

[60] Tanaka H, Doesburg K, Iwasaki T, Mierau I. Screening of lactic acid bacteria for bile salt hydrolase activity. Journal of Dairy Science. 1999;**82**(12):2530-2535

[61] Higgins JP, Higgins SE, Vicente JL, Wolfenden AD, Tellez G, Bm H. Temporal effects of lactic acid bacteria probiotic culture on *Salmonella* in neonatal broilers. Poultry Science. 2007;**86**(8):1662-1666

[62] Mead GC. Prospects for 'competitive exclusion' treatment to control *Salmonella* and other foodborne pathogens in poultry. Veterinary Journal. 2000;**159**(2):111-123

[63] Vicente JL, Torres-Rodriguez A, Higgins SE, Pixley C, Tellez G, Donoghue AM, et al. Effect of a selected *Lactobacillus* spp.-based probiotic on *Salmonella* enterica serovar enteritidis- infected broiler chicks. Avian Diseases. 2008;**52**(1):143-146

[64] Pascual M, Hugas M, Badiola JI, Monfort JM, Garriga M. *Lactobacillus salivarius* CTC2197 prevents *Salmonella* enteritidis colonization in chickens. Applied and Environmental Microbiology. 1999;**65**(11):4981-4986

[65] Wolfenden AD, Pixley CM, Higgins JP, Higgins SE, Vicente J, Torres-Rodriguez A, et al. Evaluation of spray application of a *Lactobacillus*- based probiotic on *Salmonella* enteritidis colonization in broiler chickens. International Journal of Poultry Science. 2007;**6**(7):493-496

[66] Teague KD, Graham LE, Dunn JR, Cheng HH, Anthony N, Latorre JD, et al. In ovo evaluation of FloraMax®-B11 on Marek's disease HVT vaccine protective efficacy, hatchability, microbiota composition, morphometric analysis, and *Salmonella* enteritidis infection in broiler chickens. Poultry Science. 2017;**96**(7):2074-2082

[67] De Oliveira JE, Van der Hoeven-Hangoor E, Van de Linde IB, Montijn RC, Van Der Vossen J. In ovo inoculation of chicken embryos with probiotic bacteria and its effect on posthatch *Salmonella* susceptibility. Poultry Science. 2014;**93**(4):818-829

[68] Tellez G, Rodríguez-Fragoso L, Kuttappan VA, Kallapura G, Velasco X, Menconi A, et al. Probiotics for human and poultry use in the control of gastrointestinal disease: A review of real-world experiences. Alternative and Integrative Medicine. 2013;**2**:1-6

[69] Guillermo T, Andrea L, Juan DL, Xochitl H-V, Billy MH, Todd C. Food- producing animals and their health in relation to human health. Microbial Ecology in Health and Disease. 2015;**26**(1):25876

[70] Praepanitchai O-A, Noomhorm A, Anal AK. Survival and behavior of encapsulated probiotics (*Lactobacillus plantarum*) in calcium-alginate-soy protein isolate-based hydrogel beads in different processing conditions (pH and temperature) and in pasteurized mango juice. BioMed Research International. 2019;**2019**:1-8

[71] Iravani S, Korbekandi H, Mirmohammadi SV. Technology and potential applications of probiotic encapsulation in fermented milk products. Journal of Food Science and Technology. 2015;**52**(8):4679-4696

[72] Samedi L, Charles AL. Viability of 4 probiotic bacteria microencapsulated with arrowroot starch in the simulated gastrointestinal tract (GIT) and yoghurt. Food. 2019;**8**(5):175

[73] Gbassi GK, Vandamme T. Probiotic encapsulation technology: From microencapsulation to release into the gut. Pharmaceutics. 2012;**4**(1):149-163

[74] Evivie SE, Huo G-C, Igene JO, Bian X. Some current applications, limitations and future perspectives of lactic acid bacteria as probiotics. Food & Nutrition Research. 2017;**61**(1): 1318034

[75] Chávarri M, Marañón I, Villarán MC. Encapsulation technology to protect probiotic bacteria. In: Probiotics. IntechOpen; 2012

[76] Montel Mendoza G, Pasteris SE, Otero MC, Fatima Nader-Macías ME. Survival and beneficial properties of lactic acid bacteria from raniculture subjected to freeze-drying and storage. Journal of Applied Microbiology. 2014;**116**(1):157-166

[77] Vreeland RH, Rosenzweig WD, Powers DW. Isolation of a 250 million- year-old halotolerant bacterium from a primary salt crystal. Nature. 2000;**407**(6806):897

[78] Cartman ST, La Ragione RM, Woodward MJ. Bacterial spore formers as probiotics for poultry. Food Science & Technology Bulletin Functional Foods. 2007;**4**(217):30

[79] Shivaramaiah S, Pumford NR, Morgan MJ, Wolfenden RE, Wolfenden AD, Torres-Rodriguez A, et al. Evaluation of *Bacillus* species as potential candidates for direct-fed microbials in commercial poultry. Poultry Science. 2011;**90**(7):1574-1580

[80] Wolfenden RE, Pumford NR, Morgan MJ, Shivaramaiah S, Wolfenden AD, Pixley CM, et al. Evaluation of selected direct- fed microbial candidates on live performance and *Salmonella* reduction in commercial Turkey brooding houses. Poultry Science. 2011;**90**(11):2627-2631

[81] Menconi A, Morgan MJ, Pumford NR, Hargis BM, Tellez G. Physiological properties and *Salmonella* growth inhibition of probiotic *Bacillus* strains isolated from environmental and poultry sources. International Journal of Bacteriology. 2013;**2013**:1-8

[82] Monisha R, Uma MV, Murthy VK. Partial purification and characterization of *Bacillus pumilus* xylanase from soil source. Kathmandu University Journal of Science, Engineering and Technology. 2009;**5**:137-148

[83] Mazotto AM, Coelho RRR, Cedrola SML, de Lima MF, Couri S, Paraguai de Souza E, et al. Keratinase production by three *Bacillus* spp. using feather meal and whole feather as substrate in a submerged fermentation. Enzyme Research. 2011;**2011**:1-7

[84] ArpanaM, GulabS,VarshaG, AnitaY, Aneja KR, Gautam SK, et al. Isolation and biochemical characterization of acido-thermophilic extracellular phytase producing bacterial strain for potential application in poultry feed. Jundishapur Journal of Microbiology. 2011;**4**(4):273-282

[85] Shah K. Purification and characterization of lipase from *B. subtilis* Pa2. Journal of Biochemical Technology. 2012;**3**(3):292-295

[86] Jani SA, Chudasama CJ, Patel DB, Bhatt PS, Patel HN. Optimization of extracellular protease production from alkali thermo tolerant actinomycetes: *Saccharomonospora viridis* SJ-21. Bulletin of Environment, Pharmacology and Life Sciences. 2012;**1**(6):84-92

[87] Latorre JD, Hernandez-Velasco X, Kuttappan VA, Wolfenden RE, Vicente JL, Wolfenden AD, et al. Selection of *Bacillus* spp. for cellulase and xylanase production as direct-fed microbials to reduce digesta viscosity and *Clostridium perfringens* proliferation using an in vitro digestive model in different poultry diets. Frontiers in Veterinary Science. 2015;**2**:25

[88] Zhao S, Deng L, Hu N, Zhao B, Liang Y. Cost-effective production of *Bacillus licheniformis* using simple netting bag solid bioreactor. World Journal of Microbiology and Biotechnology. 2008;**24**(12):2859-2863

[89] Wang ZR, Qiao SY, Lu WQ, Li DF. Effects of enzyme supplementation on performance, nutrient digestibility, gastrointestinal morphology, and volatile fatty acid profiles in the hindgut of broilers fed wheat-based diets. Poultry Science. 2005;**84**(6):875-881

[90] Shivaramaiah S, Wolfenden RE, Barta JR, Morgan MJ, Wolfenden AD, Hargis BM, et al. The role of an early *Salmonella* typhimurium infection as a predisposing factor for necrotic enteritis in a laboratory challenge model. Avian Diseases. 2011;**55**(2):319-323

[91] Zhou M, Zeng D, Ni X, Tu T, Yin Z, Pan K, et al. Effects of *Bacillus licheniformis* on the growth performance and expression of lipid metabolism- related genes in broiler chickens challenged with *Clostridium perfringens*- induced necrotic enteritis. Lipids in Health and Disease. 2016;**15**(1):48

[92] Jadamus A, Vahjen W, Simon O. Growth behaviour of a spore forming probiotic strain in the gastrointestinal tract of broiler chicken and piglets. Archives of Animal Nutrition. 2001;**54**(1):1-17

[93] Jin LZ, Ho YW, Abdullah N, Jalaludin S. Probiotics in poultry: Modes of action. World's Poultry Science Journal. 1997;**53**(4):351-368

[94] Prescott JF, Smyth JA, Shojadoost B, Vince A. Experimental reproduction of necrotic enteritis in chickens: A review. Avian Pathology. 2016;**45**(3):317-322

[95] Thibodeau A, Fravalo P, Yergeau É, ArsenaultJ, LahayeL, LetellierA. Chicken caecal microbiome modifications induced by *Campylobacter jejuni* colonization and by a non- antibiotic feed additive. PLoS One. 2015;**10**(7):e0131978

[96] Stanley D, Wu S-B, Rodgers N, SwickRA,MooreRJ. Differentialresponses of cecal microbiota to fishmeal, Eimeria and *Clostridium perfringens* in a necrotic enteritis challenge model in chickens. PLoS One. 2014;**9**(8):e104739

[97] Wilson KA, Kung RW, Wetmore SD. Toxicology of DNA adducts formed upon human exposure to carcinogens: Insights gained from molecular modeling. In: Advances in Molecular Toxicology. Elsevier; 2016;**10**:293-360

[98] Rawal S, Kim JE, Coulombe R Jr. Aflatoxin B1 in poultry: Toxicology, metabolism and prevention. Research in Veterinary Science. 2010;**89**(3):325-331

[99] Oguz H, Kurtoglu V. Effect of clinoptilolite on performance of broiler chickens during experimental aflatoxicosis. British Poultry Science. 2000;**41**(4):512-517

[100] Galarza-Seeber R, Latorre JD, Hernandez-Velasco X, Wolfenden AD, Bielke LR, Menconi A, et al. Isolation, screening and identification of *Bacillus* spp. as direct-fed microbial candidates for aflatoxin B1 biodegradation. Asian Pacific Journal of Tropical Biomedicine. 2015;**5**(9):702-706

[101] Galarza-Seeber R, Latorre JD, Wolfenden AD, Hernandez-Velasco X, Merino-Guzman R, Ledoux DR, et al. Evaluation of *Bacillus* spp. as direct fed microbial (DFM) candidates for aflatoxin B1 biodegradation in broiler chickens. International Journal of Probiotics and Prebiotics. 2016;**11**(1):29-37

[102] Guo M, Hao G, Wang B, Li N, Li R, Wei L, et al. Dietary administration of *Bacillus subtilis* enhances growth performance, immune response and disease resistance in Cherry Valley ducks. Frontiers in Microbiology. 2016;**7**:1975

10

Prebiotics and the Modulation on the Microbiota-GALT-Brain Axis

Elena Franco-Robles, Joel Ramírez-Emiliano, José Sergio López-Briones and Cristina Doriany Balcón-Pacheco

Abstract

It is well known that there exists a bi-directional communication system between the enteric nervous system and central nervous system. Recent research has attempted to understand the influence of intestinal bacteria on the brain and behavior. In this manner, it has been observed that pathogenic bacterial products such as lipopolysaccharides (LPSs) can induce behavioral changes such as acute anxiety, depressive symptoms, cognitive deficits, and increased sensitivity to visceral pain. The modulation of LPS production through probiotics, prebiotics, and symbiotics can prevent these changes. In addition to the neuronal, endocrine, and metabolic pathways, it has been observed that the immune mechanism also exerts an influence on the gut-brain axis. The cells of the immune system can undergo phenotypic changes by the induction of certain bacterial species, which can have an important participation in the development of brain disorders. Although the main effect of prebiotics is through the stimulation of probiotic bacteria, in this chapter, we review the indirect therapeutic potential of prebiotics on the brain through the intestinal microbiota, the gut-associated lymphoid tissue (GALT), and other components of the intestinal lumen. Thus, the objective is to elucidate the mecha- nisms underlying its effects on the gut-brain axis. Here, we will summarize the possible therapeutic effect of prebiotics on intestinal microbiota, the gut-associated lymphoid tissue (GALT), and brain.

Keywords: prebiotics, gut microbiota, central nervous system, enteric nervous system, GALT

Introduction

The intestinal microbiota contributes significantly to metabolic, trophic, and protective functions. In this regard, intestinal bacteria are responsible for metabolism of many complex substances into simple components; thus, intestinal microbi- ota contributes to the digestion of nutrients and has a key role in the nutrition of the host, to the control of certain pathogens and to the improvement of the functions of the local immune system, preventing or participating in some pathologies such as colon cancer [1]. In addition, intestinal bacteria are involved in vitamin synthesis and also in ion absorption [2, 3]. Through their trophic effect, the intestinal bacteria stimulate the proliferation and differentiation of cells on intestinal epithelium [4]; also, these bacteria may contribute to the maturation of immune cells, regulating the proliferation of pathogenic microorganisms and their toxins [1]. In addition

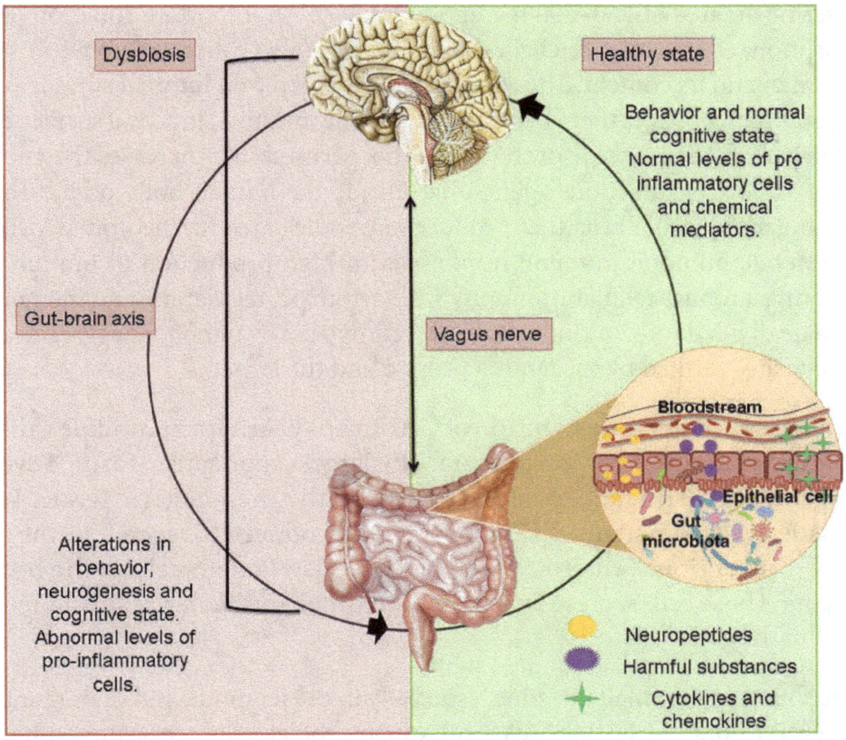

Figure 1. *Microbiota-GALT-brain axis. The vagus nerve makes the connection of the intestine to the brain and vice versa. Dysbiosis causes local alterations in the GALT and in the brain. When there is no dysbiosis, the bacterial metabolites participate in the state of local and systemic health and even more so over the brain.*

to the functions above described, it has been proposed that intestinal microbiota exerts indirect functions in other organs such as the liver and brain. Studies per- formed in humans and in animal models suggest that intestinal dysbiosis has an important role in the development of mood disorders such as depression, anxiety, and Parkinson's disease [5–7]. For these reasons, the interest in exploring the inter- actions between immune system, intestinal microbiota, and central nervous system (CNS) has increased (**Figure 1**).

On the other hand, when bacterial probiotics are administered in adequate amounts, they confer benefits on host health. The main functions of probiotics are to prevent and ameliorate several digestive and allergic disorders. Also, the micro- biota modulates ontogeny and immune system functions, as well as the interactions of the intestine-brain axis to regulate some neurological functions. However, the microbiota effects are not only in intestine but also in peripheral tissues, such as in immune system modulation and interacting with the gut-brain axis to regulate some neurologic functions.

The gut-associated lymphoid tissue (GALT)

Anatomy and physiology

The gut-associated lymphoid tissue (GALT) is a specialized component of mucosal-associated lymphoid tissue (MALT) or mucosal immune system that protects the individual's intestine from invading pathogens. The intestine-associated tissue extends throughout the small and large intestine, covering an area of 260–300 m2 approximately. An important function of intestine is nutrient absorption, where an epithelial cells monolayer (also called enterocytes) separates the GALT from lumen and its content. The enterocytes monolayer on luminal surface is coated by a glucocalix layer, which protects them from acidic pH. The intestinal mucosal surface can function as a permeable barrier to the inside of the body. This permeability increases the vulnerability to infections by a variety of infectious agents that invade the human body orally. Therefore, the largest populations of plasma cells that produce antibodies are enriched on GALT, generating a local and systemic humoral immune response with high production of immunoglobulin A (IgA), pro- moting a robust cellular immunity with cytotoxic, regulatory, and memory functions [8]. GALT can be divided into: (a) inductive sites, composed by lymphoid aggregates or follicles, and (b) effectors sites formed by the lamina propria and the lumen.

In **Figure 2** there is a complete description of GALT, thus GALT forms nodules disseminated into the submucosa and the lamina propria. The largest aggregates form "Peyer's patches," which in the small intestine are located in front of the mes- enteric tissue. Locally, new epithelial cells derived from stem cells are constantly produced to regenerate the epithelium. In addition to conventional enterocytes, there are also Paneth cells at the bottom of the epithelial crypts. These cells secrete lysozyme and other antibacterial substances to control the growth of pathogens.

These cells are found in the small intestine, especially in the jejunum, and their granules become visible after several hours of fasting. Also, the mucus goblet cells are scattered between other cell types. On the other hand, enteroendocrine cells produce polypeptides and are distributed diffusely throughout the gastrointestinal tract. On the surface of Peyer's patches, "M cells" also called "caveolate cells" are located, which capture antigens and function as intestinal chemoreceptors.

Beneath the epithelial lining is an underlying layer of connective tissue called lamina propria, which is connected to the lymphatic circulation and mesenteric

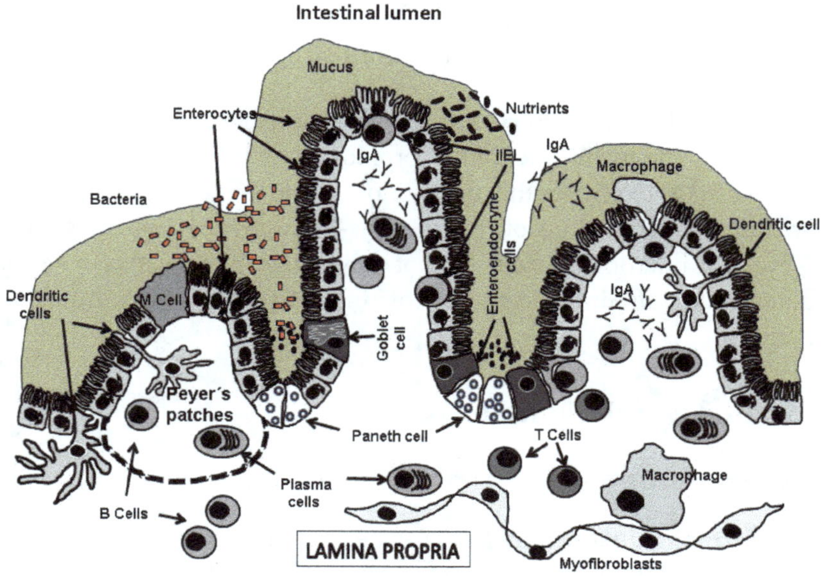

Figure 2. *Anatomy of gut-associated lymphoid tissue (GALT).*

lymph nodes. The intestine is constantly exposed to several antigens derived from the diet, microbiota, and a wide variety of bacterial, viral, fungal, and other pathogens. The immunity of this compartment is highly specialized because it is capable of both functions: (a) triggering an immune response and (b) inducing tolerance by suppressing immune response through the interaction of epithelial cells and microbiota [9, 10].

The induction of intestinal mucosa tolerance depends on several factors, such as: (a) nature of antigen (mainly protein antigens) [11]; (b) dose and frequency of antigen exposition [12–14]; (c) kinetics of antigen uptake, because antigenic exposition time with the immune system is key to induce tolerance [15]; and (d) genetic background and age of the host, because there is different susceptibility to infection, depending on age and genetic inheritance. The main antigens reaching GALT are a mixture of free amino acids and short oligopeptides, generated by gastric, pancreatic, or protease action on large proteins, which are absorbed by intestinal epithelial cells [16]. However, some intact proteins or incomplete proteolysis products can reach inductive sites through the following three non-exclusive pathways:

a. The epithelium responds to stimuli from different antigens expressing chemokine and cytokine genes [17]. Consequently, different subsets of cells carrying antigens are recruited to inductive sites [18].

b. M cells, specialized cells in both internalizing and transportation of intestinal antigens [19]. The antigens are endocytosed or phagocytosed, transported in vesicles through the M cells cytoplasm, and released on the basal surface where they are captured by antigen-presenting cells and transported to inductive sites, which will be presented to T and B cells.

c. Lamina propria (LP) dendritic cells (CDs) can go through the epithelial cells to capture antigens directly from the intestinal lumen, preserving the epithelial barrier integrity [20].

It is well known that the intestinal epithelium provides a physical barrier that separates the trillions of commensal bacteria present into intestinal lumen from both underlying lamina propria and deeper intestinal layers. The intestinal epithelium is composed of four cellular subsets derived from a common pluripotent stem cell progenitor: (1) enterocytes which constitute the majority of intestinal epithelial cells (IECs), (2) goblet cells producing mucus, (3) enteroendocrine cells producing hormones, and (4) Paneth cells that produce antimicrobial peptides and lectins. In addition, below the intestinal epithelium, stromal cells, B cells (especially plasma IgA producing cells), T cells, macrophages, and dendritic cells are found in lamina propria. In addition, strategically positioned are intraepithelial lymphocytes or iIELs (specialized T cells) and some dendritic cells, which are located between the IECs to sample the luminal content [20–22]. Thus, intraepithelial lymphocytes are considered the first line of cellular defense against any antigen that enters orally.

The iIELs belong to the lymphoid tissue associated to intestine and they are found in a ratio of 1:10 with respect to the epithelial cells along 300 m2 of intestinal surface [23]. Most iIELs contain abundant cytoplasmic granules for cytotoxic activity; also, effector cytokines such as interferon gamma (IFN-γ), interleukin-2 (IL-2), IL-4, and IL-17 can be secreted [24, 25]. The iIELs are cells that provide an immediate and efficient immune protection to prevent the spread of pathogens. However, to avoid excessive or unnecessary inflammatory responses on intestinal barrier, the iIELs also have regulatory functions [26].

Microbial ecosystem of the gastrointestinal tract

The human intestine contains a wide variety of microorganisms, approximately 500–1000 different species, of which the Bacteria, Archaea and Eukarya are the principal ones [1, 27]. The predominant bacteria populations in the intestine are the Bacteroidetes and Firmicutes, constituting about 90% [28]. This diverse ecosystem is called "gut microbiota," which has a symbiotic and mutualism relationship with the host [29, 30]. The intestinal microbiota exerts its own functions and has both direct and indirect influence on host's physiology and health, especially on metabolism. However, several pathologies, including neurological disorders such as irritable bowel syndrome, depression, anxiety, and Parkinson's disease have been associated with alteration of the intestinal microbiota known as "dysbiosis" [5–7].

With respect to dysbiosis, the wide diversity and abundance of gut microbiota population can be modified importantly by host's diet and age, as well as by other factors. The newborn is colonized by bacteria from birth and initially, there are no differences between bacteria population localized on different parts of body.

Infants who are born via vaginal delivery are mainly colonized by Lactobacillus and Prevotella, microbial populations closely related to maternal vaginal bacteria populations [31]. In contrast, infants born by cesarean-section (C-section) are exposed to *Staphylococcus* and *Corynebacterium*, which are skin microbes [32]. Thus, vaginal delivery or cesarean section as well as lactation or weaning are important factors that influence intestinal microbiota establishment. On the other hand, initially it was proposed that the prenatal environment is sterile; however, the presence of several bacterial species has been detected on placentas, amniotic fluid, and in meconium of healthy mothers, which suggests that, in the intrauterine stage there is already contact with microbes [33]. The diversity and functionality of the bacterial ecosystem is modified and increased in subsequent years of childhood [33].

Thus, in adult life, the predominant populations are both Bacteroidetes and Firmicutes, while the phylum Actinobacteria, Proteobacteria, Verrucomincrobia, archaea, and eukaryotes decrease importantly [34, 35]. Commonly, intestinal microbiota is very stable in adulthood, although a greater proportion of both Bifidobacteria and Clostridia has been found in the gut of young adults in comparison with older adults [36]. Important changes in composition and function of intestinal microbiota occur on aging. Aging has been associated with changes in intestinal microbiota composition, inducing alterations of multiple physiological functions, including intestine and immune system malfunctioning. An increased proportion of facultative anaerobes bacteria as well as an imbalance of Bacteroidetes/Firmicutes ratio in microbiota are age-related differences. Also, in people over 60 years of age, when the immune system function begins to decline, a significant decrease in Bifidobacteria has been found [37]. These previous findings were also supported by studies performed in intestinal mucosal tissue of aged and young mice, where a reduction in *Akkermansia muciniphila* proportion as well as decrease of antimicrobial factors Ang4 and lysozyme were detected in aged mice. Moreover, an important decrease in genes expression related to immunity was found, including T cell activation and other gene signaling pathways [38].

The high-carbohydrate and high-fat diet composition may produce dysbiosis.

It was described that in mice, a Western diet (WD: high-carbohydrate and high-fat diet) intake caused dysbiosis and dysregulated bile acids (BA) synthesis with reduced endogenous ligands for BA receptors, that is, farnesoid X receptor and G-protein- coupled bile acid receptor in the liver and brain [39]. More relevantly, a ketogenic low- carbohydrate high-fat diet induced changes in the oral microbiome of elite endurance athletes; the relative abundances of *Haemophilus*, *Neisseria*, and *Prevotella* spp. were decreased, and the relative abundance of *Streptococcus* spp. was increased [40].

Functions of the gut microbiota in the host

The main biological functions regulated by the gut microbiota are related to the efficiency to metabolize food and obtain energy. Polysaccharides are the main source of energy in bacterial metabolism, which are transformed into short-chain fatty acids (SCFAs). Bacterial metabolism is not limited only to SCFAs production and obtaining energy. The intestinal microbiota can synthesize several vitamins, aryl hydrocarbon receptor (AHR) ligands on host cells, polyamines [41], folate [42], indole [43], serotonin [44], and other compounds. In addition, intestinal microbiota also produces bacterial toxins called bacteriocins. To date, 13 species of bacteriocins have been found in human feces [45].

The major SCFAs produced by the gut microbiota are acetate, propionate, and butyrate, which are found at 80–130 mM [46, 47]. In this way, the SCFAs repre- sent approximately 70% of the total energy captured by the intestinal epithelial cells. Interestingly, butyrate produced by *Butyrivibrio fibrisolvens* protects against autophagy and energy starvation in the epithelium of gnotobiotic mice [48, 49]. While acetate and propionate have an important role in lipid metabolism, activation of the Gpr43 receptor promotes adipogenesis [50]. Therefore, using antibiotics at subtherapeutic doses, as commonly used in animal production, a dysbiosis is gener- ated by increasing SCFA levels, which consequently induces lipogenesis and hepatic triglycerides synthesis [50].

It is well known that microbiota strongly impacts on the expression of genes and proteins on host intestinal epithelial cells. In axenic mice, it has been found that in intestinal colonization by *Bacteroides thetaiotaomicron*, an important gene expression was induced. Expression of these genes is involved in protection, intestinal barrier function regulation, epithelium vascularization, and digestion/absorption of nutri- ents by increasing amino acid metabolism [50, 51]. It also participates in the regula- tion of endocrine, neurological, and bone density functions [33], as well as in the metabolism and absorption of phytochemicals such as polyphenols and drugs [50].

Additionally, several studies have shown that microbiota has an important role in peripheral and intestinal immune system ontogeny, as well as intestinal epithelium renewal [52, 53]. Also, microbiota-epithelial cell interaction indirectly controls the expenditure and storage of energy in the host [54]. Dysbiosis has been associated with a several pathologies affecting directly the digestive tract, including chronic inflammatory bowel diseases, colorectal cancer, constipation, and diarrhea; but in peripheral organs, it can also induce allergies, arthritis, or neurological disorders [55]. In this way, correcting the dysbiosis could improve the symptoms of diseases like irritable bowel syndrome (IBS) and functional diarrhea [56].

Intestinal bacteria are importantly involved in development and regulation of the immune system [57–59]. On this regard, mice grown under germ-free conditions exhibited several abnormalities, including hypoplastic Peyer's patches, IgA-producing cells reduction, relatively poorly structured spleen and lymph nodes, and decrease in proportion of Treg cells in colon [53]. Interestingly, when mice were exposed to intestinal bacteria for several weeks, the structure and function of immune system cells were restored [60]. Moreover, it has been shown that lipopolysaccharides from gram negative bacteria, as well as peptidoglycans from gram positive bacteria, activate Toll-like receptors (TLRs), inducing different immune responses [61]. Also, the expression of angiogenin 4 by *Bacteroides thetaiotaomicron* is induced. Angiogenin 4 is an important immune response regulator with microbicidal activity against a wide range of intestinal microbes, including bacterial and fungal pathogens [62]. A zwitterionic capsular polysaccharide of *Bacteroides fragilis* is an antigen related to T CD4 + effector cells function [63]; it also protects mice from *Helicobacter hepaticus* infections by suppressing IL-17 production and other immunological mechanisms [64]. Moreover, the genus Bifidobacteria is a producer of acetate that inhibits the translocation of Shiga toxin from *E.coli* 0157 suppressing colon inflammation [65]. In addition, *Bacteroides*, *Turicibacter*, and *Barnesiella* bacteria strains interact with T CD8 + cytotoxic cells in the mucosal compartment of both the small intestine and the colon [66]. Finally, the gut microbiota also regulates the interaction of dendritic cells with regulatory T cell through TLR 2 signaling, which induces an increased susceptibility to chronic inflammatory diseases such as colitis [67].

Host's immunomodulatory activity on the gut microbiota

The interactions between environmental signals and intestinal immune system are necessary to maintain a stable equilibrium and regulate the protective function of the intestinal barrier. Thus, in order to prevent microbial colonization and tran- sepithelial migration, several chemical substances are produced by the intestinal epithelial cells, such as gastric acid, enzymes (lactoferrin, lysozyme), antimicrobial peptides (defensins), mucins, and nitric oxide [68]. On the other hand, signaling through Toll-like receptors (TLRs) is very important for the activation of innate immune system. TLRs recognize a wide range of common antigens present in pathogens, activating the adaptive immune system for the generation of multiple highly specific and immunocompetent clones [69].

Connection between enteric nervous system (ENS) and central nervous system (CNS)

The enteric nervous system (ENS) extends from the esophagus to the anal region. The main functions are: (1) stimulation of glandular secretions, (2) motor functions such as peristalsis, and (3) ions and water exchange. The neurons found in CNS are subdivided into two main plexuses: the myenteric plexus and the submu- cosal plexus [70]. The former is responsible for peristalsis and second regulates the glandular secretions and control of blood flow. There are extrinsic fibers connecting these two plexuses, which are stimulated by both sympathetic and parasympathetic nervous system, communicating directly to spinal cord, part of the vagus nerve, and pre-vertebral ganglia of the sympathetic nervous system, although the ENS is able to function independently [71].

Generally, there are different types and subtypes of neurons throughout the ENS: excitatory neurons of the intestine, secretomotor, vasodilator, and non-vaso- dilator; some of them innervate whole endocrine cells and others intrinsic visceral afferent neurons. Most of the different types of neurons participate in reflexes corresponding to each plexus (myenteric and submucosal) (**Figure 3**).

The neurotransmitters secreted by enteric neurons are varied, the acetyl- choline (excitatory effect) and noradrenaline (inhibitory effect) being the most studied [72]. In addition, the communication of the ENS with the CNS is not only through the secretion of these neurotransmitters. Several studies have shown that bacterial metabolites generated in the intestinal ecosystem have a direct impact on the brain. Thus, it is well known that the intestinal microbiota has an important effect on CNS, because the homeostasis and intestinal functions can be regulated by the CNS [73]. The CNS and intestine connection may occur through several pathways, including: neuronal, hormonal, immune system, and intestinal bacterial metabolites [74]. Regarding the neuronal connection, the intestine is directly connected to the brain through the vagus nerve; thus, intestinal microbiota may stimulate the enteric nervous system [75]. In addition, several reports have shown that a defective communication between the brain and intestine microbiota is associated to anxiety, depression, inflammatory bowel disease, and other diseases

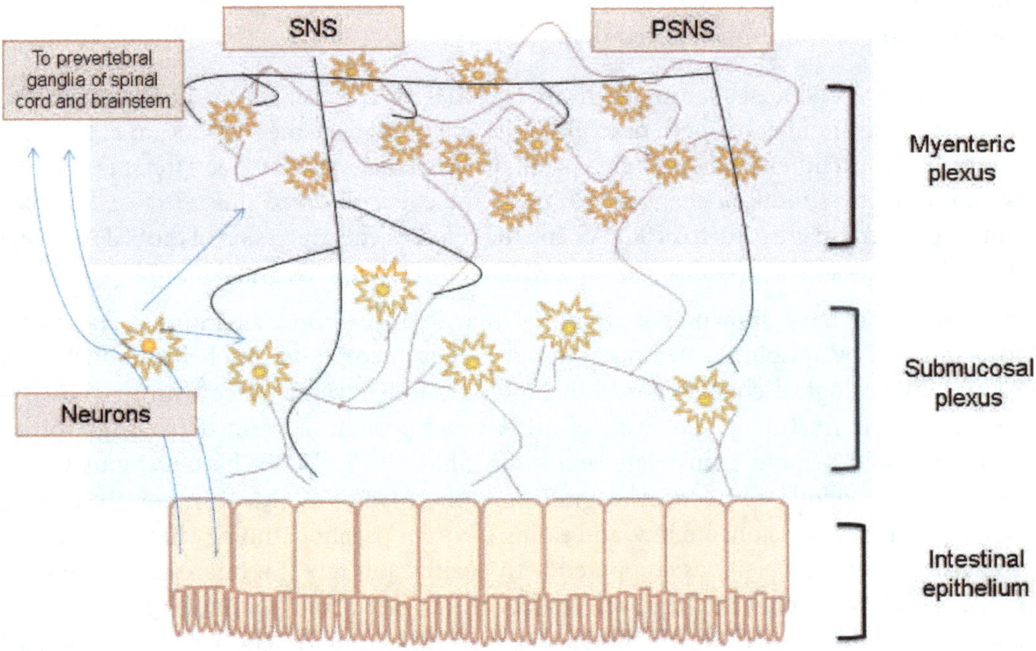

Figure 3. *Structure of the enteric nervous system. Both plexuses can be identified (myenteric and submucosal). The extension goes from the intestinal wall to the enteric plexuses and from there to the pre-vertebral ganglia of the spinal cord and brainstem. SNS, sympathetic nervous system; PSNS, parasympathetic nervous system.*

[76–78]. Interestingly, behavior changes in elderly have been associated with decreased immune function, resulting in alteration of intestinal microbiota-brain connection [79].

Dysbiosis and its effect on the neuroplasticity and behavior

In the first weeks of life, microbiota diversifies into a microbial community in which anaerobic microorganisms predominate [80]. This early colonization coincides with hypothalamic-pituitary-adrenal (HPA) axis activation, which has an important role in the innervation of the gastrointestinal (GI) tract and enteric nervous system (ENS) function. Likewise, the production of 5HT by enterochro- maffin cells is regulated by the intestinal microbiota, inducing de *novo* synthesis of 5HT [44].

Several reports have shown the relation between intestine and the CNS through metabolic, neuroendocrine, and immunological pathways, impacting neuronal plasticity and cognition. Production of proinflammatory cytokines (IL-1) in intes- tinal lumen may affect the brain through vagus nerve [81]. Also, bacterial products like lipopolysaccharide (LPS) can increase cytokines production, as well as induce both neuroinflammation and neurodegeneration [82]. In addition, it was shown that WD-fed mice had intestinal dysbiosis, which was accompanied by inflammatory signaling in the brain, microglial activation, and reduced neuroplasticity [83]. Therefore, dysbiosis of gut microbiota may increase the cytokines production and neuroinflammation, affecting mood, or it could induce psychiatric disorders such as depression and anxiety as was described in animal models [56] as well as in comparative studies performed in humans [5, 84]. Moreover, in maternal immune activation (MIA) mouse model, intestinal dysbiosis induced both higher production of both 4-ethylphenylsulfate (4EPS) and indolepyruvate, leading to autism spec- trum disorder (ASD). Interestingly, with *Bacteroides fragilis* administration, these behavioral symptoms were ameliorated [85].

A mechanism by producing substrates from bacterial metabolism such as SCFAs from the fermentation of dietary fiber has been described. The SCFAs regulate the metabolism of glucose and cholesterol [86]. Most importantly, the SCFAs (acetate, propionate, and butyrate) treatment alleviated the psychosocial stress-induced alterations in mice; the effect observed was selective, because the stress-induced body weight gain, fecal SCFAs, and the colonic gene expression of the SCFAs receptors free fatty acid receptors 2 and 3 remained unaffected by SCFAs supplementation [87].

Other studies also have shown that dysbiosis may induce mood alterations. For example, dysbiosis induced by antibiotics treatment in mice leads to cognitive and behavioral alterations as well as to neurological changes [88, 89]. In addition, in juvenile mice with dysbiosis, after a 2-week antibiotic treatment, the levels of mRNA and protein of brain-derived neurotrophic factor (BDNF) and tropomyosin-related receptor kinase B (TrKB) in hippocampus CA3 and dentate gyrus subregions, respectively, significantly increased [89]. Most importantly, patients with disorders like depression, anxiety, and eating disorder psychopathology have a significantly lower microbial alpha diversity as compared with healthy subjects [5]. Interestingly, depression and anxiety symptoms may be improved by fecal microbiota transplantation in patients with irritable bowel syndrome (IBS), functional diarrhea (FDR), or functional constipation (FC). Thus, the increase of microbiota diversity may improve the patient's mood [56].

Prebiotics

Since its initial description in 1995, the concept of prebiotic has been in constant evolution. Currently, according to Gibson and Roberfroid, a probiotic has been defined as "any substance present in diet, which specifically stimulates the growth and/or the fermentative activity of one or a limited number of bacteria species of intestinal microbiota, generating beneficial effects on health of host as a consequence of changes on either bacterial composition or metabolic activity" [90].

Generally, a food ingredient is considered as a prebiotic when it has the fol- lowing characteristics: (a) it must be kept in good condition until reaching the distal portions of the intestine; that is, it is not absorbed in the anterior part of the gastrointestinal tract and resists the hydrolysis of digestive enzymes [91] such as α-glucosidase, maltase, isomaltase, and sucrase [92]; (b) it must act as a selective substrate in the growth and/or metabolism of one or a limited number of beneficial bacterial species, such as Lactobacillus spp. and Bifidobacterium spp.; and finally (c) it must positively stimulate the microbiota, by increasing beneficial microorgan- isms and reducing pathogenic bacteria [93]. It is well known that proliferation of bifidobacteria and lactobacilli is favored by prebiotics; moreover, the proliferation of bacterial pathogenic strains such as *Clostridium*, *Escherichia*, *Campylobacter*, *Enterobacterium*, or *Salmonella* is inhibited. It has been proposed that intestinal microbiota is in involved in inhibition prebiotics mechanisms, by either competition for adhesion sites to the mucosa, or changes in the intestinal environment, such as (a) a reduction in pH as result of the synthesis of SCFA and (b) production of metabolites inhibiting pathogens proliferation, such as bacteriocins [94].

Types and sources

As already mentioned, prebiotics are normally ingested in the diet; however, only some carbohydrates (poly and oligosaccharides), whose chemical structure has β-type bonds, some peptides, some proteins, and certain lipids such as esters and ethers are food ingredients qualified as prebiotics [95].

Different types of oligosaccharides considered as prebiotics have been reported, among them are: fructooligosaccharides (FOS), oligofructose (OF), the inulin type fructans (ITFs), galactooligosaccharides (GOS), transgalactooligo- saccharides (TOS), and lactulose. However, there are others less known such as isomaltooligosaccharides (IMOS). Xylooligosaccharides (XOS) and mananooligo- saccharides (MOS) also have a probiotic potential [90, 95]. It is well known that sucrose and starch are the main carbohydrates found in higher plants, followed by glucomannans and fructans are the main reserve sources in the vegetable kingdom [96, 97].

Interaction of prebiotics in gut microbiota

The non-digestible carbohydrates or prebiotics are selectively fermented by the microbiota that produces important metabolites for the health of the host, the SCFAs acetate, propionate, and butyrate mainly. Likewise, prebiotics have a direct impact on the gut-associated lymphoid tissue (GALT) with immunomodulatory functions [98]. The symbiotic association between the host and the microbiota is fundamental in its physiology. The increase of the beneficial populations in the intestinal lumen affects the establishment of opportunistic pathogens, contributing to the strengthening of the GALT.

Potential effects on microbiota-GALT-brain axis

The stimulation of probiotic bacteria by prebiotics contributes to the increase of the production of beneficial metabolites for the host, such as the SCFAs. However, prebiotics not only have an impact on the intestinal microbiota. Oligofructose, an inulin type fructan, can bind to cellular receptors of pathogenic bacteria and block adhesion to the surface of enterocytes, helping to prevent colonization. In the same way, β (2→1)-fructans are ligands of TLR2, TLR4, TLR5, TLR7, TLR8, and NOD2.

Moreover, levan (2→6)-fructans appears to be a recognized of TLR4 to reduce IgE serum levels and Th2 responses [98]. Bacterial metabolites travel through the vagus nerve to reach specific brain regions such as cerebellum and hippocampus and modify gene expression [99]. It has been described that alterations in intestinal microbiota due to exposure to chronic stress are reversed with the administration of prebiotics and/or probiotics [100].

According to the classification of microbiota obtained by statistical analysis of sequence data, an alpha diversity has been identified, which describes the diversity of bacteria within a single individual; while beta diversity describes the diversity of specimens between different individuals. It has been shown that different types of prebiotics have an effect on the microbiota-GALT-brain axis in mice and human models. Prebiotics treatments (isolated from acorn and sago) induced increased β-diversity in heart failure patient's fecal microbiome, while no significant change in β-diversity was seen in healthy fecal microbiome. Alpha diversity was significantly higher in both healthy and diseased fecal microbiome, which was accompanied with an increase of the beneficial bacteria and SCFAs. Moreover, prebiotics treatment ameliorated HFD-induced glucose intolerance and insulin resistance in diabetic mice. Feeding both prebiotics treatments and inulin increased SCFAs levels in the mouse gut, and decreased the gut hyperpermeability and mucosal inflammatory markers in HFD-fed mice. The expression of pro-opiomelanocortin was also modulated by prebiotics administration, suggesting an important role in the hypotha- lamic energy signaling in the mice [101]. In diabetic db/db mice, the administration of oligofructose increases the expression of tight junction proteins occludin and ZO-1, which improves the integrity of the BBB in the hypothalamus and normalizes the expression of mRNA of IL-6 in the hippocampus; however, it does not improve alterations in behavior or in neurogenesis [102].

Moreover, the administration of prebiotic chitosan oligosaccharides (COSs) in Male Sprague-Dawley rats increases cognitive function and reduces levels of TNF-α and IL-1β, both pro-inflammatory cytokines [103]. In a mouse model of amyo- trophic lateral sclerosis, the oral administration of GOS reduced the motor neuron death and muscular atrophy and increased the levels of serum folate, vitamin B12, and homocysteine [104]. In the same way, in a mouse model with lipopolysac- charide-induced anxiety, the administration of bimuno-galacto-oligosaccharides (B-GOSs) reduced the pro-inflammatory cytokine IL-1β and expression of cortical 5-HT2A receptors [78]. In a mouse model of vascular dementia, the β-glucan from barley and arabinoxylan from the yeast *Triticum aestivum* demonstrated a protective effect [105].

Interestingly, our research group previously demonstrated that the administra- tion of oligofructose and *Agave* fructans decreased TBARS levels and carbonyls in learning and memory regions of the brain of overweight mice [106]. Also, in high- fat diet-induced obese mice, prebiotics not only reduces the oxidative damage in the same regions but also increases the levels of BDNF and GDNF [107].

On the other hand, the administration of synbiotics (probiotics and prebiotics in combination) shows interesting effects. In infants with cow's milk allergy, the treatment with based formula of amino-acid and symbiotics (a combination of fructo-oligosaccharides and *Bifdobacterium breve* M-16V) increased *Bifidobacterium* spp. and *Veillonella sp* bacteria substantially. Additionally, the lactate levels were increased, but the valerate and SCFAs levels were decreased [108]. Prolonged consumption of ADR-159 diet (fermentate generated from *Lactobacillus fermentum* and *Lactobacillus delbrueckii*) had no effect on anthropometrics or general health, but mice fed with ADR-159 presented increased sociability and lower baseline corticosterone levels (stress hormone). The diet also induced significant changes in the microbiota [109].

In patients with major depressive disorder (MDD), probiotics treatment (≥10 × 109 CFU of freeze-dried *Helveticus* R0052 and *B. longum* R0175 bacteria) for 8 weeks improved depression symptoms, but the serum inflammatory cytokines marker (TNF-α, IL-1β, IL-6, and IL-10) levels were not improved, and the urinary cortisol levels decreased by 20% of baseline. However, the prebiotics treatment (galactooligosaccharides, GOS) had no effect on depression symptoms nor inflam- matory marker levels [110]. In a like manner, oral GOS administration reduced the human stress hormone cortisol and increased attentional vigilance to positive versus negative stimuli [111].

Conclusions

The effect of the probiotics on the intestinal microbiota is quite important not only for the functions on intestine, but also for the development and function of immune system, metabolism, and central nervous system. Moreover, these systems are closely related, so that any alteration will impact their functionality. Therefore, the balance on microbiota-gut-central nervous system axis is very important to maintain the adequate functions of these systems.

Intestinal dysbiosis leads to alterations in development and function of central nervous system, which is significantly improved upon intestinal colonization with normal microbiota and probiotics treatment. To date, the interaction of microbiota- intestine-brain axis is not completely clear. For this reason, it is interesting to redirect the investigation of CNS diseases whose pathological mechanism is unknown.

Acknowledgements

We are thankful to the University of Guanajuato for their financial support (220/2019) to J R-E and 191/2019 to SLB. Cristina D. Balcón thanks CONACYT for the scholarship with support number 712500.

Conflict of interest

The authors declare no conflict of interest.

Author details

Elena Franco-Robles[1*], Joel Ramírez-Emiliano[2], José Sergio López-Briones[3] and Cristina Doriany Balcón-Pacheco1

1Department of Veterinary, University of Guanajuato, Irapuato, Guanajuato, Mexico

2Department of Medical Sciences, University of Guanajuato, Leon, Guanajuato, Mexico

3Department of Medicine and Nutrition, University of Guanajuato, Leon, Guanajuato, Mexico

*Address all correspondence to: e.francorobles@ugto.mx

References

[1] Shreiner AB, Kao JY, Young VB. The gut microbiome in health and in disease. Current Opinion in Gastroenterology. 2015;**31**(1):69-75. DOI: 10.1097/MOG.0000000000000139

[2] Guarner F, Malagelada JR. Gut flora in health and disease. The Lancet. 2003;**361**(9356):512-519. DOI: 10.1016/ S0140-6736(03)12489-0

[3] Bik EM. Composition and function of the human-associated microbiota. Nutrition Reviews. 2009;**67**(1):S164-S171. DOI: 10.1111/j.1753-4887.2009.00237.x

[4] Gordon JI, Hooper LV, McNevin MS, Wong M, Bry L. Epithelial cell growth and differentiation. III. Promoting diversity in the intestine: Conversations between the microflora, epithelium, and diffuse GALT. The American Journal of Physiology. 1997;**273**(3):G565-G570. DOI: 10.1152/ajpgi.1997.273.3.G565

[5] Kleiman SC, Watson HJ, Bulik-Sullivan EC, Huh EY, Tarantino LM, Bulik CM, et al. The intestinal microbiota in acute anorexia nervosa and during Renourishment: Relationship to depression, anxiety, and eating disorder psychopathology. Psychosomatic Medicine. 2015;**77**(9):969-981. DOI: 10.1097/PSY.0000000000000247

[6] Sampson TR, Debelius JW, Thron T, Janssen S, Shastri GG, Ilhan ZE, et al. Gut microbiota regulate motor deficits and Neuroinflammation in a model of Parkinson's disease. Cell. 2016;**167**:1469-1480

[7] Zhou ZL, Jia XB, Sun MF, Zhu YL, Qiao CM, Zhang BP, et al. Neuroprotection of fasting mimicking diet on MPTP-induced Parkinson's disease mice via gut microbiota and metabolites. Neurotherapeutics. 2019;**16**(3):741-760. DOI: 10.1007/ s13311-019-00719-2

[8] Mowat A. Anatomical basis of tolerance and immunity to intestinal antigens. Nature Reviews. Immunology. 2003;**3**(4):331-341

[9] MayerL.Mucosalimmunity.Pediatrics. 2003;**111**(Supplement_3):1595-1600

[10] MacDonald TT, Monteleone G. Immunity, inflammation, and allergy in the gut. Science. 2005;**307**(5717):1920-1925

[11] Peng H, Chang Z, Han S, Won M, Huang B. Chemical denaturation of ovalbumin abrogates the induction of oral tolerance of specific IgG antibody and DTH responses in mice. Scandinavian Journal of Immunology. 1995;**42**(3):297-304

[12] Mowat A, Strobel S, Drummond H, Ferguson A. Immunological responses to fed protein antigens in mice. I. Reversal of oral tolerance to ovalbumin by cyclophosphamide. Immunology. 1982;**45**(1):105-113

[13] Friedman A, Weiner HL. Induction of anergy or active suppression following oral tolerance is determined by antigen dosage. PNAS. 1994;**91**(14):6688-6692

[14] Meyer A, Benson J, Gienapp I, Cox K, Whitacre C. Suppression of murine chronic relapsing experimental autoimmune encephalomyelitis by the oral administration of myelin basic protein. Journal of Immunology. 1996;**157**(9):4230-4238

[15] Garside P, Mowat AM, Khoruts A. Oral tolerance in disease. Gut. 1999;**44**(1):137-142

[16] Erickson R, Kim Y. Digestion and absorption of dietary protein. Annual Review of Medicine. 1990;**41**:133-139

[17] Nagler-Anderson C. Man the barrier! Strategic defences in the intestinal mucosa. Nature Reviews. Immunology. 2001;**1**(1):59-67

[18] Clayburgh D, Shen L, Turner J. A porous defense: The leaky epithelial barrier in intestinal disease. Laboratory Investigation. 2004;**84**(3):282-291

[19] Neutra M, Mantis N, Kraehenbuhl J. Collaboration of epithelial cells with organized mucosal lymphoid tissues. Nature Immunology. 2001;**2**(11):1004-1009

[20] Rescigno M, Urbano M, Valzasina B, Francolini M, Rotta G, Bonasio R, et al. Ricciardi-Castagn... P: Dendritic cells express tight junction proteins and penetrate gut epithelial monolayers to sample bacteria. Nature Immunology. 2001;**2**(4):361-367

[21] Jabri B, Ebert E. Human CD8+ intraepithelial lymphocytes: A unique model to study the regulation of effector cytotoxic T lymphocytes in tissue. Immunological Reviews. 2007;**215**:202-214

[22] Niess JH, Brand S, Gu X, Landsman L, Jung S, McCormick BA, et al. CX3CR1-mediated dendritic cell access to the intestinal lumen and bacterial clearance. Science. 2005;**307**(5707):254-258

[23] Petit A, Ernst P, Befus A, Clark D, Rosenthal K, Ishizaka T, et al. Murine intestinal intraepithelial lymphocytes I. relationship of a novel Thy-1-,Lyt- 1-,Lyt-2+, granulated subpopulation to natural killer cells and mast cells. European Journal of Immunology. 1985;**15**(3):211-215

[24] Shires J, Theodoridis E, Hayday A. Biological insights into TCRgammadelta+ and TCRalphabeta+ intraepithelial lymphocytes provided by serial analysis of gene expression (SAGE). Immunity. 2001;**15**(3):419-434

[25] Guy-Grand D, Malassis-Seris M, Briottet C, Vassalli P. Cytotoxic differentiation of mouse gut thymodependent and independent intraepithelial T lymphocytes is induced locally. Correlation between functional assays, presence of perforin and granzyme transcripts, and cytoplasmic granules. The Journal of Experimental Medicine. 1991;**173**(6):1549-1552

[26] Cheroutre H, Lambolez F, Mucida D. The light and dark sides of intestinal intraepithelial lymphocytes. Nature Reviews. Immunology. 2011;**11**(7):445-456

[27] Farooqi I, O'Rahilly S. Monogenic obesity in humans. Annual Review of Medicine. 2005;**56**:443-458

[28] Hugenholtz P, Goebel BM, Pace NR. Impact of culture-independent studies on the emerging phylogenetic view of bacterial diversity. Journal of Bacteriology. 1998;**180**(18):4765-4774

[29] Xu J, Gordon JI. Inaugural article: Honor thy symbionts. PNAS. 2003;**100**(18):10452-10459

[30] Xu J, Mahowald M, Ley R, Lozupone C, Hamady M, Martens E, et al. Evolution of symbiotic bacteria in the distal human intestine. PLoS Biology. 2007;**5**(7):e156

[31] Dominguez-Bello MG, Costello EK, Contreras M, Magris M, Hidalgo G, Fierer N, et al. Delivery mode shapes the acquisition and structure of the initial microbiota across multiple body habitats in newborns. PNAS. 2010;**107**(26):11971-11975. DOI: 10.1073/pnas.1002601107

[32] Bäckhed F, Roswall J, Peng Y, Feng Q , Jia H, Kovatcheva-Datchary P, et al. Dynamics and stabilization of the human gut microbiome during the first year of life. Cell Host & Microbe. 2015;**17**(5):690-703. DOI: 10.1016/j. chom.2015.04.004

[33] Lynch SV, Pedersen O. The human intestinal microbiome in health and disease. The New England Journal of Medicine. 2016;**375**:2369-2379. DOI: 10.1056/NEJMra1600266

[34] Eckburg PB, Bik EM, Bernstein CN, Purdom E, Dethlefsen L, Sargent M, et al. Diversity of the human intestinal microbial Flora. Science. 2005;**308**(5728):1635-1638. DOI: 10.1126/science.1110591

[35] Reyes A, Haynes M, Hanson N, et al. Viruses in the faecal microbiota of monozygotic twins and their mothers. Nature. 2010;**466**:334-338

[36] Agans R, Rigsbee L, Kenche H, Michail S, Khamis H, Paliy O. Distal gut microbiota of adolescent children is different from that of adults. FEMS Microbiology Ecology. 2011;**77**(2):404-412

[37] Mariat D, Firmesse O, Levenez F, Guimaraes V, Sokol H, Dore J, et al. The Firmicutes/Bacteroidetes ratio of the human microbiota changes with age. BMC Microbiology. 2009;**9**:123

[38] Sovran B, Hugenholtz F, Elderman M, Van Beek A, Graversen K, Huijskes M, et al. Age- associated impairment of the mucus barrier function is associated with profound changes in microbiota and immunity. Scientific Reports. 2019;**9**:1437

[39] Jena PK, Sheng L, Di Lucente J, Jin LW, Maezawa I, Wan YY. Dysregulated bile acid synthesis and dysbiosis are implicated in Western diet-induced systemic inflammation, microglial activation, and reduced neuroplasticity. The FASEB Journal. 2018;**32**(5):2866-2877. DOI: 10.1096/ fj.201700984RR

[40] Murtaza N, Burke LM, Vlahovich N, Charlesson B, O'Neill HM, Ross ML, et al. Analysis of the effects of dietary pattern on the Oral microbiome of elite endurance athletes. Nutrients. 2019;**11**(3):E614. DOI: 10.3390/nu11030614

[41] Rooks MG, Garrett WS. Gut microbiota, metabolites and host immunity. Nature Reviews. Immunology. 2016;**16**(6):341-352. DOI: 10.1038/nri.2016.42

[42] Rossi M, Amaretti A, Raimondi S. Folate production by probiotic bacteria. Nutrients. 2011;**3**:118-134

[43] Shimada Y et al. Commensal bacteria-dependent indole production enhances epithelial barrier function in the colon. PLoS One. 2013;**8**:e80604

[44] Yano JM, Yu K, Donaldson GP, et al. Indigenous bacteria from the gut microbiota regulate host serotonin biosynthesis. Cell. 2015;**161**:264-276

[45] Lakshminarayanan B, Guinane CM, O'Connor PM, et al. Isolation and characterization of bacteriocin-producing bacteria from the intestinal microbiota of elderly Irish subjects. Journal of Applied Microbiology. 2013;**114**:886-898

[46] Cummings JH. Fermentation in the human large intestine: Evidence and implications for health. Lancet. 1983;**1**:1206-1209

[47] Cummings JH, Pomare EW, Branch WJ, Naylor CP, Macfarlane GT. Short chain fatty acids in human large intestine, portal, hepatic and venous blood. Gut. 1987;**28**:1221-1227

[48] Donohoe DR et al. The microbiome and butyrate regulate energy metabolism and autophagy in the mammalian colon. Cell Metabolism. 2011;**13**:517-526

[49] Asanuma N, Kawato M, Ohkawara S, Hino T. Characterization and transcription of the genes encoding enzymes involved in butyrate production in Butyrivibrio fibrisolvens. Current Microbiology. 2003;**47**:203-207

[50] Lee WJ, Hase K. Gut microbiota– generated metabolites in animal health and disease. Nature Chemical Biology. 2014;**10**:416-424

[51] Hooper LV, Wong MH, Thelin A, Hansson L, Falk PG, Gordon JI. Molecular analysis of commensal host-microbial relationships in the intestine. Science. 2001;**291**(5505):881-884

[52] Slack E, Hapfelmeier S, Stecher B, Velykoredko Y, Stoel M, Lawson MAE, et al. Innate and adaptive immunity cooperate flexibly to maintain host- microbiota mutualism. Science. 2009;**325**(5940):617-620

[53] Lee YK, Mazmanian SK. Has the microbiota played a critical role in the evolution of the adaptive immune system? Science. 2010;**330**(6012):1768-1773

[54] Backhed F, Ding H, Wang T, Hooper LV, Koh GY, Nagy A, et al. The gut microbiota as an environmental factor that regulates fat storage. PNAS. 2004;**101**(44):15718-15723

[55] Neish A. Microbes in gastrointestinal health and disease. Gastroenterology. 2009;**136**(1):65-80

[56] Kurokawa S. The effect of fecal microbiota transplantation on psychiatric symptoms T among patients with irritable bowel syndrome, functional diarrhea and functional constipation: An open-label observational study. Journal of Affective Disorders. 2018;**235**:506-512

[57] Round J, Mazmanian S. The gut microbiota shapes intestinal immune responses during health and disease. Nature Reviews. Immunology. 2009;**9**(5):313-323

[58] Brestoff JR, Artis D. Commensal bacteria at the interface of host metabolism and the immune system. Nature Immunology. 2013;**14**:676-684

[59] Fulde M, Hornef MW. Maturation of the enteric mucosal innate immune system during the postnatal period. Immunological Reviews. 2014;**260**:21-34

[60] Macpherson A, Harris N. Interactions between commensal intestinal bacteria and the immune system. Nature Reviews. Immunology. 2004;**4**(6):478-485

[61] Kamada N, Chen GY, Inohara N, Nuñez G. Control of pathogens and pathobionts by the gut microbiota. Nature Immunology. 2013;**14**:685-690

[62] Hooper L, Stappenbeck T, Hong C, Gordon J. Angiogenins: A new class of microbicidal proteins involved in innate immunity. Nature Immunology. 2003;**4**(3):269-273

[63] Mazmanian S, Liu C, Tzianabos A, Kasper D. An immunomodulatory molecule of symbiotic bacteria directs maturation of the host immune system. Cell. 2005;**122**(1):107-118

[64] Mazmanian S, Round J, Kasper D. A microbial symbiosis factor prevents intestinal inflammatory disease. Nature. 2008;**453**(7195):620-625

[65] Fukuda S et al. Bifidobacteria can protect from enteropathogenic infection through production of acetate. Nature. 2011;**469**:543-547

[66] Presley LL, Wei B, Braun J, Borneman J. Bacteria associated with Immunoregulatory cells in mice. Applied and Environmental Microbiology. 2010;**76**(3):936-941

[67] Garrett W, Lord G, Punit S, Lugo-Villarino G, Mazmanian S, Ito S, et al. Communicable ulcerative colitis induced by T-bet deficiency in the innate immune system. Cell. 2007;**131**(1):33-45

[68] Otte JM, Kiehne K, Herzig KH. Antimicrobial peptides in innate immunity of the human intestine. Journal of Gastroenterology. 2003;**38**(8):717-726. DOI: 10.1007/s00535-003-1136-5

[69] Sanz Y, Nadal I, Sanchez E. Probiotics as drugs against human gastrointestinal infections. Recent Patents on Anti-Infective Drug Discovery. 2007;**2**(2):148-156

[70] Carabotti M, Scirocco A, Maselli MA, Severi C. The gut-brain axis: Interactions between enteric microbiota, central and enteric nervous systems. Annals of Gastroenterology. 2015;**28**(2):203-209

[71] Hall JE. Guyton and Hall Textbook of Medical Physiology: Enhanced E-Book. Philadelphia, PA: Saunders Elsevier; 2010

[72] Furness JB. Types of neurons in the enteric nervous system. Journal of the Autonomic Nervous System. 2000;**81**(1-3):87-96. DOI: 10.1016/ S0165-1838(00)00127-2

[73] Moloney RD, Dinan TG, Cryan JF. P.2.031 the CBA/J mouse as a genetic model of visceral hypersensitivity with co-morbid anxiety and depression: Role of glutamate transport. European Neuropsychopharmacology. 2013;**23**:S51-S52

[74] Ochoa-Repáraz J, Kasper LH. Current Obesity Reports. 2016;**5**:51. DOI: 10.1007/ s13679-016-0191-1

[75] Forsythe P, Bienenstock J, Kunze WA. Vagal pathways for microbiome-brain-gut Axis communication. In: Lyte M, Cryan J, editors. Microbial Endocrinology: The Microbiota-Gut-Brain Axis in Health and Disease. Advances in Experimental Medicine and Biology. Vol. 817. New York, NY: Springer; 2014

[76] Collins SM, Surette M, Bercik P. The interplay between the intestinal microbiota and the brain. Nature Reviews Microbiology. 2012;**10**:735-742

[77] Cryan JF, Dinan TG. Mind-altering microorganisms: The impact of the gut microbiota on brain and behavior. Nature Reviews Neuroscience. 2012;**13**:701-712

[78] Savignac HM. Prebiotic administration normalizes lipopolysaccharide (LPS)-induced anxiety and cortical 5-HT2A receptor and IL1-β levels in male mice. Brain, Behavior, and Immunity. 2016;**52**:120-131

[79] Erny D. Host microbiota constantly control maturation and function of microglia in the CNS. Nature Neuroscience. 2015;**18**:965-977

[80] Mitsou E, Kirtzalidou E, Oikonomou I, Liosis G, Kyriacou A. Fecal microflora of Greek healthy neonates. Anaerobe. 2008;**14**(2):94-101

[81] Hickman RA et al. Consequences of gut Dysbiosis on the human brain. INTECH. 2016;**3**:41-64

[82] Mohammadi G et al. The effects of probiotic formulation pretreatment (lactobacillus helveticus R0052 and Bifidobacterium longum R0175) on a lipopolysaccharide rat model. Journal of the American College of Nutrition. 2019;**2019, 38**(3):209-217

[83] Jena PK et al. Dysregulated bile acid synthesis and dysbiosis areimplicated in Western diet– induced systemicinflammation, microglial activation, andreduced neuroplasticity. The FASEB Journal. 2018, 2018;**32**:2866-2877

[84] Heym N, Heasman BC, Hunter K, et al. The role of microbiota and inflammation in self-judgement and empathy: Implications for understanding the brain-gut-microbiome axis in depression. Psychopharmacology. 2019;**236**(5):1459-1470

[85] Hsiao EY, McBride SW, Hsien S, et al. Microbiota modulate behavioral and physiological abnormalities associated with neurodevelopmental disorders. Cell. 2013;**155**:1451-1463

[86] Leung K, Thuret S. Gut microbiota: A modulator of brain plasticity and cognitive function in ageing. Healthcare. 2015;**3**:898-916

[87] van de Wouw M. 2018 short-chain fatty acids: Microbial metabolites that alleviate stress-induced brain-gut axis alterations. The Journal of Physiology. 2018;**596**(20):4923-4944

[88] Bercik P, Denou E, Collins J, Jackson W, Lu J, Jury J, et al. The intestinal microbiota affect central levels of brain-derived neurotropic factor and behavior in mice. Gastroenterology. 2011;**141**(2):599-609. e3. DOI: 10.1053/j.gastro.2011.04.052

[89] Bistoletti M et al. Antibiotic treatment-induced dysbiosis differently affects BDNF and TrkB expression in the brain and in the gut of juvenile mice. PLoS One. 2019;**14**(2):e0212856

[90] Hernández HA, Coronel RC, Monge ZM, Quintana HC. Microbiota, Probióticos, Prebióticos y Simbióticos. Pediatría Integral. 2015;**9**(5):337-354

[91] Walton GE, Swann JR, y Gibson GR. Chapter 2: Prebiotics. Rosenberg E, Delong EF, Lory S, Stackebrandt E. y Thompson F. (Eds). The Prokaryotes: New York: Springer Science Business Media; 25-43

[92] Huazano-García A, López MG. Applied Biochemistry and Biotechnology. 2018;**184**:25. DOI: 10.1007/s12010-017-2526-0

[93] García Y, López MG, Bocourt R, Rodríguez Z, Urías SJ. Herrera M. Fermentación in vitro del extracto de Agave fourcroydes (henequén) por bacterias ácido lácticas. Revista Cubana de Ciencia Agrícola. 2012a;2(46):203-209

[94] Morales KD. y Vélez RJF. Prebióticos: Su importancia en la salud humana y propiedades funcionales en tecnología de alimentos. Temas Selectos de Ingeniería de Alimentos. 2013;1(7):12-24

[95] Pérez C. D., López G. y Ros G. 2004. Principales prebióticos y sus efectos en la alimentación humana. Anales de Veterinaria de Murcia. 20:5-20

[96] Mancilla-Margalli NA, López MG. Water-soluble carbohydrates and fructan structure patterns from agave and Dasylirion species. Journal of Agricultural and Food Chemistry. 2006;**54**(20):7832-7839. DOI: 10.1021/jf060354v

[97] Urías-Silvas J, Cani P, Delmée E, Neyrinck A, López M, Delzenne N. Physiological effects of dietary fructans extracted from Agave tequilana Gto. And Dasylirion spp. British Journal of Nutrition. 2008;**99**(2):254-261. DOI: 10.1017/ S0007114507795338

[98] Franco-Robles E, López MG. Implication of Fructans in health: Immunomodulatory and antioxidant mechanisms. The Scientific World Journal. 2015;**2015**:289267. DOI: 10.1155/2015/289267

[99] Mayer EA, Knight R, Mazmanian SK, Cryan JF, Tillisch K. Gut microbes and the brain: Paradigm shift in neuroscience. Journal of Neuroscience. 2014;**34**(46):15490-15496. DOI: 10.1523/JNEUROSCI.3299-14.2014

[100] Sarkar A, Lehto SM, Harty S, Dinan TG, Cryan JF, Burnet P. Psychobiotics and the manipulation of bacteria-gut-brain signals. Trends in Neurosciences. 2016;**39**(11):763-781. DOI: 10.1016/j.tins.2016.09.002

[101] Ahmadi S, Nagpal R, Wang S, et. al. Prebiotics from acorn and sago prevent high-fat-diet-induced insulin resistance via microbiome–gut–brain axis modulation. 2019;**67**:1-13. DOI: 10.1016/j.jnutbio.2019.01.011

[102] Fernández de Cossío L, Fourrier C, Sauvant J, Everard A, Capuron L, Cani PD, et al. Impact of prebiotics on metabolic and behavioral alterations in a mouse model of metabolic syndrome. Brain, Behavior, and Immunity. 2017;**64**:33-49. DOI: 10.1016/j.bbi.2016.12.022

[103] Jia S, Lu Z, Gao Z, An J, Wu X, Li X, et al. Chitosan oligosaccharides alleviate cognitive deficits in an amyloid-β1-42-induced rat model of Alzheimer's disease. International Journal of Biological Macromolecules. 2016;**83**:416-425. DOI: 10.1016/j. ijbiomac.2015.11.011

[104] Song L. Galactooligosaccharide improves the animal survival and alleviates motor neuron death in SOD1 G93A mouse model of amyotrophic lateral sclerosis. Neuroscience. 2013;**246**:281-290

[105] Han HS, Jang JH, Jang JH, Choi JS, Kim YJ, Lee C, et al. Water extract of Triticum aestivum L. and its components demonstrate protective effect in a model of vascular dementia. Journal of Medicinal Food. 2010;**13**(3):572-578

[106] Franco-Robles E, Ramírez-Emiliano J, López MG. Agave fructans and oligofructose decrease oxidative stress in brain regions involved in learning and memory of overweight mice. Natural Product Research. 2018;**33**(10):1527-1530. DOI: 10.1080/14786419.2017.1423297

[107] Franco-Robles E, López MG. Agavins increase Neurotrophic factors and decrease oxidative stress in the brains of high-fat diet-induced obese mice. Molecules. 2016;**21**:998

[108] Wopereis H, van Ampting MTJ, Cetinyurek-Yavuz A, Slump R, Candy DCA, Butt AM, et al. A specific synbiotic-containing amino acid-based formula restores gut microbiota in non- IgE mediated cow's milk allergic infants: A randomized controlled trial. Clinical and Translational Allergy. 2019;**9**:27. DOI: 10.1186/s13601-019-0267-6

[109] Warda AK et al. Heat-killed lactobacilli alter both microbiota composition and behaviour. Behavioural Brain Research. 2019;**362**:213-223

[110] Kazemi A et al. Effect of prebiotic and probiotic supplementation on circulating pro-inflammatory cytokines and urinary cortisol levels in patients with major depressive disorder: A double-blind, placebo-controlled randomized clinical trial. Journal of Functional Foods. 2019;**52**:596-602

[111] Schmidt K, Cowen PJ, Harmer CJ, Tzortzis G, Errington S, Burnet PW. Prebiotic intake reduces the waking cortisol response and alters emotional bias in healthy volunteers. Psychopharmacology. 2015;**232**(10):1793-1801

11

Probiotics in Health and Immunity: A First Step toward Understanding the Importance of Microbiota System in Translational Medicine

Ciro Gargiulo Isacco, Andrea Ballini, Danila De Vito, Angelo Michele Inchingolo, Stefania Cantore, Gregorio Paduanelli, Kieu Cao Diem Nguyen, Alessio Danilo Inchingolo, Gianna Dipalma and Francesco Inchingolo

Abstract

There are mounting evidences showing the relation of chronic inflammatory and autoimmune diseases with the uncontrolled intensification of gut dysbiosis. This position asserts that an elevated presence of pathogens and bacterial, fungal, and viral components is directly involved in inflammatory metabolic diseases with a strong alteration of autoimmune components such as in inflammatory bowel syndrome (IBS), ulcerative colitis, and Crohn's disease. Furthermore, the increase of unbalanced enteric microbiota is also connected to other types of conditions of metabolic origin such as diabetes mellitus, atherosclerosis, and osteo-degenerative conditions. As a matter of fact, evidence confirmed that gut damages histologically inspected revealed a situation with high expression of pro-inflammatory cytotumor necrosis factor-alpha (TNF-α), and IL1β, IL-2, IL-4, IL-6, and IL-17 together with high level of mucin-2. This chapter focuses on diverse topics related to microbiota dysfunction and systemic health condition and regenerative capacity and the therapeutic role of probiotics in gut health and disease emphasizing the potential beneficial role of probiotics in idiopathic inflammatory metabolic diseases. In brief, outcomes demonstrate that an intimate relationship between microbiota, metabo- lism, tissue/cellular damages, and regeneration is standing. Within this scenario, the gut certainly plays a big part of the regenerative mechanisms in translational medicine.

Keywords: inflammatory bowel syndrome, ulcerative colitis, Crohn's disease, gut dysbiosis, short-chain fatty acids, central nervous system, experimental autoimmune encephalopathy, vagus nerve, tumor necrosis factor-alpha, IL1β, IL-2, IL-4, IL-6, IL-17, mucin-2, probiotics

Introduction

The oral-gastro-intestinal-sex-skin can be classified as unique large and heterogeneous apparatus populated by a huge variety of microorganisms, bacteria, virus, fungi, and other single-celled creatures, that compose the totality of human microbiota that contributes together with bone/skeleton system, to maintain the body energy homeostasis. The human body hosts something like 10–100 trillion microbial cells that coexist in a strict fruitful symbiotic relation- ship that persists as long as the body is kept in a balanced healthy state [1, 2].

The gut plays an important role in regulating metabolic immune activities. The gut's essential task is the absorption of nutrients and the synthesis of important micromolecules obtained from food that cannot be assimilated by the stomach and small intestine [1–3]. Xyloglucans and fructo-oligosaccharides from vegetables and fruits, protein, and lipids; the assimilation of essential vitamins like vitamins B-12, D, and K; and the synthesis of hormones like serotonin from tryptophan amino acid take place right in the gut, thanks to the constant activity of its entire microbiota. The microbiota are able to produce 50–100 mmol·L—1 per day of extremely important short-chain fatty acids (SCFAs), such as acetic, propionic, and butyric acids—and serve as an energy source to the host intestinal epithelium and skeleton [1–4].

The importance of SCFAs has been well described by several studies during the last decade; the activity of acetic acid, for instance, has been found to be essen- tial against infections, in blood pressure regulation and against sclerotic plaque deposition in arterial walls. The presence of butyric acid is an essential anti-IBS agent due to its immune-modulator properties and anti-inflammatory action, while propionic acid has been found to be important in preventing obesity and diabetes 2 [1–4].

Although bacteria, viruses, and fungi might be very harmful and dangerous, they are indispensable for life as well. This symbiotic coexistence throughout the millennia made a deep crucial biological impact on human species, and it has become essential not only for survival but for evolution as well. Accumulating evidences have clearly demonstrated how part of these specific microorganisms can resume specific immunomodulatory roles and the way they affect either composition function or migration of various immune cell subpopulations from one site to a different location. For instance, oral macrophages may migrate under the influence of specific signal induction of local microbiota from oral either to the lungs or even the brain passing through the blood brain barrier (BBB) [5–10].

The outcomes from experiments performed on germ-free (GF) mice confirmed the great role of gut microbiota in the upsurge of immune system deficiencies. GF animals were shown to have compromised Paneth cells and low levels of natural killer (NK) cells, dendritic cells (DCs), and $\alpha/\beta +$ and $\gamma/\delta +$ T cell populations that play an important role in defense and pathogenesis during inflammation and infec- tion, especially against certain types of malignancies. In addition, GF animals were highly susceptible to frequent infections due to a decline in angiogenin-4 (Ang4), a powerful antimicrobial part of the class of microbicide proteins in Paneth cells [5–10].

The alteration of the gut microbiota may contribute to open up the invasion of exogenous pathogens that may destabilize the whole intestinal mucosa. The patho- gen systematic overgrowth will trigger a cascade of strong inflammatory responses making intestinal mucosa highly susceptible and motile. The chronic inflammation will weaken the endothelial tide junction to the point that the walls become highly relaxed and permeable causing the phenomenon known as "leaky gut" that allows the free, uncontrolled passage of microorganism into the system via the blood-stream and tissues where they start allocating. In fact, the presence of these typical gut residents could be found in eroded, inflamed, and degenerated joints and organs such as the lungs, heart, brain, and liver [11–13].

Gut dysbiosis: a modality to understand neurodegenerative diseases: the disruption of blood-brain barrier (BBB) may explain the gut-oral-brain axis relationship

The high and uncontrolled levels of pathogenic microorganism colonizing the gut contribute to a condition known as dysbiosis [14]. Since few years the dysbiosis has been associated with a variety of degenerative patterns that tend to subvert the metabolic/neuro/hormonal/immune axis contributing to a variety of disorders that round to different body systems ranging from skeleton, cardiovascular, to neuro system. There are several mechanisms proposed that are able to trigger this state of systemic disorders; one of the possibilities is linked to bacterial metabolites and immune-modulating mediators that contribute to the high permeability of intestinal mucosa allowing local pathogens to get through the mucosal barriers triggering a huge variety of immune responses. A second and though related mechanism is the sabotage of SCFAs' production; the consequences of this mechanism are the abrupt breakdown of energy balance mechanism, a reduction of cell-bacteria signaling pathway, and the worsening of epithelial cell layer integrity due to the decreased production of tight junction proteins which allows the translocation of LPS into the submucosa as well. The significant presence of pro-inflammatory cytokines and interleukins such as TNFα, IFN-γ, IL-1β, IL-2, IL-4, IL-5, and IL-6 is the peculiar trait of a dysbiotic gut (**Figure 1**) [14–16]. A third way of dysbiosis transmission is through the vagus nerve (VN), the main component of parasympathetic nervous system (PSN) which also constitutes an effective bridge of the gut/CNS axis. This hypothesis, today supported by a concrete line of evidences, proposes the existence of a reciprocal interference way between the CNS and gut through the VN. In this view the VN is able to perceive microbiota movement, grade of activity, and therefore degradation; on the other hand, pathogens once out from the gut mucosa barrier are able to communicate and move to the CNS through the VN pathway [14–18].

These essential structural alterations are at the base of neurodegenerative pathologies. Though it is a unique pathological aspect, we may see the presence of a common configuration indeed, which is a shared neurological chronic inflammatory pattern. In all these cases, the chronic neuro-inflammatory condition is characterized by an abnormal hyperactive behavior of neural immune cells, the microglia, known as macrophages of the brain [18, 19]. The chronic inflammatory state that from the gut opens up the pathway of pathogenic microbiota invasion all the way through oral and brain compartment, which is the hallmark of neuro-degenerative disorders' dynamic pathogenesis. Patients with Parkinson's disease (PD), Alzheimer's (AD), multiple sclerosis (MS), or amyotrophic lateral sclerosis all present a variety of disturbances in intestinal microbial compared to healthy individuals. Neurodegenerative-affected patients' intestinal and fecal analysis showed a clear clinical picture of microbiota dysbiosis. The test outcomes showed high level of coliform and gram-negative bacteria from *Ralstonia* genus concomitantly with low critical level of anti-inflammatory strains related to *Blautia*, *Coprococcus*, and *Roseburia* genera. Another indicator was also noted; it was the low presence of *Prevotella* generally seen as beneficial bacteria, involved in the metabolism of plant polysaccharides and vitamins strictly associated with the production of neuroactive

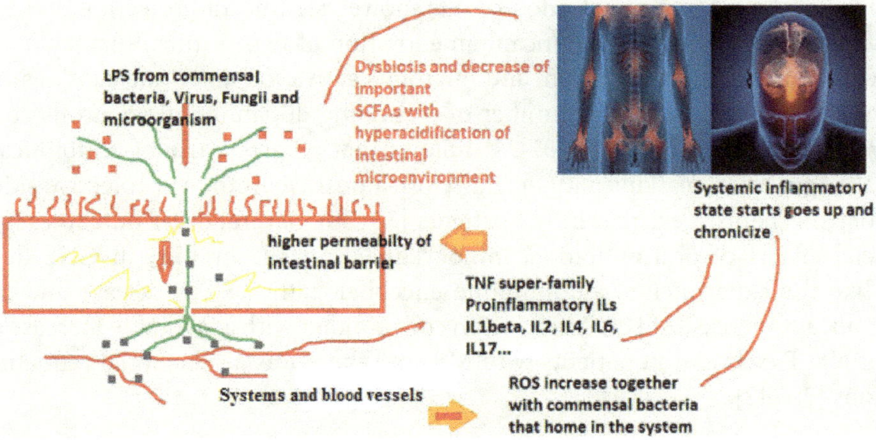

Figure 1. *The gut microbiota is a very dynamic ecosystem. The entire gut microbiota is composed of different sub-environments with unique features like niches with specific microbes and tissue interactions. The large intestine represents the more populated area and performs the highest variety of biotransformation under the guide of specific gene expression in charge of enzymes necessary for highly specific biotransformation necessary of the SCFAs. The local flora is crucial for the local microbiome homeostasis, and the whole chain of bio-reaction takes place in spaces with a specific mean pH of 6.5–7. The changes in local balance homeostasis and in pH negatively impact on the mucosa shield that repair the outside and inside permeability gradient. Once the stability and the equilibrium between all the components are broken, the gut walls become fragile under the constant attack of local immune cells that start to deteriorate the integrity of both endothelial wall and mucosa shield that induce on medium long term and accumulation of pro-inflammatory endotoxins, bacteria free passage into the system, and low antimicrobial peptide production with a consequent high gastrointestinal motility.*

SCFAs, such as GABA [20–31]. The *Prevotella* spp. is associated with mucin-type O-glycan production which is extremely important in the integrity of gut epithelial barrier; the absence of this mucin type (mucin-2 specifically) tends to compromise the correct homeostatic balance of the local microbiota, increasing intestinal permeability, a clinical feature associated with both microbial dysbiosis and neuro- degenerative diseases [32–34].

Disruption of the BBB is a hallmark in individuals with neurodegenerative dis- eases that contributes to a steady and progressive death of dopaminergic neurons in the CNS. The BBB is a part of a systemic condition that eventually allows the invasion of pathogens and immune agents from a dysbiotic gut into the CNS. However, damages are also due to a series of changes that weaken the integrity of microvascu- lature and blood vessels; these modifications are mainly due to nutritional impairment as a consequence of gut microbiota disturbances that cause low-level intake of important nutrients. Deficiencies in vitamins like C, K, D, and folates responsible for low hydroxylation for the formation of chondro-sulfate necessary for healthy microvessel endothelial walls, the augmentation of free radicals, and depletion in oxygen contents and nitrogen, matrix metalloproteinases (MMPs), cyclooxygenases (COXs), and hypoxia-inducible factor-1α (HIF-1α) are all linked with BBB disruption as neuro-inflammatory responses tend to increase and evolve [35].

Thus the scenario existing in the great majority of neurodegenerative pathologies presents a combination of higher permeability of the intestinal barriers and the BBB, inducing a greater access between gut microbiota and the CNS compartment. Experiments conducted with the use of high dose of minocycline antibiotic are well known to have an impact on specific gut and oral invasive bacteria; the post-administration results showed significant protection on LPS-induced PD in mice data confirmed by a significant amelioration of neuro-inflammatory markers such as TNF-α expression, IL-1α expression, and microglia activation and a substantial amelioration of astrocyte loss with an increased number of surviving dopaminergic neurons compared to control LPS only-injected mice [36, 37]. It follows that a correct use of antibiotics generally known to alter gut microbial diversity may disclose a positive immune protective side effect on inflammatory mechanism existent in PD patients [36–38]. Several other outcomes have shown the beneficial effects of oral antibiotic, minocycline, and tetracycline, in CNS degenerative condition like the experimental autoimmune encephalopathy (EAE) disease and MS. It was found a significant increase of IL-10 expression concomitant with a favorable increase of a subset of invariant NK T cells and in patients with MS, and there was a substantial reduction of CNS deteriorations [38–42].

Crosslink between microbiota dysbiosis and osteo-decay

Aging brings generally substantial physiological alterations—hormonal, humoral, and physical—that involves the entire homeostatic organization of the human body. Of course the GI tract and its microbiota as well undergo through profound changes that under the variations of dietary influences bring to a general decline of cognitive and immune activities. With aging, the gut microbiota lost bacterial balanced diversity with an increase of "pathogenic Proteobacteria" vs. a continuing, steady, and progressive lower level of "friend bacteria" such as Firmicutes, *Faecalibacterium prausnitzii*, and Actinobacteria (mainly bifidobacteria) [42–45].

Another important feature of gut microbiota is the ability to modulate genes that can be seen either on regulation or variation; this is one of the main factors that may explain the influences that gut microbiota eventually exert on bone development and on bone-related diseases such as osteoporosis, osteopenia, or the different types of arthritis. The delicate homeostatic balance that regulates bone formation and resorption is partly played by the activity of intestinal microorganisms. This activity is basically performed through the interaction with endocrine/nervous system axis; thereby the hormonal activity such as serotonin, cortisol, and sex hormones and several growth factors affect bone mass in mice and humans. In addition, bone marrow stem cells, circulatory stem cells, and stem cells from bone marrow niche are highly sensitive to gut microenvironment condition which eventually affects the differentiation process toward either osteoblasts or osteoclasts. In this case it has been proven that the metabolic pathway compartment which involves the ribosome activity, glycolysis, oxidative phosphorylation, carbon metabolism, and mitochondria ATP are fully responsible of regulating MSCs' functionality, growth, proliferation, and differentiation [45, 46].

This important connection has also been confirmed by Xiao and colleagues; they were able to highlight through the single-cell RNA-sequencing analysis the existing connection between the gut microbiota, BM-MSCs, and bone metabolic functional- ity. The presence of several factors such as the HIF-1 together with the expression of infection/inflammatory signaling pathways could be the scattering patterns that influence MSC mobility and immunomodulation. These outcomes showed how HIF-1 signaling is involved in BM-MSC immunomodulation. In fact, the HIF-1 is notoriously known as a triggering factor of inflammatory transcription factor NFκB and an active regulator of specific cytokine and chemokine recruitment in inflammation and infection situations. The chronic presence of an inflammatory response under the triggering activity of pro-inflammatory cytokines such as the TNF superfamily IL-1β, IL-2, IL-4, IL-6, and IL-17 can deeply disturb the osteoclast and osteoblast balance, typically resulting in a net hyper-osteoclast activity and thus bone loss. While there is an evident mechanical effect on the bone where these cytokines stimulate osteoclast differentiation with a consequent upregulation of RANKL expression in progenitor osteoblasts together with a higher RANKL expres- sion, a concomitant nonmechanical effect under the downregulation of specific deficit due to a metabolic inability of vitamin K and vitamin D synthesis in the intestinal lumen should be mentioned (**Figure 2**) [47–54].

Probiotic efficacy in therapy and clinically use

The potential impact of the therapeutic effect of probiotic on a dysbiotic situation could not be seen without taking in consideration a change of lifestyle. Diet habits, stress, poor healthy conditions, and lack of exercise can significantly impact the gut microbiota stability [55–57]. There are many evidences that nutritional habits based on "Western diet" composed of huge additive and animal fats, processed glucose and excessive quantities of hypercaloric nutritional facts, low contents of fresh food, and low level of vitamins and minerals, essential for our body, all negatively affect the correct balance of gut microbiota, which eventually lead to the insurgence of metabolic dysfunctions. It is also well known that these bad behaviors are associated with an increased risk of developing several chronic diseases that may attack oral microbiota and vaginal microbiota that recent study findings have indicated as an independent risk factor for severe neurodegenerative conditions [9, 58–60].

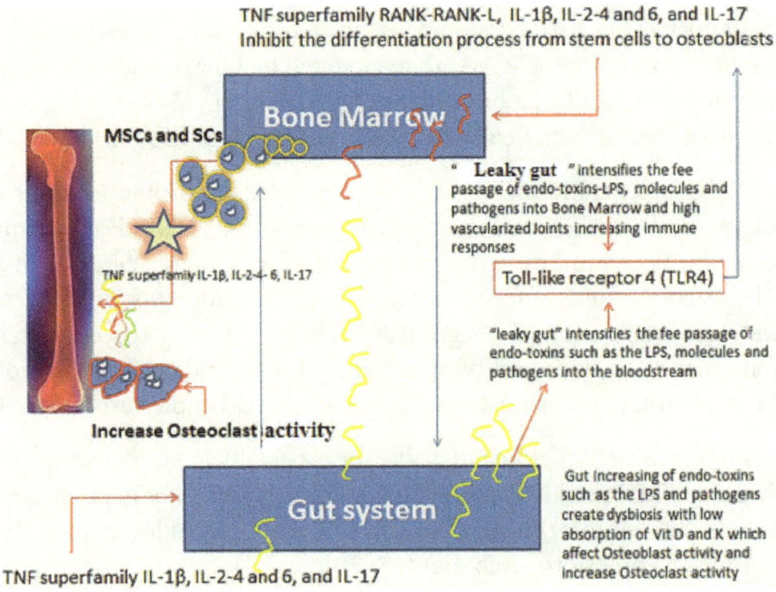

Figure 2. *The gut dysbiosis is one of the main contributors in osteo-degenerative conditions. The dysbiotic microenvironment increase the viability of systemic pro-inflammatory cytokines and interleukins generating three main anti-regenerative patterns, the increase of pH acidic level, decrease of the differentiating pathway from MSCs and SCs toward osteoblasts, and hyper-expression of osteoclast activity. The dysbiosis generates a defective absorption mechanism of important nutrients for bone homeostasis like vitamins, among them K and D, and hormones such as serotonin, testosterone, and estrogen. The prerogative of this condition is a cascade of events that will involve systemically and progressively the whole vital activity of cells, tissues, and systems of the organism.*

By definition, probiotic refers to large and diverse types of microbes both com- mensal that normally reside in the gut and exogenous that may migrate through the intestine following food or diet and supplement consumption. Probiotics might be composed of different microbial strains, the most common include species of *Lactobacillus*, *Bifidobacterium*, *Streptococcus*, and yeast *Saccharomyces* species [61].

As previously mentioned, currently there is a great interest on the use of specific probiotics as therapeutic tool to be associated as clinical approach toward immune system pathologies that may include autoimmune conditions that may attack nerves, bone, and bowel. Given the prevalence of probiotic use, the effects of probiotics on bone health is of significant interest.

Significant positive clinical outcomes have been obtained in numerous studies conducted on CNS inflammatory condition that have therapeutically used different types of probiotic strains. The results showed a reduction of CNS inflammatory level and progression; these outcomes were eventually explained by the capacity of certain strains (*Lactobacillus* species including bacteria like the *Pediococcus acidilac- tici*, *Bifidobacterium bifidum*, *Bifidobacterium animalis*, *Streptococcus thermophiles*, and *Bifidobacterium infantis* 35,624) to modulate the expression of T-regs, B-regs, and IL-10 production such as [62–64].

In addition, an experiment with genetically engineered microbial strains such as *Lactococcus lactis* capable of expressing heat shock protein 65 obtained from another strain like the *Mycobacterium leprae* was seen highly efficient in reducing EAE symptoms and disease progression [63, 64]. The beneficial outcomes in this study were associated with a decrease in IL-17 pro-inflammatory interleukin with a parallel growth of IL-10 evaluated in the mesenteric lymph node and spleen cell cultures. Furthermore, mice showed a significant higher level of endogenous CD4 + Foxp3+ regulatory T-regs and CD4 + LAP+ (latency-associated peptide). These results might be also sustained by a higher production of SCFAs that, as stated by Opazo and colleagues, were seen to induce either a decrease in RORγt, a biomarker of IL-17, or IL-23 with a higher production of IL-10 and IL-12 with a similar beneficial effect on both EAE and IBS [61–66].

Therefore beneficial homeostatic-metabolic effect of specific probiotic strains can be seen on different systems such as the cardiovascular, immune, and CNS. For instance, few strains conserve a natural ability of inhibiting the insurgence of hypercholesteremia in both mice and human. In fact the use of *Lactobacillaceae* strains such as *L. acidophilus*, *Bifidobacterium bifidum*, and *L. plantarum* Lp9 strain showed a significant role in lowering the cholesterol level under in vivo conditions thanks to their ability of secreting functional bile salt hydrolase (BSH), an enzyme crucial in the protection against the insurgence of bad cholesterol in the host.

Genomic analysis has indicated that *Lactobacillaceae* especially *L. plantarum* contain the highest presence of *BSH* genes. Intriguingly, milk fermented by *L. plantarum* NTU 102 revealed to have a high significant efficacy on total cholesterol and LDL cholesterol levels though in presence of individuals undergone a cholesterol-rich diet [67–70].

Major depressive disorders have been seen even today as a consequence of decreased serotonin level; therefore, the therapeutic strategy has mainly concentrated on producing medication, which focused on serotonin only. The major treatments are based on a class of drugs known as selective serotonin reuptake inhibitors (SSRIs). These SSRIs stimulate the serotonin uptake between neurons, and, though it has been seen some improvements, the medium long-term use has produced serious side effects on gut homeostatic balance with severe metabolic disturbances.

Nowadays, as above mentioned, following the fact that current researches have established associations between gut microbes, digestive function, and mental well-being especially under the fact that serotonin is synthesized in the gut by the combined activity of different microbiota strain such as *Lactobacillus*. The connection was firstly seen in IBS patient who also manifests clear clinical signs of depressive disorders; the analysis of gut microbiota from these patients showed a very low level of *Lactobacillus* strains versus healthy subjects that might be explained by the increased expression of serotonin transporter (SERT) [71–74].

Overall the data on this specific topic all have evidenced the positive effects of probiotics in CNS health. These effects are explained by the ability of probiotics to directly interact with fundamental metabolic agents either within the gut or outside that eventually explain the Gut-CNS axis. The use of probiotics and in specific the *Lactobacillus* strains showed that mice fed with these probiotics revealed a better capacity of reabsorbing tryptophan amino acid the precursor or serotonin, the re-established normal level of those hormones strictly related to stress deviances and depression such as the adrenocorticotropic hormone (ACTH), corticosterone, adrenaline, noradrenaline and the re-increase expression of brain-derived neu- rotrophic factor (BDNF) a marker that indicate a neuronal health and memory functionality [74–79].

To conclude, the higher permeability of gut or "leaky gut" intensifies the fee passage of endotoxins such as the LPS and other forms of molecules and pathogens to leak into the bloodstream and thus in the entire system. The upsurge of these endotoxins, pathogens, and waste molecules eventually trigger the activation of a cascade of immune responses through switching on the Toll-like receptor 4 (TLR4) that mediates the recruitment of T and B lymphocytes together with a huge number of pro-inflammatory cytokines, interleukins, and IgA (**Figure 3**) [80, 81]. The current position therefore considers the use of probiotics as a therapeutic tool that

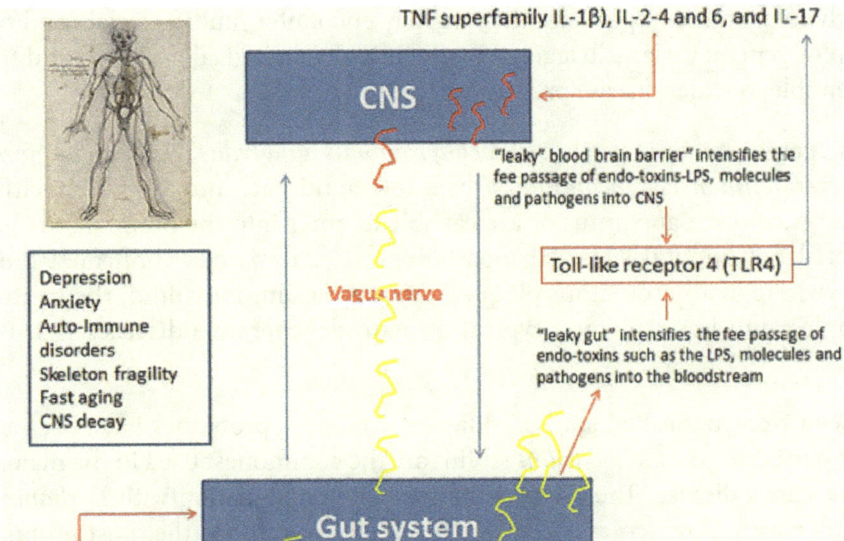

Figure 3. *There is a strict connection between CNS and gut system. The connection takes place through the afferent and efferent pathways of vagus nerve that physically connects both CNS and gut. Both CNS and gut may undergo leaky phenomena; in both cases, the barriers of either CNS or gut become extremely permeable under the chronic attack of both pathogens and immune agents overexpressed on the site. This event may eventually explain degenerative condition of both systems including IBS, ulcerative colitis, depression, PD, AD, and MS.*

may exert beneficial effects on the CNS by improving the stability homeostasis and integrity of gut microbiota, decreasing systemic inflammation.

Probiotics in the treatment of skin and oral mucosa dysbiosis

The skin represents another system where an immense variegation of micro- biota environment can be found. Skin diseases caused by disturbances at the level of local microbiota that also showed to have strict connection with gut dysbiosis are quite exhaustive in explaining these malevolence patterns. Psoriatic lesions show a very specific histopathological conformation which present highly infiltrated immune cells like the CD3+ T cells and dendritic cells (DCs). Psoriasis showed to have a genetic family trait prevalent in twins; researchers have spotted 36 genetic loci associated with PS susceptibility 1 (*PSORS1*) locus on chromosome 6p21.3 [82, 83]. Data confirmed that most of them are directly involved in the overexpression of those genes that regulate part of pro-inflammatory innate immune responses such as the NFkB activation and interleukin (IL)-23 signaling pathway. Intriguingly a 2018 study performed on mice proved the use of two specific probiotic *Lactobacillus* strains, the *L. salivarius* L305 and *L. rhamnosus* L307, in alleviating the clinical symptoms of psoriasis through inhibiting the aggressive effect of pro-inflammatory cytokines and interleukins like TNF-α, IL-1β, IL-6, IL-17, and IL-22 and promoting the anti-inflammatory/modulatory activity of IL-4 and IL-10 [84].

In oral dysbiosis, we are facing a similar inflammatory arrangement; oral diseases manifest with high-grade inflammatory patterns that spread from the gums to the adjacent structures gradually destroying the supporting tissues of the teeth, both ligaments and alveolar bones, causing early loss. Similarly to psoriasis in periodontitis, we may encounter multifactorial condition due to a combination of genetic variants triggered by the initial subgingival dysbiosis and then become highly susceptible to wider disease progression [85].

The gram-negative bacteria such as *Porphyromonas gingivalis*, *Tannerella forsythia*, and *Aggregatibacter actinomycetemcomitans* are be able to migrate into the system either down to the heart, lungs, and sex apparatus or are capable to enter into the brain via the bloodstream or via infected periodontal sites. Histopathological analysis has confirmed these bacteria almost everywhere in atheromatous plaques, the vagina, amniotic fluid, rheumatoid arthritis bioptic samples, and brain plaques typical of neurodegenerative diseases AD, PD, and MS [86–90].

As can be seen from published studies, different strains of probiotics have been used for the treatment of periodontitis. *Lactobacillus* strains are the commonest used in the majority of high-grade inflammatory disease. The use of *L. salivarius* in combination with *L. rhamnosus* and *B. subtilis* together with *L. reuteri* and *L. brevis* probiotics has shown the most promising results. High-positive results were also obtained by Laleman and colleagues in using *Streptococcus oralis* KJ3, *Streptococcus uberis* KJ2, and *Streptococcus ratti* JH145 [91, 92].

Therefore an associated altered gut microbiota may lead to chronic gut dysbiosis and propagation of systemic injuries that involve cells, tissues, system, and the intrinsic dysfunction of the regenerative mechanism.

Conclusion

In summary, the present chapter reveals that gut microbiota and a correct use of probiotic may play important pleiotropic functions on several levels and systems. It is now clear that there is a bidirectional interaction between micro- biota and nervous system, microbiota and immunity, microbiota and bones, and eventually microbiota and mitochondria. Probiotics are getting more and more attention due to the increase of evidence of their benefits in many degenerative disorders. It shows the capacity of microbiota to restore gut and vaginal and oral microbiota, thus attenuating various severe inflammatory responses. All these findings suggest that probiotics could play a role in clinical procedure and therapy approaches to decrease the risk of morbidity and mortality related to CNS diseases, cardiovascular diseases, and bone degenerations. The shared information presented on this chapter may also demonstrate that the traditional view on gut microbiota and microbiome has changed and may be eventually useful as a prospective medium for the delivery of superior, more precise, and personalized treatments in the achievement of better protective health benefits for a more and more aging society.

Conflictsofinterest

The authors declare no conflicts of interest.

Author contribution

The authors contributed equally to this work.

Author details

Ciro Gargiulo Isacco[1,2,3,4], Andrea Ballini[5,6*], Danila De Vito[6],

Angelo Michele Inchingolo[2], Stefania Cantore[2], Gregorio Paduanelli[2], Kieu Cao Diem Nguyen[1,2], Alessio Danilo Inchingolo[2], Gianna Dipalma[2] and Francesco Inchingolo[2]

1 Human Stem Cells Research Center, Ho Chi Minh City, Vietnam

2 Department of Interdisciplinary Medicine (DIM), School of Medicine, University Aldo Moro, Bari, Italy

3 Pham Chau Trinh University of Medicine, Danang, Vietnam

4 International Institute of Gene and Immunology, Ho Chi Minh City, Vietnam

5 Department of Biosciences, Biotechnology and Biopharmaceutics, University of Bari Aldo Moro, Bari, Italy

6 Department of Basic Medical Sciences, Neurosciences and Sense Organs, University of Bari Aldo Moro, Bari, Italy

*Address all correspondence to: andrea.ballini@uniba.it; andrea.ballini@me.com

References

[1] Baohong W, Mingfei Y, Longxian L, Zongxin L, Lanjuan L. The human microbiota in health and disease. Engineering. 2017;**3**(1):71-82

[2] Moya A, Ferrer M. Functional redundancy-induced stability of gut microbiota subjected to disturbance. Trends in Microbiology. 2016;**24**(5):402-413

[3] Dumas ME, Barton RH, Toye A, Cloarec O, Blancher C, Rothwell A. Metabolic profiling reveals a contribution of gut microbiota to fatty liver phenotype in insulin-resistant mice. Proceedings of the National Academy of Sciences of the United States of America. 2006;**103**:12511-12516

[4] Borycka-Kiciak K, Banasiewicz T, Rydzewska G. Butyric acid – A well- known molecule revisited. Przegląd Gastroenterologiczny. 2017;**12**(2):83-89

[5] Al-Lahham SH, Peppelenbosch MP, Roelofsen H, Vonk RJ, Venema K. Biological effects of propionic acid in humans; metabolism, potential applications and underlying mechanisms. Biochimica et Biophysica Acta. 2010;**1801**(11):1175-1183

[6] Acheson DW, Luccioli S. Microbial- gut interactions in health and disease. Mucosal immune responses. Best Practice & Research. Clinical Gastroenterology. 2004;**18**(2):387-404

[7] Thomas S, Izard J, Walsh E, Batich K, Chongsathidkiet P, Clarke G, et al. The host microbiome regulates and maintains human health: A primer and perspective for non-microbiologists. Cancer Research. 2017:1-31

[8] Round L, Mazmanian SK. The gut microbiome shapes intestinal immune responses during health and disease. Nature Reviews Immunology. 2009;**9**(5):313-323

[9] Holly MK, Smith JG. Paneth cells during viral infection and pathogenesis. Viruses. 2018;**10**(5):225

[10] Ballini B, Santacroce L, Cantore S, Bottalico L, Dipalma G, Topi S, et al. Probiotics efficacy on oxidative stress values in inflammatory bowel disease: A randomized double-blinded placebo- controlled pilot study. Endocrine, Metabolic & Immune Disorders Drug Targets. 2019;**19**(12):1-12

[11] Lannes N, Eppler E, Etemad S, Yotovski P, Filgueira L. Microglia at center stage: A comprehensive review about the versatile and unique residential macrophages of the central nervous system. Oncotarget. 2017;**8**(69):114393-114413

[12] Sonnenburg JL, Bäckhed F. Diet- microbiota interactions as moderators of human metabolism. Nature. 2016;**535**:56-64

[13] Round JL, Mazmanian SK. The gut microbiota shapes intestinal immune responses during health and disease. Nature Reviews. Immunology. 2009;**9**:313-323

[14] Cani PD, Bibiloni R, Knauf C, Waget A, Neyrinck AM, Delzenne NM, et al. Changes in gut microbiota control metabolic endotoxemia-induced inflammation in high-fat diet-induced obesity and diabetes in mice. Diabetes. 2008;**57**:1470-1481

[15] Spielman LJ, Gibson DL, Klegeris A. Unhealthy gut, unhealthy brain: The role of the intestinal microbiota in neurodegenerative diseases. Neurochemistry International. 2019;**120**:149-163

[16] Barrett E, Ross RP, O'Toole TW, Fitzgerald GF, Stanton C. Gamma- Aminobutyric acid production by culturable bacteria from the human intestine. Journal of Applied Microbiology. 2005;**113**:411-417

[17] Bergstrom KS, Xia L. Mucin- type O-glycans and their roles in intestinal homeostasis. Glycobiology. 2013;**23**:1026-1037

[18] Bonaz B, SinnigerV, Pellissier S. Anti- inflammatory properties of the vagus nerve: Potential therapeutic implications of vagus nerve stimulation. The Journal of Physiology. 2016;**594**:5781-5790

[19] Russo R, Cristiano C, Avagliano C, De Caro C, La Rana G, Raso GM, et al. Gut-brain axis: Role of lipids in the regulation of inflammation, pain and CNS diseases. Current Medicinal Chemistry. 2018;**25**(32):3930-3952

[20] Block ML, Hong JS. Microglia and inflammation-mediated neurodegeneration: Multiple triggers with a common mechanism. Progress in Neurobiology. 2005;**76**:77-98

[21] Zhang YG, Wu S, Yi J, Xia Y, Jin D, Zhou J, et al. Target intestinal microbiota to alleviate disease progression in amyotrophic lateral sclerosis. Clinical Therapeutics. 2017;**39**:322-336

[22] Zhang R, Miller RG, Gascon R, Champion S, Katz J, Lancero M, et al. Circulating endotoxin and systemic immune activation in sporadic amyotrophic lateral sclerosis (sALS). Journal of Neuroimmunology. 2008;**206**(1-2):121-124

[23] Xu R, Wang Q. Towards understanding brain-gut-microbiome connections in Alzheimer's disease. BMC Systems Biology. 2016;**10**(3):63

[24] Wu S, Yi J, Zhang YG, Zhou J, Sun J. Leaky intestine and impaired microbiome in an amyotrophic lateral sclerosis mouse model. Physics Reports. 2015;**3**(4):e12356

[25] Wang D, Ho L, Faith J, Ono K, Janle EM, Lachcik PJ, et al. Role of intestinal microbiota in the generation of polyphenol-derived phenolic acid mediated attenuation of Alzheimer's disease beta-amyloid oligomerization. Molecular Nutrition & Food Research. 2015;**59**:1025-1040

[26] Villumsen M, Aznar S, Pakkenberg B, Jess T, Brudek T. Inflammatory bowel disease increases the risk of Parkinson's disease: A Danish Nationwide cohort study, 1977-2014. Gut. 2018

[27] Villarán RF, Espinosa-Oliva AM, Sarmiento M, De Pablos RM, Argüelles S, Delgado-Cortés MJ, et al. Ulcerative colitis exacerbates lipopolysaccharide-induced damage to the nigral dopaminergic system: Potential risk factor in Parkinson's disease. Journal of Neurochemistry. 2010;**114**(6):1687-1700

[28] Toepfer M, Folwaczny C, Klauser A, Riepl RL, Müller-Felber W, Pongratz D. Gastrointestinal dysfunction in amyotrophic lateral sclerosis. Amyotrophic Lateral Sclerosis. 1999;(1):1, 15-19

[29] Rowin J, Xia Y, Jung B, Sun J. Gut inflammation and dysbiosis in human motor neuron disease. Physiological Reports. 2017;**5**(18):e13443

[30] Holmqvist S, Chutna O, Bousset L, Aldrin-Kirk P, Li W, Björklund T, et al. Direct evidence of Parkinson pathology spread from the gastrointestinal tract to the brain in rats. Acta Neuropathologica. 2014;**128**(6):805-820

[31] Cersosimo MG, Raina GB, Pecci C, Pellene A, Calandra CR, Gutiérrez C, et al. Gastrointestinal manifestations in Parkinson's disease: Prevalence and occurrence before motor symptoms. Journal of Neurology. 2013;**260**(5):1332-1338

[32] Barrett E, Ross RP, O'Toole PW, Fitzgerald GF, Stanton C. γ-Aminobutyric acid production by culturable bacteria from the human intestine. Journal of Applied Microbiology. 2012;**113**(2):411-417

[33] Arpaia N, Campbell C, Fan X, Dikiy S, van der Veeken J, deRoos P. Metabolites produced by commensal bacteria promote peripheral regulatory T-cell generation. Nature. 2013;**504**:451-455

[34] Tailford LE, Crost EH, Kavanaugh D, Juge N. Mucin glycan foraging in the human gut microbiome. Frontiers in Genetics. 2015;**6**:81

[35] Rausch P, Rehman A, Kunzel S, Hasler R, Ott SJ, Schreiber S. Colonic mucosa-associated microbiota is influenced by an interaction of Crohn disease and FUT2 (secretor) genotype. Proceedings of the National Academy of Sciences of the United States of America. 2011;**108**:19030-19035

[36] Rosenberg GA. Neurological diseases in relation to the blood-brain barrier. *Journal of Cerebral Blood Flow and Metabolism*. 2012;**32**(7):1139-1151

[37] Zaura E, Brandt BW, Teixeira de Mattos MJ, Buijs MJ, Caspers MP, Rashid MU, et al. Same exposure but two radically different responses to antibiotics: Resilience of the salivary microbiome versus long- term microbial shifts in feces. MBio. 2015;**6**:e01693-e01615

[38] Tomas-Camardiel M, Rite I, Herrera AJ, de Pablos RM, Cano J, Machado A, et al. Minocycline reduces the lipopolysaccharide-induced inflammatory reaction, peroxynitrite- mediated nitration of proteins, disruption of the blood-brain barrier, and damage in the nigral dopaminergic system. Neurobiology of Disease. 2004;**16**:190-201

[39] Sampson TR, Debelius JW, Thron T, Janssen S, Shastri GG, Ilhan ZE, et al. Gut microbiota regulate motor deficits and neuroinflammation in a model of Parkinson's disease. Cell. 2016;**167**:1469-1480

[40] Yokote H, Miyake S, Croxford JL, Oki S, Mizusawa H, Yamamura T. NKT cell-dependent amelioration of a mouse model of multiple sclerosis by altering gut flora. The American Journal of Pathology. 2008;**173**:1714-1723

[41] Popovic N, Schubart A, Goetz BD, Zhang SC, Linington C, Duncan ID. Inhibition of autoimmune encephalomyelitis by a tetracycline. Annals of Neurology. 2002;**51**:215-223

[42] Colpitts SL, Kasper LH. Influence of the gut microbiome on autoimmunity in the central nervous system. Journal of Immunology. 2017;**198**(2):596-604

[43] Odamaki T, Kato K, Sugahara H, Hashikura N, Takahashi S, Xiao JZ, et al. Age-related changes in gut microbiota composition from newborn to centenarian: A cross-sectional study. BMC Microbiology. 2016;**16**:90

[44] Salazar N, López P, Valdés L, Margolles A, Suárez A, Patterson AM, et al. Microbial targets for the development of functional foods accordingly with nutritional and immune parameters altered in the elderly. Journal of the American College of Nutrition. 2013;**32**:399-406

[45] Ticinesi A, Tana C, Nouvenne A, Prati B, Lauretani F, Meschi T. Gut microbiota, cognitive frailty and dementia in older individuals: A systematic review. Clinical Interventions in Aging. 2018;**13**:1497-1511

[46] Salazar N, Valdés-Varela L, González S, Gueimonde M, de Los Reyes-Gavilán CG. Nutrition and the gut microbiome in the elderly. Gut Microbes. 2017;**8**:82-97

[47] Xiao E, He L, Wu Q , et al. Microbiota regulates bone marrow mesenchymal stem cell lineage differentiation and immunomodulation. Stem Cell Research & Therapy. 2017;**8**(1):213

[48] Nicholson JK, Holmes E, Kinross J, Burcelin R, Gibson G, Jia W, et al. Host– gut microbiota metabolic interactions. Science. 2012;**336**(6086):1262-1267

[49] Walmsley SR, Print C, Farahi N, Peyssonnaux C, Johnson RS, Cramer T, et al. Hypoxia-induced neutrophil survival is mediated by HIF-1alpha- dependent NF-kappaB activity. The Journal of Experimental Medicine. 2005;**201**(1):105-115

[50] Scortegagna M, Cataisson C, Martin RJ, Hicklin DJ, Schreiber RD, Yuspa SH, et al. HIF-1alpha regulates epithelial inflammation by cell autonomous NFkappaB activation and paracrine stromal remodeling. Blood. 2008;**111**(7):3343-3354

[51] Spees JL, Lee RH, Gregory CA. Mechanisms of mesenchymal stem/ stromal cell function. Stem Cell Research & Therapy. 2016;**7**(1):125

[52] Szychlinska MA, Di Rosa M, Castorina A, Mobasheri A, Musumeci G. A correlation between intestinal microbiota dysbiosis and osteoarthritis. Heliyon. 2019;**5**(1):e01134

[53] Lam J et al. TNF-alpha induces osteoclastogenesis by direct stimulation of macrophages exposed to permissive levels of RANK ligand. The Journal of Clinical Investigation. 2000;**106**(12):1481-1488

[54] Walsh NC, Gravallese EM. Bone remodeling in rheumatic disease: A question of balance. Immunological Reviews. 2010;**233**(1):301-312

[55] Romas E et al. Expression of osteoclast differentiation factor at sites of bone erosion in collagen-induced arthritis. Arthritis and Rheumatism. 2000;**43**(4):821-826

[56] Huxley RR, Ansary-Moghaddam A, Clifton P, Czernichow S, Parr CL, Woodward M. The impact of dietary and lifestyle risk factors on risk of colorectal cancer: A quantitative overview of the epidemiological evidence. International Journal of Cancer. 2009;**125**:171-180

[57] Rothschild D, Weissbrod O, Barkan E. Environment dominates over host genetics in shaping human gut microbiota. Nature. 2018;**555**:210-215

[58] Warburton DE, Nicol CW, Bredin SS. Health benefits of physical activity: The evidence. CMAJ. 2006;**174**:801-809

[59] Inchingolo F, Dipalma G, Cirulli N, Cantore S, Saini RS, Altini V, et al. Microbiological results of improvement in periodontal condition by administration of oral probiotics. Journal of Biological Regulators and Homeostatic Agents. 2018;**32**(5):1323-1328

[60] Cantore S, Ballini A, De Vito D, Abbinante A, Altini V, Dipalma G, et al. Clinical results of improvement in periodontal condition by administration of oral probiotics. Journal of Biological Regulators and Homeostatic Agents. 2018;**32**(5):1329-1334

[61] Kim D, Yoo SA, Kim WU. Gut microbiota in autoimmunity: Potential for clinical applications. Archives of Pharmacal Research. 2016;**39**:1565-1576

[62] Opazo MC, Ortega-Rocha EM, Coronado-Arrázola I, et al. Intestinal microbiota influences non-intestinal related autoimmune diseases. Frontiers in Microbiology. 2018;**9**:432

[63] Kwon HK, Kim GC, Kim Y, Hwang W, Jash A, Sahoo A, et al. Amelioration of experimental autoimmune encephalomyelitis by probiotic mixture is mediated by a shift in T helper cell immune response. Clinical Immunology. 2013;**146**(3):217-227

[64] RezendeRM,OliveiraRP,MedeirosSR, Gomes-Santos AC, Alves AC, Loli FG, et al. Hsp65-producing Lactococcus lactis prevents experimental autoimmune encephalomyelitis in mice by inducing CD4+LAP+ regulatory T cells. Journal of Autoimmunity. 2013;**40**:45-57

[65] Mizuno M, Noto D, Kaga N, Chiba A, Miyake S. The dual role of short fatty acid chains in the pathogenesis of autoimmune disease models. PLoS One. 2017;**12**(2):e0173032

[66] Takata KT, Tomita T, Okuno M, Kinoshita T, Koda JA, Honorat M, et al. Dietary yeasts reduce inflammation in central nerve system via microflora. Annals of Clinical Translational Neurology. 2015;**2**:56-66

[67] Konieczna P, Groeger D, Ziegler M, Frei R, Ferstl R, Shanahan F, et al. Bifidobacterium infantis 35624 administration induces Foxp3 T regulatory cells in human peripheral blood: Potential role for myeloid and plasmacytoid dendritic cells. Gut. 2012;**61**:354-366

[68] Mohania D, Kansal VK, Shah D, Nagpal R, Kumar M, Gautam SK, et al. Therapeutic effect of probiotic dahi on plasma, aortic, and hepatic lipid profile of hypercholesterolemic rats. Journal of Cardiovascular Pharmacology and Therapeutics. 2013;**18**(5):1-8

[69] Brandvold KR, Weaver JM, Whidbey C, Wright AT. A continuous fluorescence assay for simple quantification of bile salt hydrolase activity in the gut microbiome. Scientific Reports. 2019;**9**(1359):1-7

[70] Liang L, Yi Y, Lv Y, Qian J, Lei X, Zhang G. A comprehensive genome survey provides novel insights into bile salt hydrolase (BSH) in Lactobacillaceae. Molecules. 2018;**23**(5):1157

[71] Guo Z, Liu XM, Zhang QX, Shen Z, Tian FW, Zhang H, et al. Influence of consumption of probiotics on the plasma lipid profile: A meta-analysis of randomised controlled trials. Nutrition, Metabolism, and Cardiovascular Diseases. 2011;**21**:844-850

[72] Cao YN, Feng LJ, Liu YY, et al. Effect of *Lactobacillus rhamnosus* GG supernatant on serotonin transporter expression in rats with post- infectious irritable bowel syndrome. World Journal of Gastroenterology. 2018;**24**(3):338-350

[73] Zhang ZF, Duan ZJ, Wang LX, Yang D, Zhao G, Zhang L. The serotonin transporter gene polymorphism (5-HTTLPR) and irritable bowel syndrome: A meta-analysis of 25 studies. BMC Gastroenterology. 2014;**14**:23

[74] Wheatcroft J, Wakelin D, Smith A, Mahoney CR, Mawe G, Spiller R. Enterochromaffin cell hyperplasia and decreased serotonin transporter in a mouse model of postinfectious bowel dysfunction. Neurogastroenterology and Motility. 2005;**17**:863-870

[75] Wallace CJK, Milev R. The effects of probiotics on depressive symptoms in humans: A systematic review [published correction appears in annals of general psychiatry 2017 Mar 7;16:18]. Annals of General Psychiatry. 2017;**16**:14

[76] Ait-Belgnaoui A, Durand H, Cartier C, et al. Prevention of gut leakiness by a probiotic treatment leads to attenuated HPA response to an acute psychological stress in rats. Psychoneuroendocrinology. 2012;**37**(11):1885-1895

[77] Ait-Belgnaoui A, Colom A, Braniste V. Probiotic gut effect prevents the chronic psychological stress-induced brain activity abnormality in mice. Neurogastroenterology and Motility. 2014;**26**(4):510-520

[78] Bravo JA, Forsythe P, Chew MV, et al. Ingestion of *Lactobacillus* strain regulates emotional behavior and central GABA receptor expression in a mouse via the vagus nerve. PNAS. 2011;**108**(38):16050-16055

[79] Stetler C, Miller GE. Depression and hypothalamic–pituitary–adrenal activation: A quantitative summary of four decades of research. Psychosomatic Medicine. 2011;**73**(2):114-126

[80] Sen S, Duman R, Sanacora G. Serum brain-derived neurotrophic factor, depression, and antidepressant medications: Meta-analyses and implications. Biological Psychiatry. 2008;**64**(6):527-532

[81] Soares JB, Pimentel-Nunes P, Roncon-Albuquerque R, Leite-Moreira A. The role of lipopolysaccharide/toll-like receptor 4 signaling in chronic liver diseases. Hepatology International. 2010;**4**(4):659-672

[82] Chamian F, Lowes MA, Lin SL, Lee E, Kikuchi T, Gilleaudeau P, et al. Alefacept reduces infiltrating T cells, activated dendritic cells, and inflammatory genes in psoriasis vulgaris. Proceedings of the National Academy of Sciences of the United States of America. 2005;**102**:2075-2080

[83] Lowes MA, Chamian F, Abello MV, Fuentes-Duculan J, Lin SL, Nussbaum R, et al. Increase in TNF- alpha and inducible nitric oxide synthase-expressing dendritic cells in psoriasis and reduction with efalizumab (anti-CD11a). Proceedings of the National Academy of Sciences of the United States of America. 2005;**102**:19057-19062

[84] Holowacz S, Blondeau C, Guinobert I, Guilbot A, Hidalgo S, Bisson JF. *Lactobacillus salivarius* L-307 and *Lactobacillus rhamnosus* LA-305 attenuate skin inflammation in mice. Beneficial Microbes. 2018;**9**(2):299-309

[85] Armitage GC. Development of a classification system for periodontal diseases and conditions. Annals of Periodontology. 1999;**4**:1-6

[86] PageRC,OffenbacherS,SchroederHE, Seymour GJ, Kornman KS. Advances in the pathogenesis of periodontitis: Summary of developments, clinical implications and future directions. Periodontology. 2000;**14**:216-248

[87] Leon R, Silva N, Ovalle A, Chaparro A, Ahumada A, Gajardo M, et al. Detection of porphyromonas gingivalis in the amniotic fluid in pregnant women with a diagnosis of threatened premature labor. Journal of Periodontology. 2007;**78**:1249-1255

[88] Forner L, Larsen T, Kilian M, Holmstrup P. Incidence of bacteremia after chewing, tooth brushing and scaling in individuals with periodontal inflammation. Journal of Clinical Periodontology. 2006;**33**:401-407

[89] Quirke AM, Lugli EB, Wegner N, Hamilton BC, Charles P, Chowdhury M, et al. Heightened immune response to autocitrullinated Porphyromonas gingivalis peptidylarginine deiminase: A potential mechanism for breaching immunologic tolerance in rheumatoid arthritis. Annals of the Rheumatic Diseases. 2014;**73**:263-269

[90] Stein PS, Steffen MJ, Smith C, Jicha G, Ebersole JL, Abner E, et al. Serum antibodies to periodontal pathogens are a risk factor for Alzheimer's disease. Alzheimers Dement. 2012;**8**:196-203

[91] Jayaram P, Chatterjee A, Raghunathan V. Probiotics in the treatment of periodontal disease: A systematic review. Journal of Indian Society of Periodontology. 2016;**20**(5):488-495

[92] Laleman I, Yilmaz E, Ozcelik O, Haytac C, Pauwels M, Herrero ER, et al. The effect of a streptococci containing probiotic in periodontal therapy: A randomized controlled trial. Journal of Clinical Periodontology. 2015;**42**(11):1032-1041

12

Use of Prebiotics as an Alternative to Antibiotic Growth Promoters in the Poultry Industry

Bruno Solis-Cruz, Daniel Hernandez-Patlan, Billy M. Hargis and Guillermo Tellez

Abstract

Nowadays there is a great concern about antimicrobial resistance (AMR), which has been recognized as one of the most serious global public health threats. Multilateral organizations focused on global health accept the use of antibiotics in animal production as one of the main drivers of AMR, so that many strategies to control this problem have been proposed, resulting in the total ban of antibiotics as growth-promoting agents. On the other hand, this ban has led to an increase in the incidence of bacterial infections or even to the use of antibiotics at therapeutic doses, which could cause a worse scenario of bacterial resistance. Poultry is one of the most commonly exploited species worldwide and a sector that continues to grow and industrialize in many parts of the world, so it was to be expected that a large part of the antibiotics used in animal production was destined to this industry. The reduction or complete abolition of antibiotics in poultry production would have a positive effect in the control of AMR, but this would also have negative economic and public health repercussions, caused by foodborne pathogens and the decrease of the productive parameters. For that, many specific alternatives have been evaluated and marketed, prebiotics being one of the most promising alternatives for the poultry industry.

Keywords: prebiotics, antibiotics, antimicrobial resistance, poultry, intestinal microbiota

Introduction

Over recent years, the scientific community has expressed great concern about antimicrobial resistance (AMR), which has been recognized as one of the most serious global public health threats in this century [1]. Nowadays, most multilateral organizations focused on global health accept the use of antibiotics in animal intended for food production as one of the main drivers of AMR infections in human health, adopting national action plans that commit to reduce the indiscriminate use of antibiotics by their members [2–4]. These action plans propose many strategies to control this problem, particularly by encouraging reasonable and limited use of antibiotics in food animal production, particularly those that are considered of critical importance for both human and veterinary medicine. The World Health Organization (WHO) has issued a series of guidelines and resolutions in regard to the use of antimicrobial agents in animal production, among those that stand out being the overall reduction in the use of all classes of antimicrobials, with the conditional recommendation not to use those that have been classified as critically important for human medicine, as well as the complete restriction for growth promotion and prevention of infectious diseases that have not yet been clinically diagnosed [5], for which some government regulatory agencies have taken action on the use of antibiotics for animal production, resulting in their total ban as growth-promoting agents [6, 7].

It could be expected that the total ban on the use of antibiotics as growth promoters will lead to a decrease in the levels of antibiotic resistance [8]. However, we cannot ignore some issues resulting from this ban, such as the increase in the incidence of bacterial infections which would also increase the use of other antibiotics at pro- phylactic or even therapeutic doses, accelerating the development of AMR in these pathogens and making it a worse scenario [9, 10]. In addition, some farming practices must be implemented to reduce the use of antibiotics in animal production, such as adequate animal vaccination, good hygiene and husbandry practices, higher animal welfare, and improved breeding programs, which implies an increase in production costs, and it is still not enough to completely reduce the risks of infection [11, 12].

Poultry is one of the most commonly exploited species worldwide, and a sector that continues to grow and industrialize in many parts of the world [13], so it was to be expected that a large part of the antibiotics used in animal production was destined to this industry [14, 15]. Antibiotics have been used in poultry production for therapeutic, prophylactic, or growth promotion purposes, especially in broiler chickens, which has resulted in huge profits for poultry producers [16].

Although it is a fact that the reduction or complete abolition in the use of antibiotics for poultry production would have a positive effect in the control of AMR and public health, this would also have negative economic repercussions, since production costs and, consequently, the prices of the final products, as well as the international trade of poultry products, would be affected [17, 18]. Furthermore, the antibiotic- free production of poultry could imply public health problems caused by foodborne pathogens such as e *C. perfringens*, *E. coli*, *S. aureus*, *Campylobacter spp.*, or *Salmonella spp.* [19–21] while increasing production costs caused by bacterial infections, along with the detriment in the health of the birds and the decrease of the productive parameters. Hence, the ban on the use of antibiotics for poultry production, as well as other increasingly popular trends, such as the growth of the organic products market, has forced poultry producers to find viable alternatives with similar benefits to antibiotics. For that, many specific alternatives have been evaluated and marketed, such as enzymes, prebiotics, probiotics, organic acids, dietary fiber, highly available nutrients, herbs, spices, essential oils, plant components, and vaccines [15].

Among all the available alternatives, prebiotics have proven to be promising alternatives for the poultry industry because they are able to pass through the digestive tract, which facilitates and supports the symbiotic relationship between the host and gastrointestinal tract (GIT) microbiota and results in health benefits for the birds [22–24]. Thus, this chapter exposes the use of prebiotics as feed additives in poultry, with emphasis on their beneficial effects on the microbiota composition, their ability to control pathogenic infections, positive changes in intestinal morphology, improved productive parameters, and immunomodulatory effects as possible mechanisms of action, which make them potential alternatives to avoid the use of antibiotics as growth promoters in the poultry industry.

Types of prebiotics used in the poultry industry

It is difficult to describe in a few words what a prebiotic is; nevertheless, all definitions agree that these compounds, when administered as feed ingredients, are resistant to enzymatic digestion and cannot be absorbed, and still they confer a health benefit for the host animal by selectively stimulating the growth, metabolism, and composition of beneficial native bacteria in the GIT and eliminating the pathogenic ones [25–27]. In general, prebiotics share these common properties, but there are some others that are also common among them, including resistance to gastric acidity, selective fermentability by a limited number of potentially beneficial microorganisms, alteration of the GIT microbiota toward a healthier composition, and modulation of the host animal defense system [28].

Although only carbohydrate-based compounds, such as nondigestible oligosaccharides and non-starch polysaccharides, were previously considered as prebiotic candidates, nowadays the prebiotic concept has expanded to "a substrate that is selectively utilized by host microorganisms conferring a health benefit," so that other substances might fit to it, including a diversity of oligosaccharides with varying carbon chain lengths and even polyphenols and polyunsaturated fatty acids converted to respective conjugated fatty acids [29, 30]. However, to confirm its status as a prebiotic, studies for each candidate must be performed in the target animal species for its intended use, demonstrating its beneficial health effects mediated through the microbiota.

In aviculture, a wide range of prebiotic alternatives have been evaluated, trying to improve the GIT health and resistance against pathogen colonization; nevertheless, all of them have been well characterized, indicating their source, purity, chemical composition and structure, suitable dose, and side effects, and have the status of generally recognized as safe (GRAS). The most commonly used prebiotics in poultry diets are nondigestible oligosaccharides (NDO), including fructooligosaccharides (FOS) and inulin type, mannan oligosaccharides (MOS), xylooligosaccharides (XOS), galactooligosaccharides (GOS), and isomaltooligosac- charide (IMO), as well as some structural carbohydrate components of non-starch polysaccharides (NSP), such as β-glucan [25, 31, 32]. These prebiotics are com- monly administered to poultry orally at first hours or days after hatching, either spraying them directly in the feed or by their direct addition in drinking water; but recently, the administration of in ovo prebiotics in chicken embryos has been proposed as a better route of delivery, since the doses of prebiotics used in ovo could be at least 10 times lower than after hatching, with the same beneficial effects as the oral administration [23, 33].

As mentioned above, the main purpose of prebiotics is to modify the intestinal microbiota in a favorable manner for the host animal and induce positive effects, not only in the intestinal environment but also systemically, which is reflected in positive improvements of the productive parameters such as egg production, body weight gain, feed conversion ratio, and mortality index [34–36]. Besides improved host health and productivity, prebiotics have also proven their efficacy to reduce colonization of important pathogens both for poultry production and public health, such as *Salmonella*, *Campylobacter*, *C. perfringens*, and *E. coli* [37–41]. This set of beneficial effects, along with the lower risks of undesirable side effects in the host and the fact that they are cheaper and easier to produce in a large scale than probiotics, make prebiotics an excellent option as an alternative to minimize the use of antibiotics in poultry production, thus contributing to reduce the problem of AMR [42, 43].

Mechanisms of action of prebiotics in poultry

There are many mechanisms and functions of prebiotics which have been associ- ated with the poultry GIT microbiota, and it seems that there are several bacteria involved in their use; but there is evidence that other microbiota-independent

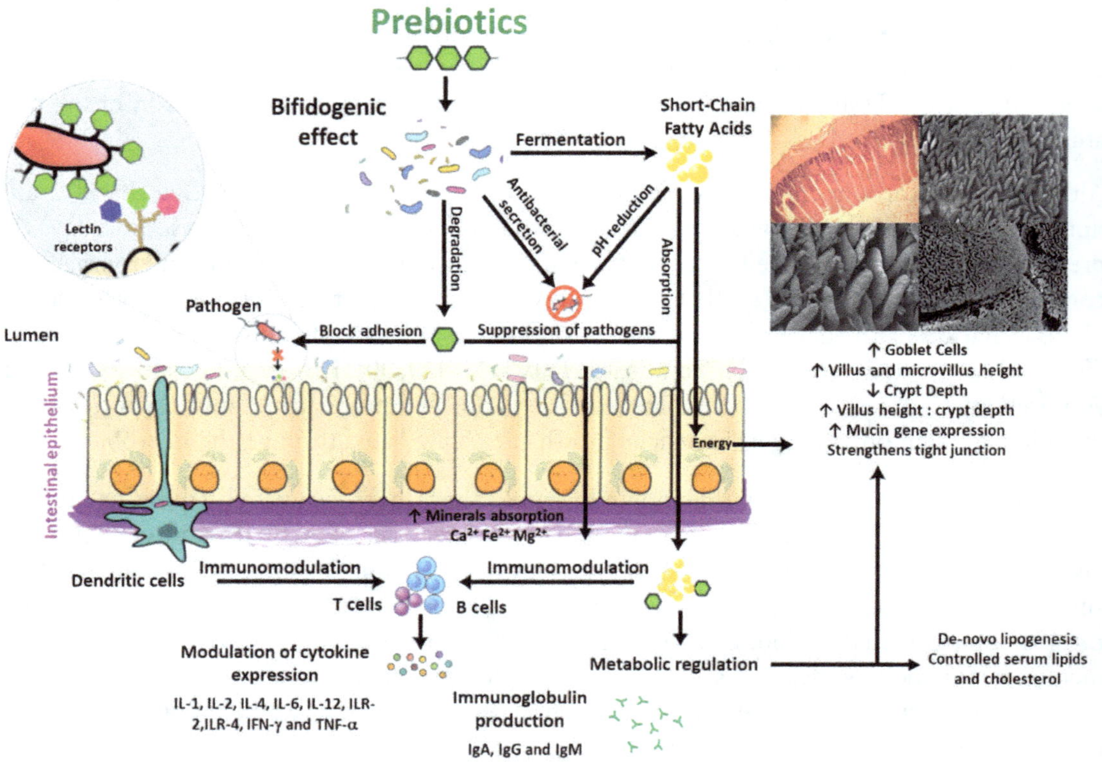

Figure 1. *Some of the potential mechanisms of action of prebiotics in poultry species, showing how they are metabolized by the host microbiota and have positive effects on immunity, gut health, metabolic activity, and pathogen colonization. Modified and adapted from Refs. [25, 45].*

interactions also exist [30]. Thus, there is no specific mechanism of how prebiotics influence the host animal health or how they can inhibit pathogen establishment, but all their beneficial effects are the result of a complex set of interactions on the host physiology, including their positive effects on intestinal morphology and microbiota balance, the suppressive effects on the enteric and systemic infections by pathogens due to the immune stimulation, the higher nutrient digestibility, and positive metabolic alterations (**Figure 1**), which result in improved productive parameters, higher quality of food products from poultry origin, general welfare of animals, and ultimately decrease in production cost [42, 44].

Intestinal microbiota balance

In poultry species, as in almost all the others, the microbiota largely determines the intestinal integrity, functionality, and health, which in turn plays a vital role in nutrient digestion and absorption, immune system development, and pathogen exclusion [46]. Since a symbiotic interaction between host and its microbiota is fundamental to poultry health and production, the alteration of the intestinal micro- biota when using prebiotic has been one of the most widely investigated mechanisms of action [25, 32, 47]. Nowadays it is well known that prebiotics alter positively the intestinal microbiota in poultry; however, the exact mechanism (s) and type of interaction involved will depend on the structure of the prebiotic and the host species. Among all the microorganisms that make up the intestinal microbiota in poultry, some particular bacteria, such as *Lactobacillus spp.* and *Bifidobacterium spp.*, have positive effects on intestinal physiology and, thus, are beneficial for the host health. These beneficial bacteria, which have been reported to have higher abundance when chickens are fed diets supplemented with prebiotics, can ferment and metabolize prebiotics, selectively stimulating their proliferation and activity [45, 48, 49], which

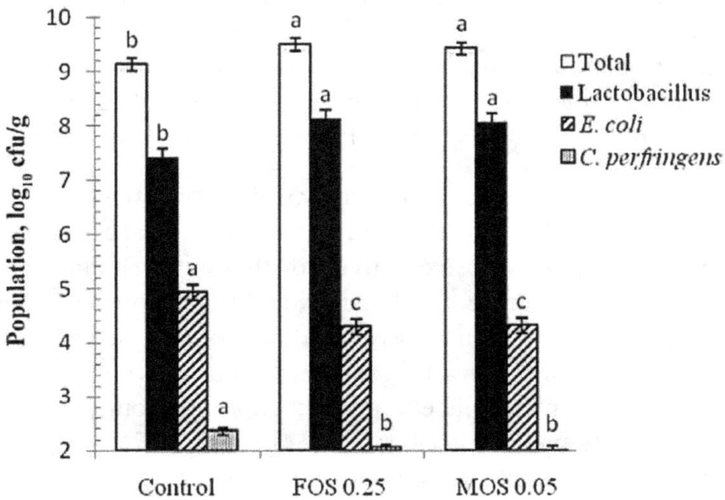

Figure 2. *Intestinal microbiota modifications in the small intestinal content of broiler chickens at 4 weeks of age. Within a bacterial species (or total), bars with different letters (a–c) are different ($P < 0.05$, $n = 8$). Modified from Ref. [40].*

can potentially influence the production of short-chain fatty acids (SCFA) and the consequent decrease in intestinal pH, improve the metabolism by increasing diges- tive enzyme activity and vitamin production and decreasing levels of triglycerides, cholesterol, and odor compounds, and stimulate the immune system that contributes to the inhibitory effects on the growth of pathogenic bacteria [50, 51].

On the other hand, several studies have shown that prebiotics influenced the beneficial intestinal microbiota of broiler chickens while maintaining low levels of potential pathogens in the small intestine and cecal digesta. Addition of FOS as prebiotic to the basal diet (4.0 g/kg) significantly increased the viable count of *Bifidobacterium* and *Lactobacillus* in the small intestinal digesta of male broiler chickens, while the number of *Escherichia coli* was significantly reduced compared to the control group [52].

It has been also reported that feeding 0.25% of FOS and 0.05% of MOS to broilers resulted in an increased diversity and population of *Lactobacillus* and decreased populations of *E. coli* and *C. perfringens* in the ileum, as shown in **Figure 2** [40]. In laying hens, dietary supplementation with different levels of inulin linearly reduced coliform bacteria counts by increasing concentrations of this prebiotic, while 2.0% of inulin achieved significantly increased cecal *Bifidobacterium* counts compared with the control group [53].

A recent study has shown that 3.5 mg of a GOS mixture delivered in ovo had a bifidogenic effect in adult chickens, since the relative abundance of *Bifidobacterium* communities was higher in four sections of intestinal content (duodenum, jejunum, ileum, and cecum), while the values of

Lactobacillus abundance resulted to be higher in the control group for most of the four sections [32].

Inhibition of pathogen colonization

The ability of prebiotics in poultry diet to reduce colonization of pathogens results from the combination of several mechanisms occurring in the GIT, from those that are directly related to the selective stimulation of the favorable micro- biota to those in which the prebiotics directly affect the pathogens or the host animal in a microbiota-independent manner. To date, it is not possible to define an exact mechanism of prebiotics to reduce pathogenic infections, so more research is required to fully elucidate their exact function and mode of action.

Various potential mechanisms have been proposed by which prebiotics can provide resistance to pathogens, one of the main ones being directly related to the beneficial bacteria in the GIT, such as *Lactobacillus* and *Bifidobacterium*, whose selective growth results in an increased concentration of SCFA, especially acetate, propionate, and butyrate, and lactate during primary fermentation process at the ceca [30]. This is mainly because these bacteria secrete several hydrolases, which monogastric animals cannot, hydrolyzing the carbohydrate-based prebiotics through a fermentation process whose metabolic end products not only contribute to the nutrition of poultry, but they have additional beneficial effects [54]. Since SCFA are the principal luminal anions, and they are relatively weak acids, their increased concentration is correlated with a lower intestinal pH, which in turn is associated with a suppression of pathogens by dissipating the proton motive force across the bacterial cell mem- brane [55], although it has also been reported that SCFA, butyrate specifically, can downregulate expression of invasion genes in *Salmonella* at low doses [56].

For instance, the effect of 14 or 19 days of 10% dietary lactose administration was evaluated in Leghorn chicks, resulting in a significantly increase of acetic, propionic, butyric, and lactic acid concentration in the cecal contents as compared with the control group; additionally, lactose decreased the pH of cecal contents, with the consequent reduction of the total number of chicks with organ cultures that were positives for this pathogen (**Figure 3**) [57].

Another study was conducted to investigate if changes in SCFA could decrease the numbers of *Enterobacteriaceae* in the ceca of broiler chickens during growth; the authors found a significant negative correlation between the log CFU of *Enterobacteriaceae* and the concentration of acetate, and the undissociated form of acetate, propionate, and butyrate, evidencing that SCFA are one of the mechanisms responsible for the decrease in numbers of these bacteria in the ceca of broiler chickens during growth, while they did not affect beneficial GIT bacteria such as *Lactobacillus* [58].

Nevertheless, there are many other mechanisms by which SCFA may be useful to avoid pathogen colonization in the GIT, such as the increased produc- tion of mucin by goblet cells that serves as a physical barrier against pathogens and contributes to their lower colonization [59, 60]. The effect of inulin dietary supplementation at different levels on mucin mRNA expression was evaluated at 21 and 42 days in broiler chickens, and it was found that dietary supplementa- tion of this prebiotic at 10 and 15 g/kg enhanced mucin mRNA expression in the jejunum both days [61]. Moreover, depending on the poultry species, the SCFA provide different levels of the of the total metabolic energy requirements, serving as the preferred energy source of colonocytes and stimulating intestinal integrity [54, 62].

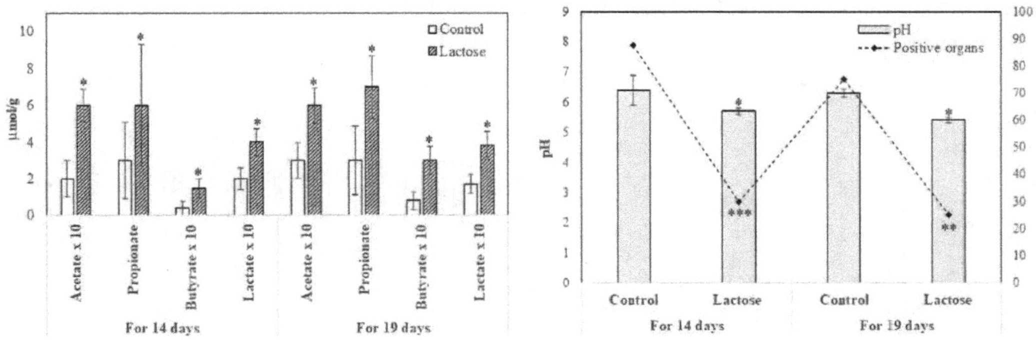

Figure 3. *Effect of dietary lactose (10%) during 14 and 19 days on cecal organic acid concentrations, pH of cecal contents, and Salmonella enteritidis organ invasion in Leghorn chicks. (*) (**) (***) significantly different from controls (P < 0.05) (P < 0.005) (P < 0.001), respectively. Data obtained from Ref. [57].*

On the other hand, the natural antipathogen activity of the intestinal microbiota in poultry has been documented by the Nurmi concept of competitive exclusion, also known as “bacterial antagonism” or “bacterial interference,” through which beneficial microorganisms compete with potentially pathogenic bacteria for limiting nutrients and attachment sites on the mucosa, or even by the production of bacteriocins like lactocin, helveticin, curvacin, nisin, or bifidocin, which may be destructive to various Gram-positive or Gram-negative intestinal pathogens, particularly *Salmonella*, *Campylobacter*, and *E. coli* [47, 63, 64]. It has been demonstrated that competitive exclusion is potentiated with prebiotics, since they promote growth of beneficial bacteria which are ubiquitous in the host animal and are capable to survive in GIT conditions.

In a study carried out in broiler chicks, the effect of treatments with dietary 7% lactose and 6.3 × 106 of anaerobic organisms, alone or in combination, on cecal colonization by *Salmonella typhimurium* (ST) after 10 and 15 days at different inoculum doses, was evaluated. The authors report that treatment with anaerobes without the addition of lactose did not effectively control cecal colonization of ST, while chick- ens treated with the combination of anaerobic organisms and lactose were resistant to cecal colonization by this pathogen, concluding that oral administration of only total anaerobes did not function well as competitive exclusion cultures [65].

In another similar study, the inhibitory effect of competitive exclusion and 0.1% concentration of FOS, singly and in combination, on *Salmonella enteritidis* SE colonization of chicks was investigated. Chicks received this pathogen at 7 or 21 days, and then birds from each group were slaughtered at 1, 7, and 14 days after for count of SE in cecal contents. Additionally, quantification of the major cecal microbiota was performed. Results from this study demonstrated the efficacy of CE on chicks 7 days post inoculation with SE, but this efficacy was not clearly demonstrated

21 days post inoculation, indicating that the efficacy of CE to reduce susceptibility to SE colonization is higher on young chicks, while FOS offered protection to chickens particularly in 21-day-old chicks (**Figure 4**). Nevertheless, when FOS was given in combination with a CE treatment, both in the 7- and 21-day-old chicks, a reduction in the number of SE per g of ceca was observed, so that low doses of FOS in the diet of chickens with a CE treatment may result in reduced susceptibility to *Salmonella* colonization. Regarding the intestinal microbiota, few

changes in *Bifidobacterium*, *Bacteroides*, and *Lactobacillus* in the cecal contents of treated groups were observed compared with the control group, both 7- and 21-day-old chicks, although when chickens were fed FOS for long times, *Bifidobacterium* and/or *Lactobacillus* of the intestinal flora may increase [39].

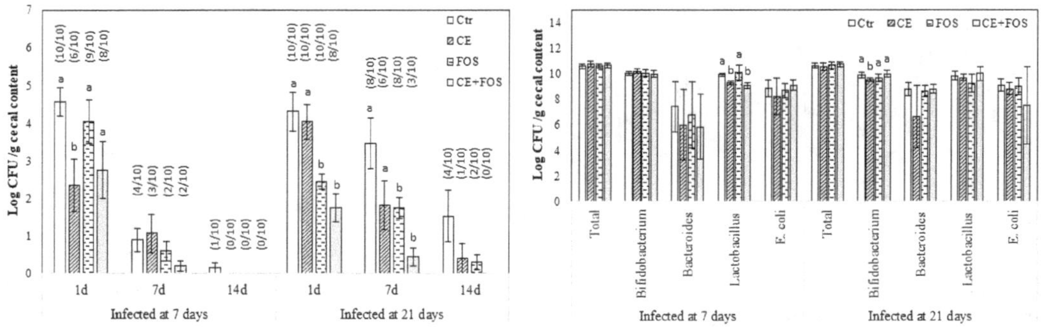

Figure 4. *Effect of competitive exclusion (CE) and FOS (0.1%) on recovery of Salmonella enteritidis from cecal contents of chicks infected at 7 and 21 days. Numbers in parentheses indicate the number of birds positive for SE/birds examined. The right graph shows the effect of these treatments on the major bacterial population of cecal microbiota. Data obtained from Ref. [39].*

Some pathogenic bacteria, such as *Salmonella spp., E. coli*, or *Vibrio cholerae*, have mannose-specific lectins (Type 1 fimbriae) on their surface, which recognize glycoprotein receptors rich in mannose on the intestinal cells of the host animal and are key to initiate attachment and colonization [45, 48]. Prebiotics, specifically MOS, can also reduce pathogen colonization by their direct union to the pathogen lectins, avoiding its attachment to the intestinal epithelial cells and, thus, passing through the GIT without colonizing.

This mechanism has been also corroborated in poultry species, both in vitro and in vivo, in two independent studies. In the first study, the in vitro effect of D-mannose, galactose, methyl-α-D-mannoside, and arabinose, on the adherence of *Salmonella typhimurium* to epithelial cells of the small intestine from 1-day-old chicks, was investigated. Authors showed that the small intestine of the chicken has receptors for bacteria with Type 1 fimbriae, and those fimbriae-positive strains of ST adhered significantly better than fimbriae-negative strains. They reported that adherence of ST to chicken small intestinal cells was inhibited more than 90% by methyl-α-D-mannoside and D-mannose and to a lesser extent by arabinose and galactose through the mechanism of blocking [66].

In the other study, the same effect of mannose was demonstrated in vivo. For that, 1-day-old broiler chickens were fed normal drinking water or drinking water supplemented with mannose (2.5% w/v) for 10 days. On day 3, birds were challenged orally with *S. typhimurium* (108 CFU), and then the cecal contents were examined on day 10. Results corroborated the blocking action of D-mannose, which could reduce the percentage of chickens colonized by ST from 78 to 28%, 82 to 21%, and 93 to 43%, in three trials [67].

In a more recent study, the ability of MOS from yeast cell walls to decrease the concentrations of enteric pathogens that express Type 1 fimbriae in poultry was evaluated. In the first part of this work, the ability of different enteric pathogens and coliforms to adhere to the MOS was measured in vitro, evaluating qualitatively if agglutination was modified with the presence of fructose, galactose, glucose, and mannose. Results of the agglutination test showed that 5 of

7 strains of *Escherichia coli* and 7 of 10 strains of *Salmonella typhimurium* and *S. enteritidis* agglutinated MOS. Other strains like *S. montevideo*, *S. give*, *S. kedougou*, and *S. dublin* also caused agglutination of MOS, but strains of *S. choleraesuis*, *S. pullorum*, and *Campylobacter* did not lead to agglutination. Nevertheless, agglutination of these Gram-negative bacteria could be inhibited by mannose and fructose, although it took much more fructose to observe the inhibitory responses than mannose. Authors reported that MOS had reduced cecal *S. typhimurium* con- centrations by about 25-fold compared to the control group; concentrations of coliforms also tended to be lower when MOS was added to the feed, while con- centrations of *lactobacilli*, *enterococci*, and anaerobic bacteria were not affected by treatment; concentration mean values are shown in the upper left graph of **Figure 5**. This tendency to reduce the salmonella concentration was observed during the time after the challenge with the pathogen, as shown in the upper right graph of **Figure 5**. The last part of the study consisted in a similar challenge using *S. dublin*, in which the percentage of prevalence this pathogen was lower in the MOS-treated groups (55.7%) than in the control group (89.8%), while concentrations of the other bacterial populations were not different. Since no changes in cecal parameters were observed with MOS addition, such as a major shift in bacterial populations or changes in pH or SCFA concentrations, which are known to affect salmonella, together with the in vitro agglutination results, authors conclude that adsorption of salmonellae by MOS could be a possible mode of action by which adhesion of these pathogens to the wall is avoided [68].

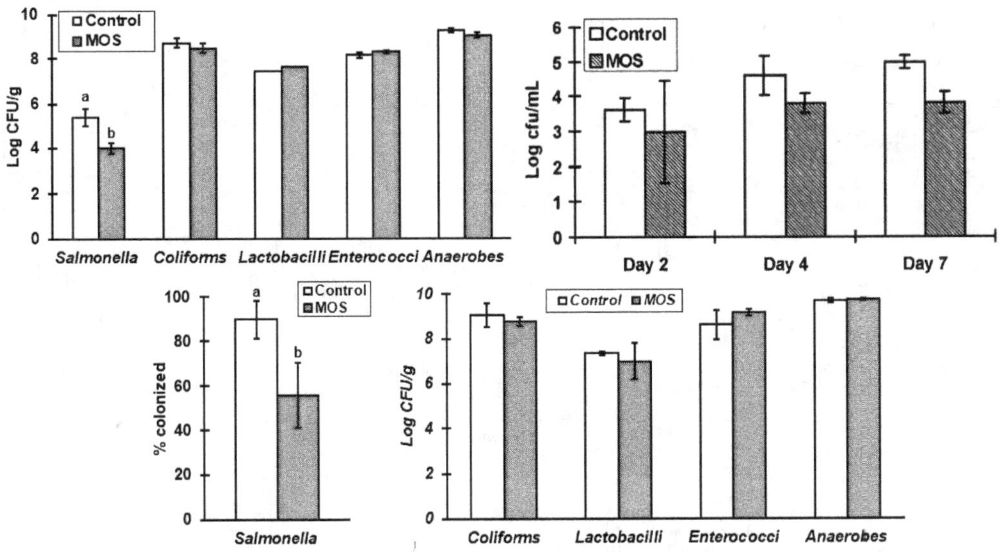

Figure 5. *Upper graphs: Effect of dietary added MOS on concentrations of different bacterial populations (left) and concentration of Salmonella typhimurium at different times after challenge in the ceca of chicks (right). Lower graphs show different bacterial populations (left) in the ceca of chicks challenged with S. dublin and (right) the percentage of birds from which S. dublin was recovered. Within bacterial populations, bars with different letters (a, b) are different (P < 0.05, n = 6). Data were obtained from Ref. [68].*

Intestinal morphology

Another proposed mechanism for health benefits of prebiotics is the improved intestinal morphological structure; several prebiotics have proven their capacity to modify positively intestinal morphology, both on macroscopic (intestinal length) and microscopic (size and density of villi and microvilli and crypt depth) structures of different sections of the intestine in poultry species [52, 69, 70]. Furthermore, an increased number of goblet cells of the intestinal villi have been reported after dietary administration of prebiotics; these specialized cells are responsible for secreting glycoprotein compounds, mainly mucins, which bind pathogenic microorganisms and reduce their adherence to the intestinal mucosa [71]. These morphological changes lead to a higher efficiency of nutrient absorption, since well-developed and functional enterocytes have been associated with increased absorptive area of the intestine [72] but also with an increased activity of the intestinal brush border enzymes and the nutrient transport systems [70, 73].

In turkeys, the dietary addition of two doses of a product based on MOS and β-glucans (1 and 2 lb./ton) on gastrointestinal tract development was evaluated through the measurement of ileal, jejunal, and duodenal morphology of turkey poults at 7 and 21 days of age. Data derived from this study suggest that feed supplemented with MOS and β-glucans could accelerate GIT maturation in turkey poults and was more pronounced in the ileum than in other portions of the small intestine. Ileum villus height, surface area, lamina propria thickness, and crypt depth were enhanced with the prebiotic treatment both on day 7 and 21, in a dose-dependent manner for many of the parameters evaluated, as it can be observed in **Figure 6**. In the jejunum results were consistently higher only for the highest dose of treatment (2 lb./ton) compared with the control group on both days, while in the duodenum results were better for the highest dose on day 7, although intestinal morphology of this intestinal section was not different on day 21. Furthermore, density of neutral, sialomucin, and sulfomucin goblet cells that were taken per intestinal section was also evaluated, showing a very similar tendency than the enteric morphometric

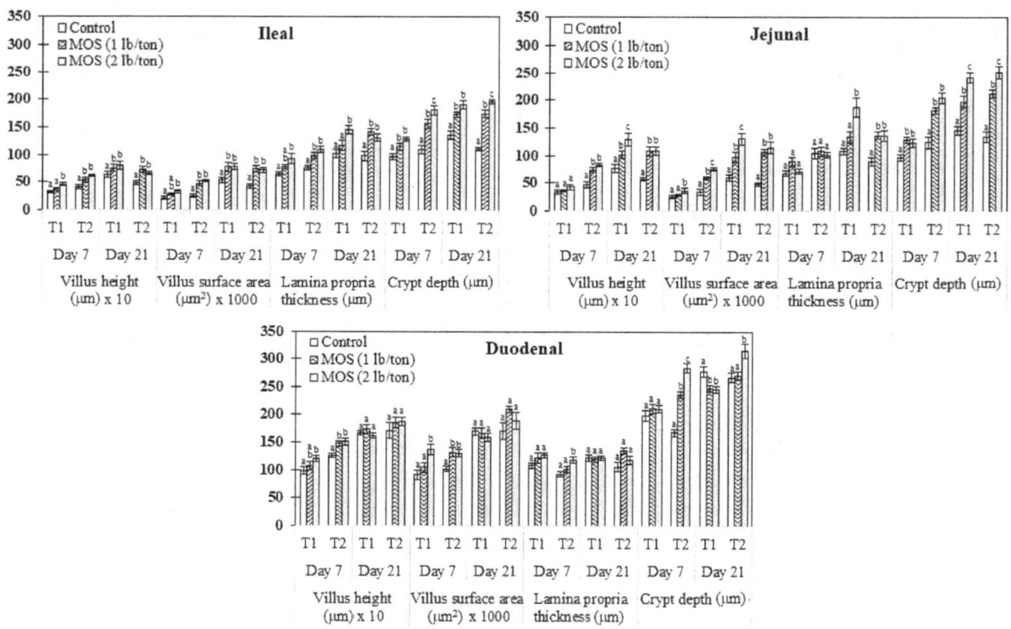

Figure 6. *Effect of MOS dietary addition at two different doses on the ileal, jejunal, and duodenal morphology of Turkey poults at 7 and 21 days of age. Within treatments, bars with different letters (a–c) are different (P < 0.05, N = 9 birds, n = 20 measurements/bird). Data were obtained from Ref. [74].*

evaluation, providing evidence of the immunostimulatory effects of this MOS- and β-glucan-based additive, because the numbers of neutral, sialomucin, and sulfomucin goblet cells in the GIT were increased in supplemented poults [74].

Studies have also been conducted whose results demonstrate the beneficial effect of prebiotics on changes at the macroscopic level. A study to evaluate and compare the effectiveness of adding inulin (1%) and oligofructose (1%) to the feed of broiler chickens was conducted, being one of the objectives to evaluate the intestinal length considering the influence of the bird sex. The experiment the experiment lasted 6 weeks, during which the productive parameters were also evaluated. Results from this study suggest that the longer the intestinal length, the better in nutrient absorption which resulted in a heavier body weight, showing correlation coefficients between intestinal length and body weight of 0.68 and 0.74 for the male and female birds, respectively, regardless of the treatments. Oligofructose-treated birds resulted to have a longer intestinal length, especially for the females, although inulin-fed birds also had a longer small intestine than control birds. There were no visible differences in villi density among the males, regardless of the treatments, while for females, the villi from inulin- and oligofructose-treated birds appeared to be denser than those of the controls [69].

Productive performance

Undoubtedly, one of the main objectives of the use of food additives in the poultry industry is the improvement of productive performance, a major indicator of poultry well-being that is directly tied to efficiency of nutrients utilization and, thus, to the profitability of production. In fact, replacement of antibiotics as growth promoters with prebiotics to observe improvements in poultry performance is the major reason for the researches [28]. As mentioned above, there is no exact mechanism of action for beneficial effects of prebiotics, so that stimulation of poultry performance results from the very complex interactions of all mechanisms previously described, for instance, by decreasing pathogen colonization, since it has been described that pathogens depress performance by interfering with nutrient digestion, absorption, and utilization; impairment of normal cellular function; negative impact on enzyme activity, epithelial integrity, and function; and diversion of energy for growth to immune response purposes [75]. Prebiotics can potentially stimulate growth performance through increased SCFA production in poultry, mostly acetate, propionate, and butyrate, since they are directly absorbed in the intestine and used as an energy source in tissues and besides that stimulate a higher metabolic activity in the intestine [76, 77]. SCFA may also serve as a potent regulator of insulin homeostasis in the chicken and carbohydrate metabolism, which stimulate the metabolic activity of striated muscle cells, possibly also having an influence on muscle protein synthesis and, thus, growth performance [78]. Beneficial effects of prebiotic on nutrient digestibility of poultry have been also reported, such as improved digestibility of crude protein, fat, dry matter, energy, and minerals [79, 80]. These results have been attributed to an increase in the beneficial microbiota, such as *Lactobacillus*, changes in the intestinal mucosal structure, and improved intestinal health, which result from the morphological changes in the intestine that lead to a higher efficiency of nutrient absorption and a better nutrient transport system, as discussed above [70, 81, 82].

Other prebiotic effects that might influence productive performance of poultry species are alterations on lipid metabolism and mineral absorption [83, 84]. Studies have demonstrated that prebiotic supplements have a positive effect on the mineral metabolism of Ca, P, Zn, Cu, and Fe [85–88], whose intake is influenced by factors such as the lower luminal pH that increases their solubility promotion and thus their passive absorption, changes in the intestinal mucosa and increased absorption surface area, elevated expression of Ca-binding proteins, release of bone-modulating factors, phytate degradation by probiotic bacteria enzymes, and improved overall intestinal health [89, 90]. On the other hand, although no exact mechanisms have been reported for the alteration on lipid metabolism caused by prebiotics, it has been demonstrated that intestinal microbiota play a role in maintaining lipid metabolism [91], so that the increase in bacterial numbers or a change in the composition of the intestinal microbiota might be related to the lipidic alterations. Studies have shown that prebiotics have a positive effect on lipid metabolism in poultry species, such as hypocholesterolemic effect both in serum and eggs, which has been attributed to many reasons. The enhanced production of SCFA results in inhibition of cholesterol biosynthesis in the liver, due to inhibition of the incorporation of colonic acetate into plasma lipids [92]. Another mechanism through which prebiotics may exert hypocholesterolemic effect is via bile acids, since they enter the small intestine and are absorbed and directed to the liver; however, during reabsorption, conjugated bile acids are exposed to intestinal microflora that hydro- lyze conjugated bile acids, making cholesterol unavailable for absorption into the circulation [53]. Although it has not been evaluated in poultry, other studies have also suggested that prebiotics may modify gene expression of lipogenic enzymes, with reduced concentration of plasma phospholipids, triacylglycerols, and lipo- proteins [93–96]. However, reports of prebiotics on the performance of poultry have been very variable, and often contradictory, as their effectiveness is strongly dependent on the type of prebiotic and the source, dose used, time of consumption, type of diet and interactions with other feed additives, administration route, animal characteristics, hygiene, husbandry conditions, and environmental stress [28, 50].

In a study carried out in White Leghorn hens, the performance parameters were measured to test two prebiotic treatments consumed for 4 weeks, oligofructose (1% w/w) and inulin (1% w/w), during the later part of the first laying cycle. Egg production, cumulative egg weight per bird, and average egg weight for each treatment were calculated weekly. Besides, body weight change, feed consumption, and feed conversion ratio were also monitored. Results showed that oligofructose and inulin increased weekly egg production by 13.35 and 10.73% and cumulative weekly egg weight per bird by 12.50 and 10.96%, respectively, as compared to the control group. Both prebiotics also improved the feed conversion ratio. Nevertheless, there were no differences in average egg weight, feed consumption, or the percentages of changes in live body weight after 4 weeks, as shown in **Figure 7** [85].

In another study, the effect of MOS at a dosage of 2 g/kg on growth performance and nutrient digestibility of two cereal-based diets (corn or wheat) in broiler chickens was evaluated, over an experimental period of 21 days. For that, body weight, feed intake, and feed conversion ration were measured at week 1 and weeks 2–3. Also, the ileal digestibility of nutrients was evaluated on day 21. Authors reported that dietary addition of MOS did not affect the body weight gain of birds but increased their feed intake during the first 7 days, while the feed conversion ratio also tended to increase with MOS, regardless of the type of cereal-based diet. Contrary, between 7 and 21 days, dietary MOS improved the growth performance of birds given the wheat-based diet compared to that of birds given the corn-based diet. Regarding the ileal digestibility of starch, the addition of MOS improved it and showed a high interaction with the type of cereal, indicating that this positive effect of MOS was more profound for the wheat diet than for the corn diet.

Enhancement of immune system

Currently, much research has focused in modulation of the immune system by the use of prebiotics, which results from the interaction of all the mechanisms mentioned above, so that it is not an isolated mechanism. A multitude of mechanisms and functions associated with the immunomodulatory effect of prebiotics have been reported, by the activation of genes and pathways implicated in immune processes [25]. It has been cited for many authors that the use of prebiotics in poultry diets improves bird's immunity through the selective growth of beneficial microbiota, resulting in an increased production of a variety of substances, such as bacteriocins and SCFA, that, in addition to being able to inhibit growth of pathogens, play a role in signaling pathway of immune system [97–99].

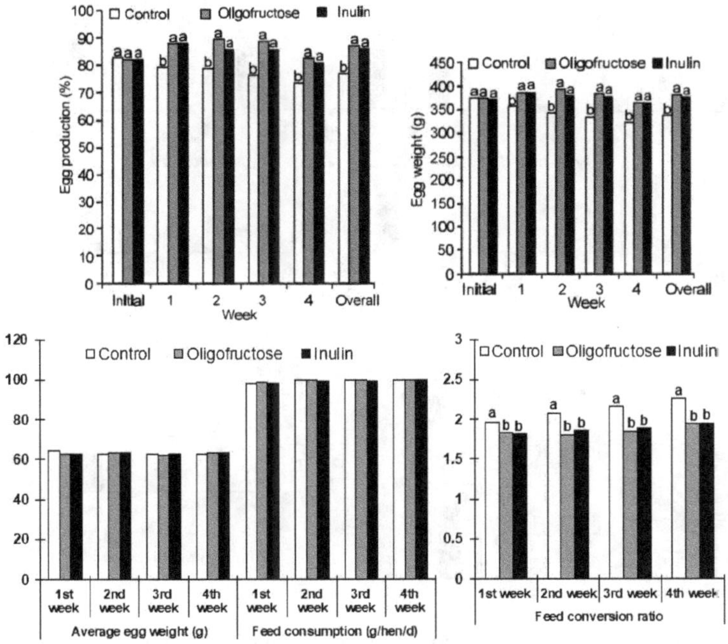

Figure 7. *Performance parameters as affected by dietary oligofructose and inulin in laying hens. Within treatments, bars with different letters (a, b) are different (P < 0.05, n = 10). Data were obtained from Ref. [85].*

The chicken gut microbiota, especially *Lactobacillus* and *Bifidobacterium*, has also been reported to modulate intestinal gene expression, T cell-mediated immu- nity, and accelerated intestinal immune system maturation, by influencing the intestinal epithelium to produce antimicrobial peptides and cytokines such as IL-12, IFN-γ, IL-10, IL-1β, and TNF-α; modulating the immune system through enhancement of phagocytosis and proliferation of immune cells such as macrophages and monocytes; enhancing production of IgA, IgM, and IgG, reactive oxygen species, and reactive nitrogen species; and proliferating natural killer cells, CD3, CD4, and CD8 T cells [25, 47, 75, 100]. Some prebiotics have shown to increase the production of secretory IgA in the intestine, which inhibits the attachment and penetration of bacteria in the lumen, increases the production of mucus, and prevents inflamma- tion that could cause epithelial tissue damage [40, 42].

On the other hand, as previously mentioned, prebiotics can inhibit pathogen colonization, decreasing detrimental molecules produced by pathogenic bacteria, which act as exogenous signals called pathogen-associated molecular patterns (PAMPs). These PAMPs can be recognized by pattern recognition receptors (PRR)

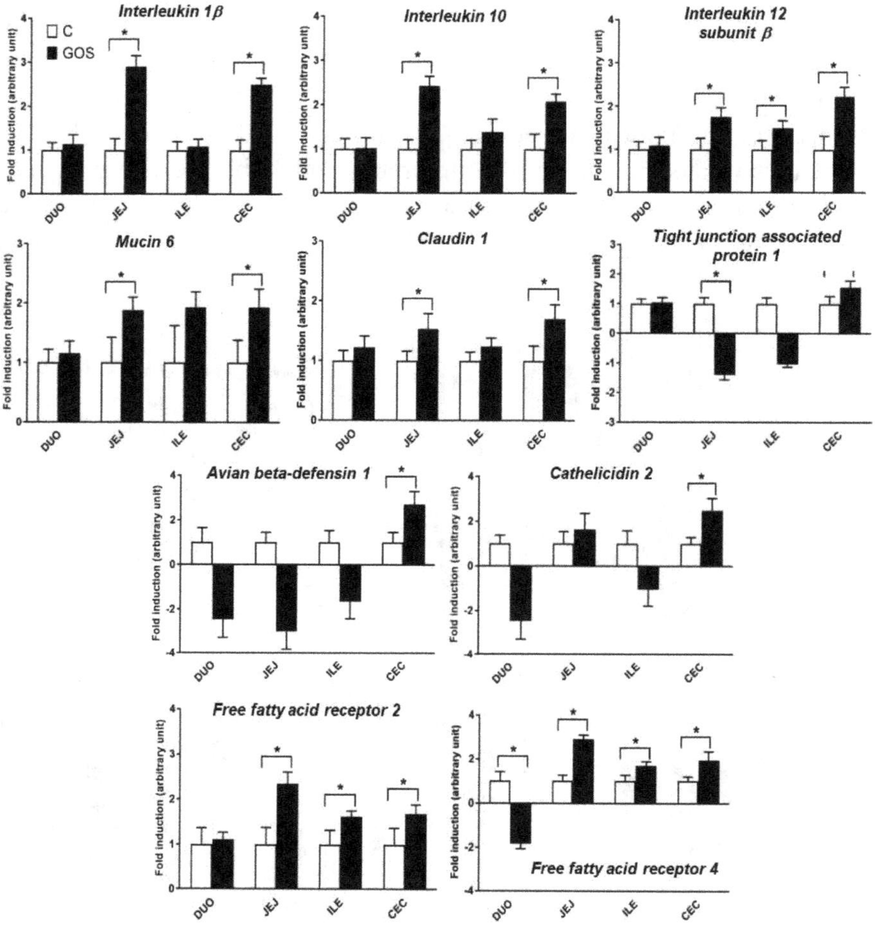

Figure 8. *Relative mRNA expression of immune-related (cytokines and host defense peptides) and barrier function (mucin and free fatty acid receptors) genes in different segments of intestinal mucosa in chickens injected in ovo with GOS. Asterisk indicates pair-wise significant differences ($P < 0.05$, $n = 10$). Graphs were obtained from Ref. [32].*

expressed on the surface of epithelial cells, macrophages, mast cells, and dendritic cells, including toll-like receptors and NOD-like receptors, and once recognized are activated, producing cytokines for the regulation of further innate immune responses [45]. Although little data show direct effects of prebiotics on immune function, some studies have indicated that prebiotics have an improved response to salmonella vaccine, which could be because prebiotics can act as nonpathogenic antigens themselves, being recognized by receptors of immune cells, which consequently modulate host immunity beneficially [45, 101].

Immunomodulatory effect of prebiotics has been evaluated in vitro and in vivo. For instance, in an in vitro study, the influence of a FOS-inulin prebiotic at 200 µg/ mL on the ability of the chicken macrophage HD11 cell line to phagocytose and kill *Salmonella enteritidis* was tested. The influence on their ability to express selected inflammatory cytokines and chemokines, such as for IL-1β, lipopolysaccharide- induced TNF factor (LITAF), C-C motif chemokine ligand 4 (CCL4) and inducible nitric oxide synthase (iNOS), and nitric oxide production, was also evaluated.

Results showed that phagocytosis of SE by macrophages was not affected with the FOS-inulin treatment, but there was a significant reduction of viable intracellular SE in macrophages treated with the prebiotic. On the other hand, prebiotic treat- ment did not influence the nitric oxide production, thus suggesting that the FOS- inulin-mediated bacterial clearance was not mediated by this compound. Similarly, prebiotic treatment has no influence on expression of LITAF, CCL4, nor iNOS; however, IL-1β expression was significantly lower in macrophages treated with FOS-inulin, suggesting that this prebiotic can modulate the innate immune system by preventing IL-1β-associated macrophage cell death [102].

In a more recent study, GOS prebiotic was in ovo administered to evaluate the modulation of chicken intestinal microflora and demonstrate the molecular responses of the host animal. The study was performed on meat-type chickens, with 3.5 mg GOS delivered by in ovo injection on day 12 of egg incubation, and the analysis of microbial communities and mucosal gene expression was performed at day 42 post-hatching. Results showed that GOS increased the relative abundance of *Bifidobacterium* in the cecum. GOS also upregulated cytokine and barrier function genes in the jejunum and cecum, host defense peptides in the cecum, and free fatty acid receptors in the jejunum, ileum, and cecum, as shown in **Figure 8**, so that it has been demonstrated that GOS prebiotics have a bifidogenic effect in adult chickens, modulating gene expression related to intestinal immune responses and gut barrier function [32].

Conclusion

Due to the great concern about AMR, it is imperative to avoid the use of antibiotics as growth promoters and look for effective alternatives that can help poultry production to improve the welfare of the poultry birds, performance, and production costs. As a result of all the studies that have been carried out, we can conclude that dietary addition of prebiotics has a positive effect on poultry production, highlighting the improvement of intestinal health, immune system, control of pathogens, and performance parameters, which are achieved through a series of interrelated mechanisms and interactions involving interactions between the organisms of the intestinal microbiota and the microbiota with the host animal. Nevertheless, effectiveness of prebiotics will depend on many factors, like the type of supplement, doses, composition of the basal diet, animal characteristics, and environmental condition, showing variable effects on poultry species, so that it is necessary to determine conditions under which prebiotics are effective and elucidate the mechanisms(s) of action involved, ensuring their effective use. Many studies have elucidated mechanisms involved in the effectiveness of prebiotics, but we believe that there is still information that remains to be discovered or that must be confirmed, including the identification of new prebiotics and their application in the poultry industry, for which we can take hold of the emerging analysis technologies.

Acknowledgements

This research was supported by the Arkansas Bioscience Institute under the project Development of an avian model for evaluation early enteric microbial colonization on the gastrointestinal tract and immune function. The authors thank the CONACyT for the doctoral grant number 270730.

Author details

Bruno Solis-Cruz[1], Daniel Hernandez-Patlan[1], Billy M. Hargis[2] and Guillermo Tellez[2*]

1 Faculty of Superior Studies Cuautitlán, Multidisciplinary Research Unit L5, National Autonomous University of Mexico, Cuautitlán Izcalli, Mexico State, Mexico

2 Department of Poultry Science, University of Arkansas, Fayetteville, Arkansas, USA

*Address all correspondence to: gtellez@uark.edu

References

[1] Prestinaci F, Pezzotti P, Pantosti A. Antimicrobial resistance: A global multifaceted phenomenon. Pathogens and Global Health, Taylor & Francis. 2015;**109**(7):309-318

[2] FAO. The FAO Action Plan on Antimicrobial Resistance 2016-2020. Rome, Italy; 2016

[3] OIE. The OIE Strategy on Antimicrobial Resistance and the Prudent Use of Antimicrobials. Paris, France; 2016

[4] WHO. Global Action Plan on Antimicrobial Resistance. Geneva, Switzerland; 2015

[5] WHO. WHO guidelines on use of medically important antimicrobials in food-producing animals: Web Annex A: Evidence base. 2017

[6] (EC) EC. Regulation No 1831/2003 of the European Parliament and Council of 22 September 2003 on additives for use in animal nutrition. Official Journal of the European Union. 2003;**268**:29-43

[7] US Food, Drug Administration D. New animal drugs and new animal drug combination products, administered in or on medicated feed or drinking water of food-producing animals: recommendations for drug sponsors for voluntarily aligning product use conditions with GFI# 209. 2014

[8] Aarestrup FM, Seyfarth AM, Emborg HD, Pedersen K, Hendriksen RS, Bager F. Effect of abolishment of the use of antimicrobial agents for growth promotion on occurrence of antimicrobial resistance in fecal enterococci from food animals in Denmark. Antimicrobial Agents and Chemotherapy. 2001;**45**(7):2054-2059

[9] Hao H, Cheng G, Iqbal Z, Ai X, Hussain HI, Huang L, et al. Benefits and risks of antimicrobial use in food-producing animals. Frontiers in Microbiology, Frontiers. 2014;**5**:288

[10] Marshall BM, Levy SB. Food animals and antimicrobials: Impacts on human health. Clinical microbiology reviews. The American Society for Microbiology. 2011;**24**(4):718-733

[11] Cervantes HM. Antibiotic-free poultry production: Is it sustainable? Journal of Applied Poultry Research, Poultry Science Association, Inc. 2015;**24**(1):91-97

[12] Siekkinen KM, Heikkilä J, Tammiranta N, Rosengren H. Measuring the costs of biosecurity on poultry farms: A case study in broiler production in Finland. Acta Veterinaria Scandinavica, BioMed Central. 2012;**54**(1):12

[13] FAO. Poultry Development Review. 2013. p. 120. Available from: http:// www.fao.org/3/i3531e/i3531e00.htm

[14] Agyare C, Boamah VE, Zumbi CN, Osei FB. Antibiotic Use in Animal Production and its Effects on Bacterial Resistance. Rijeka, Croatia: IntechOpen; 2018

[15] Mehdi Y, Létourneau-Montminy MP, Gaucher ML, Chorfi Y, Suresh G, Rouissi T, et al. Use of antibiotics in broiler production: Global impacts and alternatives. Animal Nutrition, Elsevier. 2018;**4**(2):170-178

[16] JETACAR. The use of antibiotics in food-producing animals: Antibiotic- resistant bacteria in animals and humans. In: Commonwealth Department of Health and Aged Care and the Commonwealth Department of Agriculture, Fisheries and Forestry, editor. Report of the Joint Expert Advisory Committee on Antibiotic Resistance (JETACAR). Australia: Department of Health of the Australian Government; 1999

[17] Maron DF, Smith TJ, Nachman KE. Restrictions on antimicrobial use in food animal production: An international regulatory and economic survey. Globalization and Health, BioMed Central. 2013;**9**(1):48

[18] Castanon J. History of the use of antibiotic as growth promoters in European poultry feeds. Poultry Science, Oxford University Press. 2007;**86**(11):2466-2471

[19] Diarra MS, Malouin F. Antibiotics in Canadian poultry productions and anticipated alternatives. Frontiers in Microbiology, Frontiers. 2014;**5**:282

[20] Cox LA Jr, Popken DA. Quantifying human health risks from virginiamycin used in chickens. Risk Analysis: An International Journal, Wiley Online Library. 2004;**24**(1):271-288

[21] Cox LA Jr, Popken DA. Quantifying potential human health impacts of animal antibiotic use: Enrofloxacin and macrolides in chickens. Risk Analysis: An International Journal, Wiley Online Library. 2006;**26**(1):135-146

[22] Sarangi NR, Babu L, Kumar A, Pradhan C, Pati P, Mishra J. Effect of dietary supplementation of prebiotic, probiotic, and synbiotic on growth performance and carcass characteristics of broiler chickens. Veterinary World. 2016;**9**(3):313

[23] Bednarczyk M, Stadnicka K, Kozlowska I, Abiuso C, Tavaniello S, Dankowiakowska A, et al. Influence of different prebiotics and mode of their administration on broiler chicken performance. Animal, Cambridge University Press. 2016;**10**(8):1271-1279

[24] Ganguly S. Supplementation of prebiotics, probiotics and acids on immunity in poultry feed: A brief review. World's Poultry Science Journal, Cambridge University Press. 2013;**69**(3):639-648

[25] Pourabedin M, Zhao X. Prebiotics and gut microbiota in chickens. FEMS Microbiology Letters, Oxford University Press. 2015;**362**(15):fnv122

[26] Ricke S. Potential of fructooligosaccharide prebiotics in alternative and nonconventional poultry production systems. Poultry Science, Poultry Science Association, Inc. 2015;**94**(6):1411-1418

[27] Samal L, Behura N. Prebiotics: An emerging nutritional approach for improving gut health of livestock and poultry. Asian Journal of Animal and Veterinary. 2015;**10**:724-739

[28] Adhikari PA, Kim WK. Overview of prebiotics and probiotics: Focus on performance, gut health and immunity—A review. Annals of Animal Science, De Gruyter Open. 2017;**17**(4):949-966

[29] Gibson GR, Hutkins R, Sanders ME, Prescott SL, Reimer RA, Salminen SJ, et al. Expert consensus document: The International Scientific Association for Probiotics and Prebiotics (ISAPP) consensus statement on the definition and scope of prebiotics. Nature Reviews Gastroenterology and Hepatology, Nature Publishing Group. 2017;**14**(8):491

[30] Ricke S. Impact of prebiotics on poultry production and food safety. The Yale Journal of Biology and Medicine. 2018;**91**(2):151-159

[31] Józefiak D, Kaczmarek S, Rutkowski A. A note on the effects of selected prebiotics on the performance and ileal microbiota of broiler chickens. Journal of Animal and Feed Sciences. 2008;**17**:392-397

[32] Slawinska A, Dunislawska A, Plowiec A, Radomska M, Lachmanska J, Siwek M, et al. Modulation of microbial communities and mucosal gene expression in chicken intestines after galactooligosaccharides delivery In Ovo. PloS One, Public Library of Science. 2019;**14**(2):e0212318

[33] Siwek M, Slawinska A, Stadnicka K, Bogucka J, Dunislawska A, Bednarczyk M. Prebiotics and synbiotics-in ovo delivery for improved lifespan condition in chicken. BMC Veterinary Research, BioMed Central. 2018;**14**(1):402

[34] Piray A, Kermanshahi H, Tahmasbi A, Bahrampour J. Effects of cecal cultures and aspergillus meal prebiotic (Fermacto) on growth performance and organ weights of broiler chickens. International Journal of Poultry Science. 2007;**6**(5):340-344

[35] Li X, Qiang L, Xu C, et al. Effects of supplementation of fructooligosaccharide and/or *Bacillus subtilis* to diets on performance and on intestinal microflora in broilers. Archives Animal Breeding, Copernicus GmbH. 2008;**51**(1):64-70

[36] Sims M, Dawson K, Newman K, Spring P, Hoogell D. Effects of dietary mannan oligosaccharide, bacitracin methylene disalicylate, or both on the live performance and intestinal microbiology of turkeys. Poultry Science, Oxford University Press. Oxford, UK. 2004;**83**(7):1148-1154

[37] Micciche AC, Foley SL, Pavlidis HO, McIntyre DR, Ricke SC. A review of prebiotics against salmonella in poultry: Current and future potential for microbiome research applications. Frontiers in Veterinary Science. Frontiers Media SA. 2018;**5**:1-11

[38] Bailey J, Blankenship L, Cox N. Effect of fructooligosaccharide on salmonella colonization of the chicken intestine. Poultry Science, Oxford University Press. Oxford, UK. 1991;**70**(12):2433-2438

[39] Fukata T, Sasai K, Miyamoto T, Baba E. Inhibitory effects of competitive exclusion and fructooligosaccharide, singly and in combination, on salmonella colonization of chicks. Journal of Food Protection, International Association for Food Protection. 1999;**62**(3):229-233

[40] Kim GB, Seo Y, Kim C, Paik I. Effect of dietary prebiotic supplementation on the performance, intestinal microflora, and immune response of broilers. Poultry Science. 2011;**90**:75-82

[41] Kim SA, Jang MJ, Kim SY, Yang Y, Pavlidis HO, Ricke SC. Potential for prebiotics as feed additives to limit foodborne campylobacter establishment in the poultry gastrointestinal tract. Frontiers in Microbiology, Frontiers Media SA. 2019:10

[42] Yadav AS, Kolluri G, Gopi M, Karthik K, Singh Y. Exploring alternatives to antibiotics as health promoting agents in poultry—A review. The Journal of Experimental Biology. 2016;**4**(3S-10.18006):3S

[43] Clavijo V, Flórez MJV. The gastrointestinal microbiome and its association with the control of pathogens in broiler chicken production: A review. Poultry Science, Oxford University Press. 2018;**97**(3):1006

[44] Hajati H, Rezaei M. The application of prebiotics in poultry production. International Journal of Poultry Science. 2010;**9**(3):298-304

[45] Teng PY, Kim WK. Roles of prebiotics in intestinal ecosystem of broilers. Frontiers in Veterinary Science. Frontiers Media SA. 2018;**5**:1-8

[46] Shang Y, Kumar S, Oakley B, Kim WK. Chicken gut microbiota: Importance and detection technology. Frontiers in Veterinary Science, Frontiers. 2018;**5**:254

[47] Buclaw M. The use of inulin in poultry feeding: A review. Journal of Animal Physiology and Animal Nutrition, Wiley Online Library. 2016;**100**(6):1015-1022

[48] Gaggia F, Mattarelli P, Biavati B. Probiotics and prebiotics in animal feeding for safe food production. International Journal of Food Microbiology, Elsevier. 2010;**141**:S15-S28

[49] Guillermo T, Andrea L, Juan DL, Xochitl HV, Billy MH, Todd C. Food- producing animals and their health in relation to human health. Microbial Ecology in Health and Disease, Taylor & Francis. 2015;**26**(1):25876

[50] Patterson J, Burkholder K. Application of prebiotics and probiotics in poultry production. Poultry Science, Oxford University Press. Oxford, UK. 2003;**82**(4):627-631

[51] Gibson GR, Roberfroid MB. Dietary modulation of the human colonic microbiota: Introducing the concept of prebiotics. Journal of Nutrition, Oxford University Press. 1995;**125**(6):1401-1412

[52] Xu Z, Hu C, Xia M, Zhan X, Wang M. Effects of dietary fructooligosaccharide on digestive enzyme activities, intestinal microflora and morphology of male broilers. Poultry Science, Oxford University Press. Oxford, UK. 2003;**82**(6):1030-1036

[53] Shang H, Hu T, Lu Y, Wu H. Effects of inulin on performance, egg quality, gut microflora and serum and yolk cholesterol in laying hens. British Poultry Science, Taylor & Francis. 2010;**51**(6):791-796

[54] Józefiak D, Rutkowski A, Martin S. Carbohydrate fermentation in the avian ceca: A review. Animal Feed Science and Technology, Elsevier. 2004;**113**(1-4):1-15

[55] Russell J. Another explanation for the toxicity of fermentation acids at low pH: Anion accumulation versus uncoupling. Journal of Applied Bacteriology, Wiley Online Library. 1992;**73**(5):363-370

[56] Van Immerseel F, Russell J, Flythe M, Gantois I, Timbermont L, Pasmans F, et al. The use of organic acids to combat salmonella in poultry: A mechanistic explanation of the efficacy. Avian Pathology, Taylor & Francis. 2006;**35**(3):182-188

[57] Tellez G, Dean C, Corrier D, Deloach J, Jaeger L, Hargis B. Effect of dietary lactose on cecal morphology, pH, organic acids, and *Salmonella enteritidis* organ invasion in Leghorn chicks. Poultry Science, Oxford University Press. Oxford, UK. 1993;**72**(4):636-642

[58] Van der Wielen PW, Biesterveld S, Notermans S, Hofstra H, Urlings BA, van Knapen F. Role of volatile fatty acids in development of the cecal microflora in broiler chickens during growth. The American Society for Microbiology. 2000;**66**(6):2536-2540

[59] Tellez G, Higgins S, Donoghue A, Hargis B. Digestive physiology and the role of microorganisms. Journal of Applied Poultry Research, Oxford University Press. Oxford, UK. 2006;**15**(1):136-144

[60] Willemsen L, Koetsier M, Van Deventer S, Van Tol E. Short chain fatty acids stimulate epithelial mucin 2 expression through differential effects on prostaglandin E1 and E2 production by intestinal myofibroblasts. Gut, BMJ Publishing Group. 2003;**52**(10):1442-1447

[61] Huang Q , Wei Y, Lv Y, Wang Y, Hu T. Effect of dietary inulin supplements on growth performance and intestinal immunological parameters of broiler chickens. Livestock Science, Elsevier. 2015;**180**:172-176

[62] Alloui MN, Szczurek W, 'Swikatkiewicz S. The usefulness of prebiotics and probiotics in modern poultry nutrition: A review. Annals of Animal Science, Versita 2013;13(1):17-32.

[63] Chen Y-S, Srionnual S, Onda T, Yanagida F. Effects of prebiotic oligosaccharides and trehalose on growth and production of bacteriocins by lactic acid bacteria. Letters in Applied Microbiology, Wiley Online Library. 2007;**45**(2):190-193

[64] Munoz M, Mosquera A, Almeciga-DiazC,MelendezA,SanchezO. Fructooligosaccharides metabolism and effect on bacteriocin production in lactobacillus strains isolated from ensiled corn and molasses. Anaerobe, Elsevier. 2012;**18**(3):321-330

[65] Ziprin RL, Corrier DE, Hinton A Jr, Beier RC, Spates GE, DeLoach JR, et al. Intracloacal *Salmonella typhimurium* infection of broiler chickens: Reduction of colonization with anaerobic organisms and dietary lactose. Avian Diseases, JSTOR. 1990:749-753

[66] Oyofo B, Droleskey R, Norman J, Mollenhauer H, Ziprin R, Corrier D, et al. Inhibition by mannose of in vitro colonization of chicken small intestine by *Salmonella typhimurium*. Poultry Science, Oxford University Press. Oxford, UK. 1989;**68**(10):1351-1356

[67] Oyofo B, DeLoach J, Corrier D, Norman J, Ziprin R, Mollenhauer H. Prevention of *Salmonella typhimurium* colonization of broilers with D-mannose. Poultry Science, Oxford University Press. Oxford, UK. 1989;**68**(10):1357-1360

[68] Spring P, Wenk C, Dawson K, Newman K. The effects of dietary mannaoligosaccharides on cecal parameters and the concentrations of enteric bacteria in the ceca of salmonella-challenged broiler chicks. Poultry Science, Oxford University Press. Oxford, UK. 2000;**79**(2):205-211

[69] Chen TC. Effects of adding chicory fructans in feed on broiler growth performance, serum cholesterol and intestinal length. International Journal of Poultry Science. 2003;**2**(3):214-219

[70] Rehman H, Rosenkranz C, Böhm J, Zentek J. Dietary inulin affects the morphology but not the sodium- dependent glucose and glutamine transport in the jejunum of broilers. Poultry Science, Oxford University Press. Oxford, UK. 2007;**86**(1):118-122

[71] Johansson ME, Phillipson M, Petersson J, Velcich A, Holm L, Hansson GC. The inner of the two Muc2 mucin-dependent mucus layers in colon is devoid of bacteria. Proceedings of the National Academy of Sciences. 2008;**105**(39):15064-15069

[72] Maiorka A, Silva A, Santin E, Dahlke F, Bruno L, Boleli I, et al. Effect of broiler breeder age and glutamine supplementation on the development of the intestinal mucosa of 7-day-old chicks. Brazilian Journal of Poultry Science, SciELO Brasil. 2016;**18**(1):17-22

[73] Bogucka J, Dankowiakowska A, Elminowska-Wenda G, Sobolewska A, JankowskiJ,SzpindaM,etal.Performance and small intestine morphology and ultrastructure of male broilers injected in ovo with bioactive substances. Annals of Animal Science, De Gruyter Open. 2017;**17**(1):179-195

[74] De Los Santos FS, Donoghue A, Farnell M, Huff G, Huff W, Donoghue D. Gastrointestinalmaturationisaccelerated in Turkey poults supplemented with a mannan-oligosaccharide yeast extract (Alphamune). Poultry Science, Oxford University Press. Oxford, UK. 2007;**86**(5):921-930

[75] Ajuwon K. Toward a better understanding of mechanisms of probiotics and prebiotics action in poultry species. Journal of Applied Poultry Research, Oxford University Press. 2015;**25**(2):277-283

[76] Chapman M, Grahn M, Boyle M, Hutton M, Rogers J, Williams N. Butyrate oxidation is impaired in the colonic mucosa of sufferers of quiescent ulcerative colitis. Gut, BMJ Publishing Group. 1994;**35**(1):73-76

[77] De Vadder F, Kovatcheva- Datchary P, Goncalves D, Vinera J, Zitoun C, Duchampt A, et al. Microbiota-generated metabolites promote metabolic benefits via gut- brain neural circuits. Cell, Elsevier. 2014;**156**(1-2):84-96

[78] Matis G, Kulcsar A, Turowski V, Febel H, Neogrády Z, Huber K. Effects of oral butyrate application on insulin signaling in various tissues of chickens. Domestic Animal Endocrinology, Elsevier. 2015;**50**:26-31

[79] Yang Y, Iji P, Kocher A, Mikkelsen L, Choct M. Effects of dietary mannanoligosaccharide on growth performance, nutrient digestibility and gut development of broilers given different cereal-based diets. Journal of Animal Physiology and Animal Nutrition, Wiley Online Library. 2008;**92**(6):650-659

[80] Yun W, Lee D, Choi Y, Kim I, Cho J. Effects of supplementation of probiotics and prebiotics on growth performance, nutrient digestibility, organ weight, fecal microbiota, blood profile, and excreta noxious gas emissions in broilers. Journal of Applied Poultry Research, Oxford University Press. 2017;**26**(4):584-592

[81] Wang X, Lin X, Zhang L, Du Y, Bai X, Shu S. Effect of oligo-chitosan on broiler performance, small intestine structure and muscle mineral element concentration. Journal of the Chinese Cereals and Oils Association. 2005;**20**:83-88

[82] Li X, Piao X, Kim S, Liu P, Wang L, Shen Y, et al. Effects of chito- oligosaccharide supplementation on performance, nutrient digestibility, and serum composition in broiler chickens. Poultry Science, Oxford University Press. Oxford, UK. 2007;**86**(6):1107-1114

[83] Swiatkiewicz S, Arczewska- Wlosek A. Prebiotic fructans and organic acids as feed additives improving mineral availability. World's Poultry Science Journal, Cambridge University Press. 2012;**68**(2):269-279

[84] Macfarlane GT, Steed H, Macfarlane S. Bacterial metabolism and health-related effects of galacto- oligosaccharides and other prebiotics. Journal of Applied Microbiology, Wiley Online Library. 2008;**104**(2):305-344

[85] Chen Y, Chen T. Mineral utilization in layers as influenced by dietary oligofructose and inulin. International Journal of Poultry Science, Asian Network for Scientific Information. 2004;**3**(7):442-445

[86] Scholz-Ahrens KE, Schrezenmeir J. Inulin and oligofructose and mineral metabolism: The evidence from animal trials. The Journal of Nutrition, Oxford University Press. 2007;**137**(11):2513S-2523S

[87] Ortiz L, Rodriguez M, Alzueta C, Rebolé A, Trevino J. Effect of inulin on growth performance, intestinal tract sizes, mineral retention and tibial bone mineralisation in broiler chickens. British Poultry Science, Taylor & Francis. 2009;**50**(3):325-332

[88] Swikatkiewicz S, Koreleski J, Arczewska-Wlosek A. Effect of prebiotic fructans and organic acids on mineral retention in laying hens. Acta Agriculturae Scandinavica Section A, Taylor & Francis. 2010;**60**(2):125-128

[89] Scholz-Ahrens KE, Schrezenmeir J. Inulin, oligofructose and mineral metabolism—Experimental data and mechanism. British Journal of Nutrition, Cambridge University Press. 2002;**87**(S2):S179-S186

[90] Whisner CM, Castillo LF. Prebiotics, bone and mineral metabolism. Calcified Tissue International, Springer. 2018;**102**(4):443-479

[91] Ghazalpour A, Cespedes I, Bennett BJ, Allayee H. Expanding role of gut microbiota in lipid metabolism. Current Opinion in Lipidology, NIH Public Access. 2016;**27**(2):141

[92] Wolever T, Spadafora PJ, Cunnane SC, Pencharz PB. Propionate inhibits incorporation of colonic [1, 2-13C] acetate into plasma lipids in humans. The American Journal of Clinical Nutrition, Oxford University Press. 1995;**61**(6):1241-1247

[93] Samanta A, Jayapal N, Senani S, Kolte A, Sridhar M. Prebiotic inulin: Useful dietary adjuncts to manipulate the livestock gut microflora. Brazilian Journal of Microbiology, SciELO Brasil. 2013;**44**(1):1-14

[94] Delzenne NM, Kok NN. Biochemical basis of oligofructose- induced hypolipidemia in animal models. The Journal of Nutrition, Oxford University Press. 1999;**129**(7):1467S-1470S

[95] Fiordaliso M, Kok N, Desager J-P, Goethals F, Deboyser D, Roberfroid M, et al. Dietary oligofructose lowers triglycerides, phospholipids and cholesterol in serum and very low-density lipoproteins of rats. Lipids, Wiley Online Library. 1995;**30**(2):163-167

[96] Kok N, Roberfroid M, Robert A, Delzenne N. Involvement of lipogenesis in the lower VLDL secretion induced by oligofructose in rats. British Journal of Nutrition, Cambridge University Press. 1996;**76**(6):881-890

[97] Lopetuso L, Giorgio M, Saviano A, Scaldaferri F, Gasbarrini A, Cammarota G. Bacteriocins and bacteriophages: Therapeutic weapons for gastrointestinal diseases? International Journal of Molecular Sciences, Multidisciplinary Digital Publishing Institute. 2019;**20**(1):183

[98] Vinolo MA, Rodrigues HG, Nachbar RT, Curi R. Regulation of inflammation by short chain fatty acids. Nutrients, Molecular Diversity Preservation International. 2011;**3**(10):858-876

[99] Vamanu E, Vamanu A. The influence of prebiotics on bacteriocin synthesis using the strain lactobacillus paracasei CMGB16. African Journal of Microbiology Research, Academic Journals. 2010;**4**(7):534-537

[100] Patterson JA. The Commensal Microbiota. Direct-Fed Microbials and Prebiotics for Animals. New York, USA: Springer; 2012. pp. 3-11

[101] Steed H, Macfarlane S. Mechanisms of prebiotic impact on health. In: Charalampopoulos D, Rastall RA, editors. Prebiotics and Probiotics Science and Technology. New York, NY: Springer; 2009. pp. 135-161

[102] Babu US, Sommers K, Harrison LM, Balan KV. Effects of fructooligosaccharide-inulin on salmonella-killing and inflammatory gene expression in chicken macrophages. Veterinary Immunology and Immunopathology, Elsevier. 2012;**149**(1-2):92-96

13

Functional Attributes and Health Benefits of Novel Prebiotic Oligosaccharides Derived from Xylan, Arabinan, and Mannan

Bradley A. Saville and Sandra H. Saville

Abstract

Prebiotic oligosaccharides are produced from many different sources, with sub- stantial differences in chemical structure, bonds between subunits, and degree of polymerization. These structural differences can materially affect microbial utilization and the dose required for efficacy. Most prebiotic oligosaccharides are based on subunits comprised of 6-carbon sugars such as glucose/fructose and alpha bonds.

Newer/novel oligosaccharides are derived from 5 carbon sugars and/or connected via beta bonds. Clinical trials with xylooligosaccharides, arabinoxylanoligosaccha-rides, and mannooligosaccharides have shown improvements in lipids, cholesterol, management of blood glucose, weight management, and laxation, at doses typically ranging from 1 to 4 g per day. Mannooligosaccharides are also showing promise for animal health, with the potential to reduce antibiotic use. These novel prebiotics are showing promise due to greater selectivity and their ability to deliver health benefits at a lower dose compared to conventional prebiotics.

Keywords: xylooligosaccharide, mannooligosaccharide, arabinoxylooligosaccharide, prebiotics, human health, clinical trials, animal studies

Introduction

There is increasing recognition of the beneficial role of prebiotics in modulating the microbiome of humans and animals, with the opportunity to beneficially impact health. While the initial focus has been the digestive tract, there has been increasing attention on modification of the microbiome on the skin, in the mouth, and in the urogenital tract. The efficacy of prebiotics is predicated upon selectively feeding beneficial microbes within the target region, with key metabolites, inflammatory and immune markers, etc. driving "distal" health benefits. Within this chapter, we focus on prebiotics that modulate the microbial community in the digestive tract, with a specific focus on novel prebiotics that are based upon 5-carbon sugars as well as 6-carbon sugars with less common beta bonds. When prebiotics are consumed, they are feeding a mixed community of microbes in a dynamic environment in the gastrointestinal tract, where carbohydrates (polymers, oligomers, dimers, monomers) of various types are broken down by digestive enzymes, absorbed into the bloodstream, and utilized by microbes via specific transport and enzyme systems. We elaborate on these concepts in sections 2 and 3, below.

We thus start by describing the environment within the digestive tract – digestive enzymes, systems for carbohydrate absorption, and the microbial communities therein. We then discuss various types of prebiotics, with a particular emphasis on differences in subunits and bond structure. These differences, coupled with differences in microbial enzymes and transport systems, contribute to differences in efficacy, selectivity, and dose between prebiotics. We then focus on prebiotics derived from xylan, arabinan, and mannan, differentiating them from "conventional" prebiotics that rely on subunits of common 6-carbon sugars. Finally, we discuss results from clinical and animal trials with these novel prebiotics, discussing impacts on human and animal health.

The gastrointestinal tract, gut microbiota, and carbohydrate absorption

Downstream of the stomach, the digestive tract is comprised of the small intestine and the colon. The colon is often discussed in terms of the proximal and distal regions, which is relevant in the context of prebiotics, considering both rates and locations for bacterial growth. The small intestine is the primary region for drug and nutrient absorption, although some nutrients and metabolites from microbial growth and metabolism are also absorbed from the colon, where the majority of the gastrointestinal bacteria reside. The transit time through the small intestine is very short – only a few hours, whereas the transit time through the colon may be on the order of 30–40 hours or more, depending upon dietary fiber and fluid intake, among other factors [1].

Enzymes and nutrient absorption in the digestive tract

The small intestine of the human gastrointestinal system contains amylases (from the pancreas) to break down glycogen and starch, a 6-carbon sugar with α-1,4 bonds, and brush border enzymes (lactase, maltase, dextrinase, sucrase) to break down short chain glucooligosaccharides and disaccharides such as sucrose, lactose, and maltose into glucose, fructose, and galactose [2]. Other enzymes aid in the digestion of fats and proteins. The epithelium in the jejunum and ileum of the small intestine is also specifically designed to absorb 6-carbon sugars such as glucose, fructose and galactose via passive, facilitated and active transport systems. Dimers must be broken down into monomers before absorption, and oligomers can persist further into the colon, where they feed microbes that contain enzymes and transport systems to break down complex polymers and oligomers such as xylan and inulin. Short chain fatty acids (SCFAs) can be absorbed from the small intestine, or from the colon if produced as metabolites of bacterial fermentation.

It is estimated that >90% of the SCFAs produced in the colon are absorbed, where they can influence, e.g., hepatic regulation of glucose and lipids, and hormones that regulate satiety [3].

Microbial communities in the digestive tract

The digestive tract is proposed to contain about 1013 bacteria, with 100–300 different taxa and thousands of phenotypes [4]. The dominant gut bacteria identified by 16S RNA are *Actinobacteria, Bacteroidetes, Firmicutes, Proteobacteria, Verrucomicrobia.* The small intestine, where simple carbohydrates are abundant, contains about 1–10000 bacteria per gram of intestinal content, primarily *Clostridium, Bacteroides,* and *Streptococcus* species. Zoetendal et al. [5] sampled the small intestine of four healthy subjects, and observed that *Bacteroidetes*, *Clostridium* (clusters XIVa, IV, XI, and IX) and *Proteobacteria* dominated, with some *Actinobacteria* and *Bacilli* also present. Zoetendal et al. [5] provides evidence that the microbiota, particularly *Streptococcus* species, in the small intestine adapts rapidly to changes in dietary intake, particularly carbohydrates, based upon the presence of transport systems and enzymes that quickly and efficiently utilize simple carbohydrates.

The colon, which survives on complex carbohydrates that are not digested or absorbed in the small intestine, contains about 1011 bacteria per gram of intestinal content, mainly *Prevotella, Ruminococcus*, and *Bacteroides*. Mucin degraders such as *Akkermansia mucinophila* are commonly found in the mucous layer. The GI tract, for various reasons, may also contain pathogenic bacteria at levels that may or may not be clinically significant. Furthermore, there is increasing recognition of the impact of the gut microbiota on the efficacy of drugs, since some microbes contain enzymes similar to those in the liver [6, 7], or may initiate breakdown of prodrugs. Such effects have been observed with metformin and L-DOPA, among other drugs, and may account for at least part of the interindividual variation in drug efficacy and side effects.

Most microbes evolved to process conventional 6 carbon sugars, particularly monomeric glucose, fructose and galactose. A smaller fraction is able to use man- nose, arabinose, and xylose [8].

Background on prebiotics

Definition, brief description of different types of prebiotics, their structure, and function

Historically, the definition of prebiotics was specific to oligosaccharides affecting the gut microbiome, and health impacts arising therein. Recently, the International Scientific Association for Probiotics and Prebiotics modified the consensus definition to include other types of compounds that may act as prebiotics, and also included prebiotics that could work outside of the digestive tract [9]. Even so, it is essential that that a prebiotic must *selectively* stimulate the *growth* of beneficial bacteria, and that there must be a health benefit arising from the consumption/ application of the prebiotic. Molecules such as antibiotics that modify the micro- biome by acting as antimicrobial agents against undesirable bacteria would not be considered prebiotics, since they do not act as substrates for beneficial bacteria.

Fructans such as inulin and fructooligosaccharides (FOS) are most common among prebiotics in the marketplace, although galactooligosaccharides (GOS) produced from lactose are also available, and xylooligosaccharides (XOS) were available in Japan since the 1980s [10]. Recently, forms of resistant starch (RS), isomaltooligosaccharides (IMOS), arabinoxylooligosaccharides (AXOS) and man- nooligosaccharides (MOS) have become available commercially.

Fundamentally, these prebiotics are all materially different, even though the end goal – promoting growth of beneficial bacteria, is the same. The types of bacteria that can be fed by each of these prebiotics depend upon enzymes and transport systems present in the bacteria, which can vary considerably. The selectivity of a prebiotic is also tied to these enzyme and transport systems; if a high percentage of the bacteria have the necessary enzymes and transport systems, then the prebiotic will feed a diverse array of bacteria, including beneficial bacteria along with undesirable bacteria. Conversely, a highly selective prebiotic may not feed many types of bacteria, because fewer types of microbes have the right microbial "machinery" to utilize these prebiotics. Such prebiotics are less likely to directly feed undesirable bacteria if these bacteria do not have the right transport systems and enzymes.

Below, we describe the different chemical structures of prebiotics, along with the enzymes and transport systems responsible for their utilization.

Chemical structures of prebiotics

Prebiotics are typically comprised of oligomers of 6-carbon and/or 5-carbon sugars, with different bonding structures, and different chain lengths (see **Figure 1** for examples) [11].

Fructans from inulin are typically linear oligomers primarily made up of fructose monomers connected by β-2,1 bonds. Fructans from agave tend to have a more complex structure with multiple side branches and β-2,6 linkages in addition to β-2,1 bonds [12]. FFn-type fructans such as inulin are comprised entirely of fructose subunits, whereas GFn-type fructans (typically shorter chain oligosaccharides) may have a glucose subunit connected by an α-2,1 bond onto the main fructan chain.

These short-chain GFn-type FOS molecules are typically produced by enzymatically adding fructose subunits onto sucrose (glucose-fructose) as the starting substrate. In contrast, short chain FFn FOS would be produced by hydrolysis of inulin (long chain FFn) using endoinulinases [11, 12].

Galactooligosaccharides (GOS) may also be comprised exclusively of galactose subunits, or it may have a glucose terminal subunit arising from use of lactose (glucose-galactose) as the initial substrate to produce GOS using β-galactosidases. The galactose subunits may be connected by β-1,3, β-1,4, or β-1,6 bonds, and may include branched structures, depending upon the type of β-galactosidase [11, 13].

Xylooligosaccharides (XOS) are primarily comprised of xylose subunits connected by β-1,4 bonds, although longer chain XOS may contain branches of arabinose subunits, acetyl groups, or uronic acids (originally present in the xylan source material) that can influence their functionality. XOS that includes arabi- nose sub-groups are frequently referred to as arabinoxylanoligosaccharides, or AXOS [11, 14].

Mannooligosaccharides are derived from mannan in biomass, and thus are typically made up of mannose subunits connected by β-1,4 bonds [15].

Isomaltooligosaccharides (IMOS) contain glucose subunits, but vary in their bond- ing structure (affected by manufacturing method). α-1,6 bonds are typical, with either linear or branched structures [11].

Complex starch structures that resist breakdown by pancreatic enzymes into glucose can persist into the colon – thus leading to the concept of "resistant starch" as a prebiotic. Resistant starch (RS) is available in four forms, depending upon method of manufacture and bond structure [16, 17]. Like starch, most of the bonds are of the α-1,4 variety; the presence of other types of bonds may confer "resistance". RS1 is conventional starch that may be trapped within whole grains. RS2 is typically starch with more complex branching or bond structures, rendering it less accessible to amylase. RS3 is produced when starch undergoes retrogradation, i.e., cooked starch is cooled below its gelatinization temperature. RS4 starches have undergone chemical modification, usually by acidification or cross-linking [16].

The microbial ability to utilize specific substrates is dependent upon the enzymes and transporters encoded within the cells. Some microbes can utilize a broad set of substrates, while others are more selective. This also points to the selectivity of the prebiotic; preferably, the prebiotic preferentially feeds beneficial bacteria, with limited growth of undesirable bacteria. **Table 1** summarizes the key hydrolytic enzymes and the corresponding substrates/reactions [19]. These

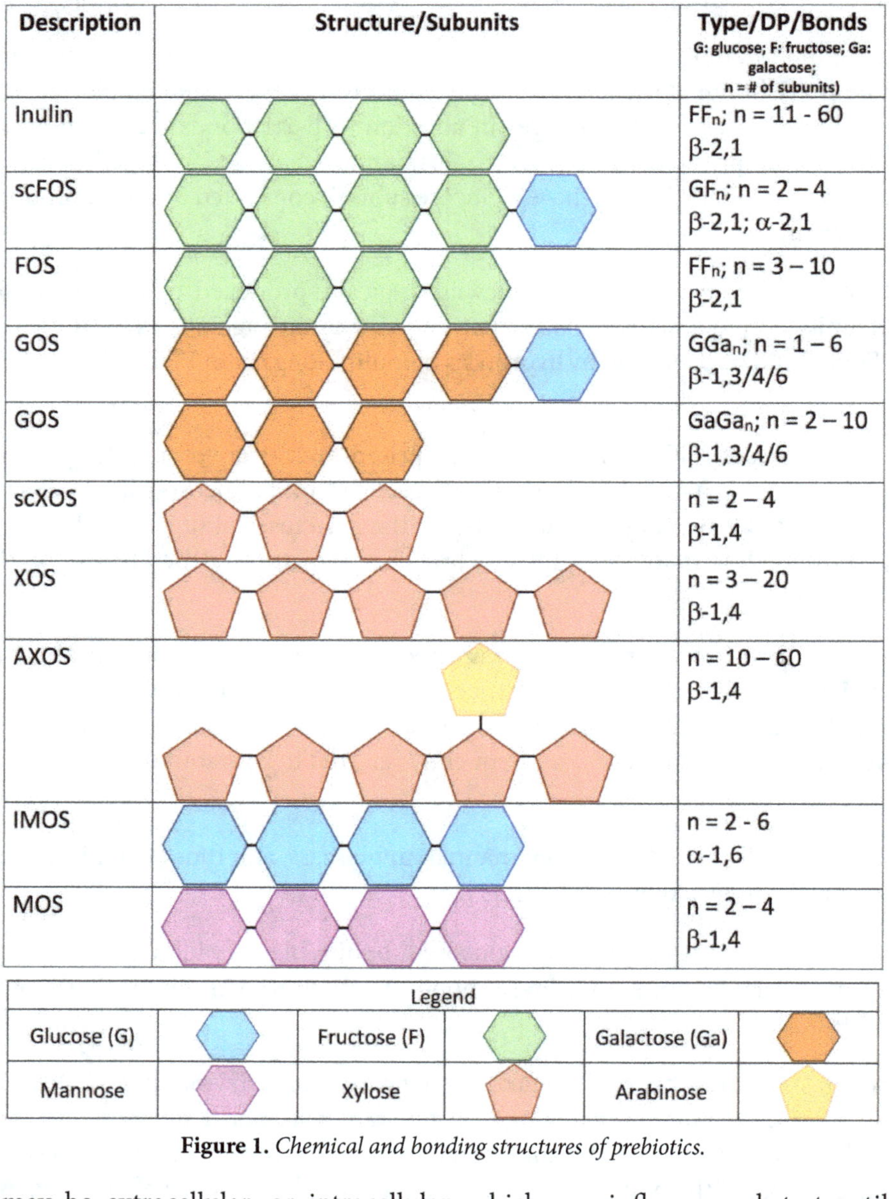

Figure 1. *Chemical and bonding structures of prebiotics.*

enzymes may be extracellular, or intracellular, which can influence substrate utili- zation; intracellular enzymes require a corresponding transporter system.

There are various types of transport systems that move prebiotics into the cell, although microbes are likely to only possess a subset, targeted towards a narrower set of substrates. Some transport systems are specific to (certain) monomers, while others target specific oligosaccharides. Chemical structures must be matched to the structure of the transport system in order to be transported into the cell. **Table 2** summarizes key membrane transport systems involved in utilization of prebiotics and other carbohydrates.

From a selectivity perspective, it is advantageous if the substrate is used intracel- lularly – thus, the requisite transport system plus intracellular fructofuranosidases

Enzyme	**Substrates, reactions**
α-amylase, glucoamylase, pullulanase	Converts starch and RS into glucose
Inulinase, β-fructofuranosidase	Converts inulin into FOS, and FOS into fructose/ glucose
β-galactosidases	Converts lactose into glucose and galactose; aids individuals with lactose intolerance [18]
β-Endoxylanases, β-xylosidases	Converts XOS into short chain XOS (sc-XOS) and xylobiose, then xylose
β-endoglucanase, cellobiohydrolase, β-glucosidase	Converts cellulose and β-gluco-oligosaccharides into cellobiose, then glucose
β-galactanase, β-galactosidase, α-galactosidase	Converts galactan into β-GOS, then galactose; β-galactosidase can also remove galactose subunits that are present as side chains in xylan/XOS
β-mannanase, β-mannosidase	Converts MOS into mannobiose, then mannose
α-arabinofuranosidase	Releases arabinose from side chains of xylan and AXOS

Table 1. *Key enzymes for carbohydrate utilization.*

would have an advantage (i.e., FOS that can be transported intracellularly would have an advantage over FOS/inulin that can only be processed via extracellular enzymes). The restricted capacity of most transporter systems precludes use of long chain fructans by many *Lactobacillus* species [11].

Types of microbes with enzyme/transport systems of various types

Microbes that have fructofuranosidase encoded extracellularly are able to use long-chain FFN-type FOS, and inulin. Some species of *Lactobacillus* (casei, paracasei) and *Streptococcus* are examples [11]. The resulting short chain fructans may be transported intracellularly for utilization. Mao et al. [23] identified 19 strains from human feces that were capable of metabolizing FOS, including multiple strains of *E. coli*, *Enterobacter cloacae*, *Bifidobacterium* spp., and *Lactobacillus* spp. Additional bacteria also proved capable of growth on FOS. This includes additional strains of *E. coli and Bifidobacteria*, along with several strains of *Streptococcus, Clostridia, Roseburia, Klebsiella*, and *Enterococcus* [23].

According to Rossi et al. [24], virtually all *Bifidobacteria* are able to grow on short-chain FOS. However, most *Bifidobacteria* grew poorly on inulin (only 8 out of 55), because most of the fructofuranosidases are intracellular, and inulin cannot be transported intracellularly. Scott et al. [25] made similar observations when inulin with a DP >25 were fed to *Bifidobacteria*. However, in a mixed culture system, inulin may be broken down into shorter chain FFn-FOS, and then used by *Bifidobacteria*. Thus, growth of *Bifidobacteria* in the presence of inulin is primarily due to cross- feeding in the presence of other microbes that act as primary degraders, rather than direct feeding by inulin.

Different species of *Bifidobacteria* contain ABC transporters, sucrose perme- ases, fructose PTS transporters, and MFS transporters. The type(s) of available transporters dictate substrate utilization, whether GFn-type short chain FOS, FFn-type FOS, or analogous substrates. Although several strains of *Bifidobacteria* can utilize GOS, there are significant differences between strains [26]. Certain strains of *B. breve* and *B. longum* contain an extracellular galactanase that breaks down long-chain GOS and galactan in plant fiber, producing di- and

Transport system	Target molecules	Examples/implications
ATP-dependent binding cassette (ABC-type) transporter system	There are many variations of the ABC transporter. Transporters in the CUT1 class work on sucrose, lactose, maltose, FOS, maltodextrins, XOS, and other oligosaccharides. Transporters in the CUT2 class generally transport monomers such as arabinose, xylose, ribose, glucose [20, 21]	Such a system is present in L. acidophilus, for utilization of short-chain FOS; Bifidobacteria have an ABC transporter specific to XOS; various enterobacteria, including E. coli, have CUT1 and CUT2 transporters for maltose and various 6-carbon sugars
Sucrose phosphoenolpyruvate phosphotransferase (PTS) transport system	Transport system is highly specific to compounds that incorporate sucrose as part of their structure.	Allows L. plantarum to utilize shortchain FOS synthesized from sucrose (GFn-type), but FFn FOS, which lacks the sucrose structure, cannot be used by L. plantarum
Major facilitator superfamily (MFS) transporter system	A major transporter system with various types, allowing intracellular transport of glucose, lactose, xylose, oligosaccharides, FFn-type FOS [11, 21]	Present in many bacteria, fungi, yeasts, plants, animals, humans; key for energy homeostasis
Fructose PTS transporters	Transport system targets the fructose component of substrates, thus allowing use of fructose, sucrose, inulin, and both FFn and GFn-type FOS.	L. rhamnosus GG contains a fructose PTS transporter, which allows growth on various types of FOS and inulin
Lactose PTS transporters	Transport system targets the lactose component of substrates	L. gasseri contains a lactose PTS transporter [22]
LacS and LacY permeases	MFS-type transport systems enabling transport of molecules with lactose module. The LacS transport system allows a microbe to use lactose, GOS (with lactose terminus), and lactitol. LacS and LacY differ based upon source family.	LacS is stated to be the sole transporter for GOS, with specificity for β-galactosides [22].

Sucrose permease	Transport of substrates with a sucrose module, such as GFntype FOS.	
CUT = Carbohydrate Uptake Transporter; FFn = FOS comprised of "n" fructose (F) subunits and a fructose terminal unit; GFn = FOS containing "n" fructose (F) subunits and a terminal glucose (G); PTS = phosphoenolpyruvate phosphotransferase.		

Table 2. *Key Transmembrane transport systems.*

tri-saccharides that can be transported into the cell and converted into galactose via intracellular β-galactosidase [27]. *B. lactis* BI-04 contains lactose permease and ABC transport systems, along with β-galactosidase [28], that enable utiliza- tion of GOS. *L. acidophilus* has the ability to utilize many different prebiotics, with various monomeric subunits and bond structures [28]. Such broad utili- zation is due to a multiplicity of molecular transport systems and hydrolytic enzymes, including up to nine different enzymes from the GH13 family that act on α-glucan [11].

Starch, owing to its high DP and complex branched structure, would be degraded in the presence of extracellular enzymes that can act on α-1,4 and α-1,6 linkages between glucose subunits. Certain *Bifidobacteria*, including *B. pseudolongum* and *B. breve* have the necessary enzymes for extracellular starch utilization [29].

Microbes such as the *L. acidophilus* cluster (including *L. johnsonii, L. helveticus, L. reuteri* and *L. plantarum*) contain LacS permease and β-galactosidase which allow these microbes to transport GOS into the cell, then break it down into glucose and galactose for metabolism [28].

Several *Bifidobacteria* contain the hydrolytic enzymes needed to break down the β-1,4 linkages present in xylooligosaccharides (XOS) and XOS with arabinose side groups (AXOS). Key enzymes include β-xylosidase and β-xylanase, the latter which breaks down longer chain XOS into shorter chains, ultimately xylobiose, that may be converted into xylose using β-xylosidase. AXOS requires arabinofuranosidase enzymes to process the arabinose side group. Some carbohydrate esterases may also be present to deal with acetyl or feruloyl side groups. The enzymes may be intracellular or extracellular; intracellular enzymes also require transporters such as an ABC transport system to act on the longer chain oligomers. Ejby et al. [30] noted that ABC transporters specific to XOS are exclusive to *Bifidobacteria. B. lactis, B. breve*, and *B. bifidum* are among the many species of *Bifidobacteria* that have the requisite enzymes and trans- port systems for utilization of short and longer chain XOS and AXOS. Crossfeeding of *Bifidobacteria* is aided by *Bacteroides* and *Prevotella*, which act as primary degraders that break down insoluble xylan in plant fiber into soluble oligosaccharides.

The wide variation in structures of prebiotics, along with the different transport and enzyme systems, ultimately dictate the selectivity of the prebiotic in a mixed culture. Monoculture systems provide some useful insights into the utilization of prebiotics by various substrates. Makelainen et al. [31] conducted a thorough study of the growth of >15 microbes, some beneficial, some pathogenic, in the presence of 11 different carbohydrate sources. Aggregate growth over 24 hours was reported as the area under the curve of DP600 measurements. **Figures 2–6** show growth of various probiotics and pathogenic microbes on glucose, FOS, GOS, scXOS, and

XOS, respectively. As expected, all bacteria grew well on glucose (**Figure 2**), consistent with the widespread ability of microbes to utilize simple 6-carbon sugars.

Low DP FFn-FOS proved to be a good substrate for several strains of *Bifidobacteria*, along with *L. paracasei* and *L. acidophilus*, but was also used by *E. coli EHEC*, *S. epidermis*, and *C. perfringens*, among pathogenic bacteria tested. GOS also grew well on *Bifidobacteria* and *Lactobacilli*, but also proved to be an excellent substrate for several pathogenic bacteria, including *E. coli* and *C. perfringens*.

Consistent with the unique chemical structure and enzyme/transporter requirements, there was less microbial growth on scXOS and XOS. Fewer strains of *Bifidobacteria* utilized XOS, along with select strains of Lactobacillus. Li et al., in a study using 29 *Lactobacillus* strains and 35 strains of *Bifidobacterium*, observed that all *Bifidobacterium* strains tested grew on a high dose of XOS, and 30 of 35 strains grew on low dose XOS [32]. They also noted that *Lactobacillus* strains were able to uti- lize XOS, albeit with fewer strains and at lower efficiency compared to *Bifidobacteria*.

However, importantly, Makelainen et al. [31] noted minimal growth of patho- genic bacteria in the presence of XOS (**Figures 5** and **6**), consistent with a much higher selectivity of XOS for beneficial bacteria. This is a key advantage in a mixed microbial environment such as the GI tract. *Bifidobacteria* and *Lactobacillus* species have to compete with many more bacteria for FOS and GOS, which thus increases the dose required for efficacy. XOS, conversely, is better targeted to *Bifidobacteria*, and in a mixed culture, could be efficacious at a lower dose.

The aggregate area under the curve data reported by Makelainen et al. [31] do not, however, capture changes in growth rates, which can vary over time. Similarly,

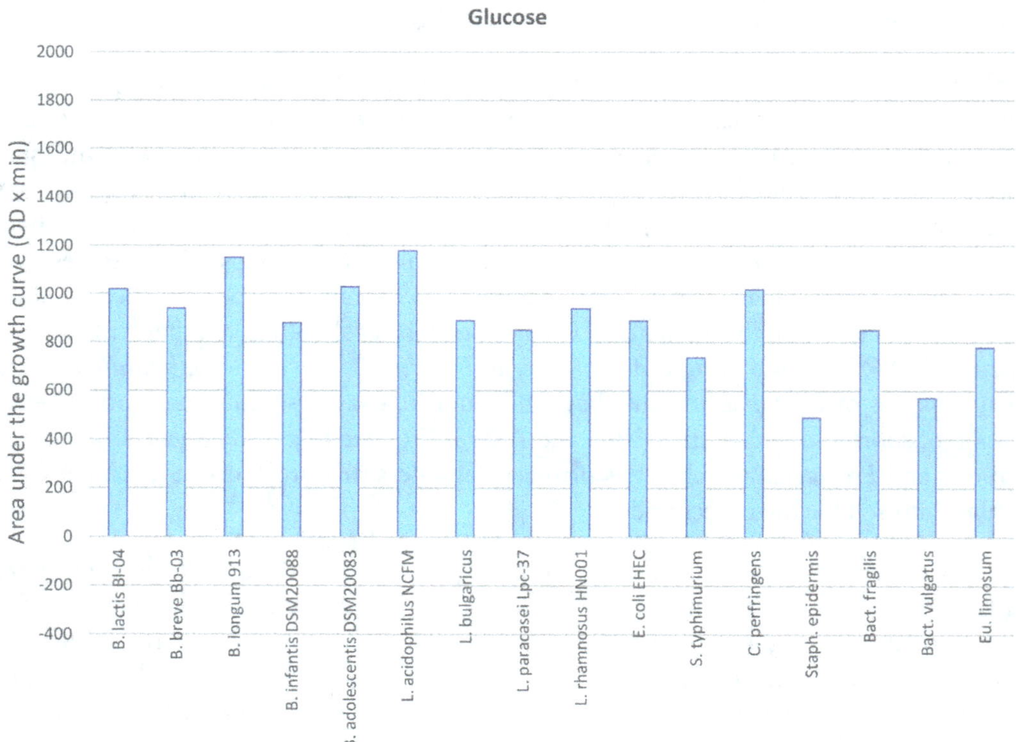

Figure 2. *Microbial growth on glucose (positive control). Data from Makelainen et al. [31].*

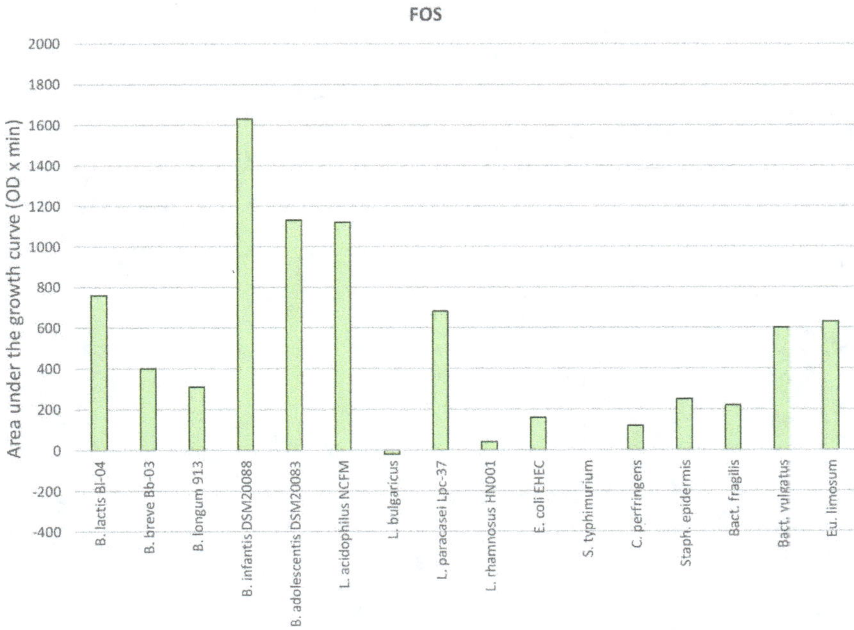

Figure 3. *Microbial growth on FFn FOS (DP 2–7). FFn = FOS comprised exclusively of fructose (F) subunits. Data from Makelainen et al. [31].*

any issues with viability of microbes could be masked by rapid early growth, which may not be sustained. **Figure 7** illustrates growth of a strain of *B. breve* on FOS, XOS, and inulin, showing temporal effects. A noteworthy observation is that viability of *B. breve* decreased significantly after ~12–16 hours if grown on FOS or inulin, whereas growth on XOS sustained *B. breve* for a longer period, even up to 48 hours.

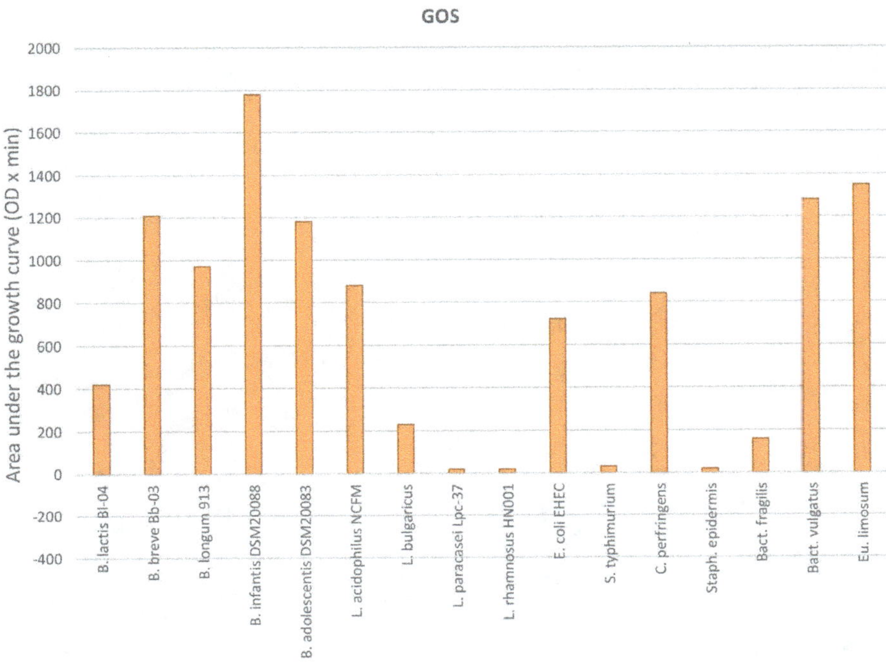

Figure 4. *Microbial growth on GGan GOS (DP 3–5). GGan GOS is comprised of n subunits of galactose (Ga) with a terminal glucose (G) subunit. Data from Makelainen et al. [31].*

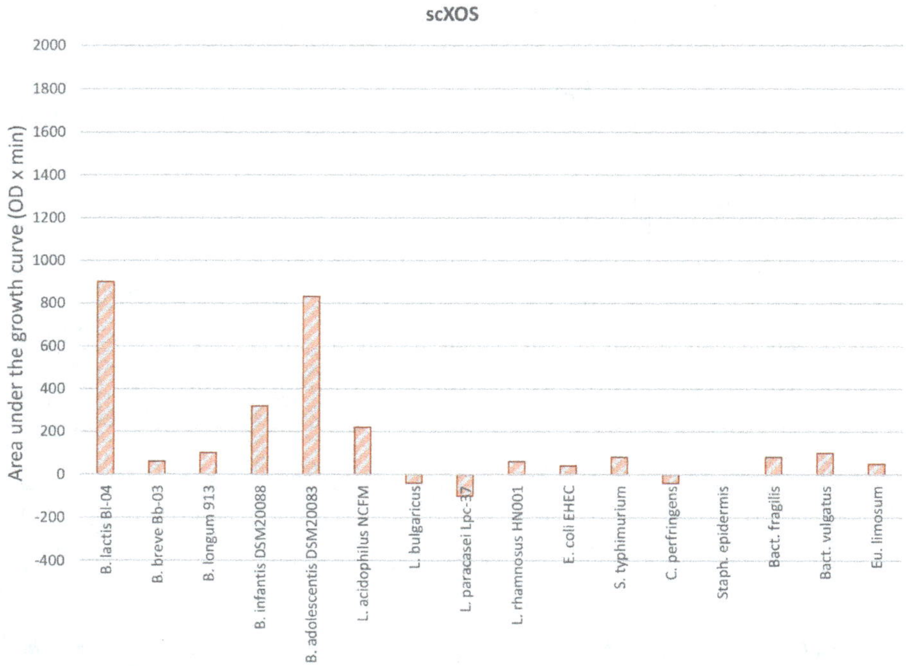

Figure 5. *Microbial growth on short chain XOS (scXOS; DP 2–5). Data from Makelainen et al. [31].*

This may have important implications in terms of sustaining key microbes in the digestive tract.

In the next section, we describe health impacts of prebiotics, with a particular emphasis on studies with XOS, AXOS, and MOS, due to their distinct chemical structures and selectivity for beneficial bacteria.

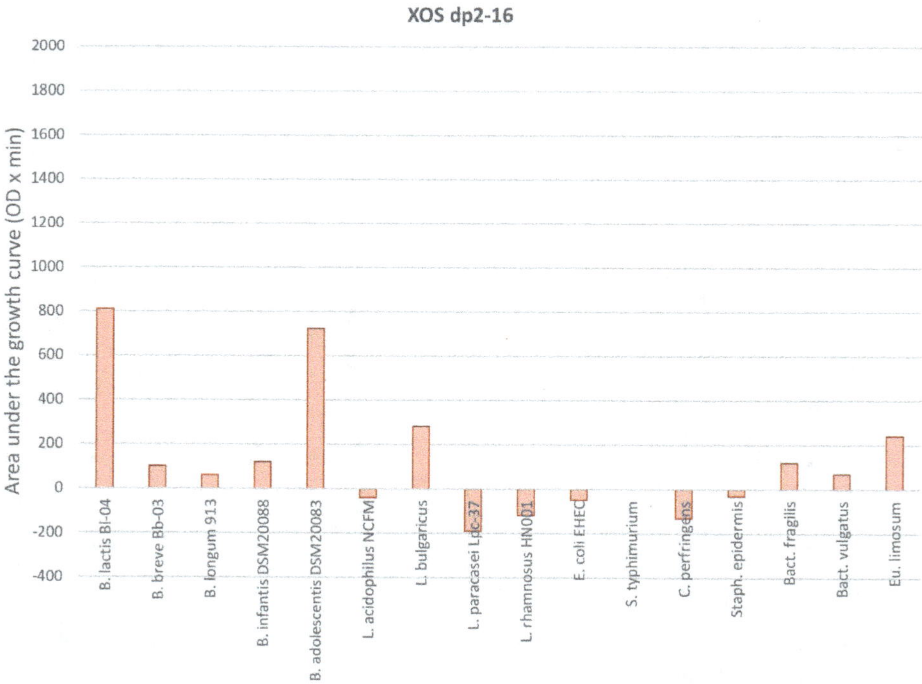

Figure 6. *Microbial growth on XOS (DP 2–10). Data from Makelainen et al. [31].*

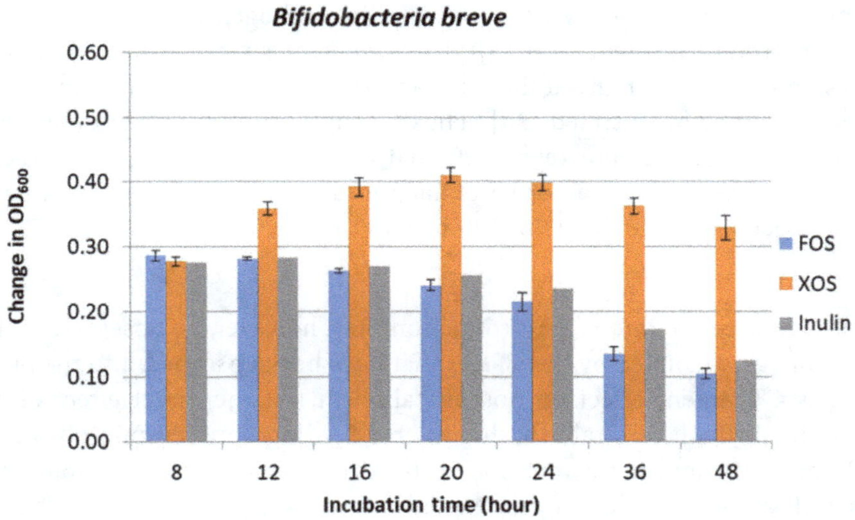

Figure 7. *Comparative growth of B. breve on FOS, XOS, and inulin.*

Health impacts of prebiotics

Stimulating the growth of beneficial bacteria with prebiotics can lead to a cascade of health effects, as illustrated in **Figure 8**. Much of the historical focus had been on *Bifidobacterium spp.* and *Lactobacillus spp.*, although more recently, beneficial health outcomes have been associated with other microbial species as well. Short chain fatty acids, primarily acetate, propionate, and butyrate, along with lactate, are produced by fermentation of non-digestible fibers and prebiotics, potentially reducing the pH within the colon. This can promote absorption of minerals such as

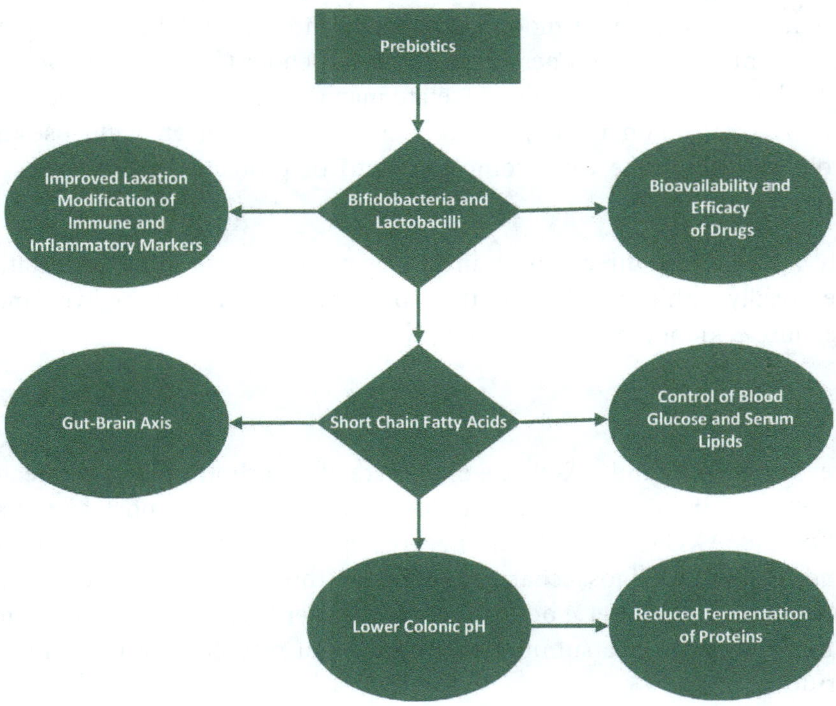

Figure 8. *Health impacts associated with prebiotic intake.*

calcium, and limit growth of yeasts, potentially pathogenic bacteria, and microbes responsible for protein fermentation [3]. The extent of such a pH reduction *in vivo* is difficult to establish, however, given that greater than 90% of SCFAs are typically absorbed, and as little as 5% may be excreted [33]. Thus, changes in fecal SCFA levels (or pH) may be difficult to detect, or may not represent what is present in the digestive tract. Autopsy samples indicate SCFA levels are an order of magnitude greater in the cecum than the ileum, and that levels decrease from about 70–140 mM in the proximal colon to 20–70 mM in the distal colon [33].

The effect of increased colonic SCFA production may, however, be observed in the form of distal health impacts mediated by the SCFAs that have been absorbed into the bloodstream. Each SCFA has a different effect on host metabolism, yet the direct effect of a prebiotic on SCFAs can be difficult to establish, due to crossfeeding between microbes that directly consume the prebiotic and other microbes within the digestive tract. Dietary fiber intake also affects SCFAs, further clouding interpretation of fecal SCFA levels. The impacts of prebiotics on SCFA levels are easier to detect via *in vitro* cultures, particularly monocultures, and in gut simulators that contain mixed microbial communities that enable cross-feeding, such as acetate production via one class of microbes, and acetate conversion into butyrate by, e.g.,

E. rectale, F. prausnitzii, or *Roseburia* spp. [3].

SCFAs provide energy for colonocytes, and, after absorption into the portal vein, are metabolized by the liver, modulating cholesterol synthesis, maintaining glucose homeostasis in peripheral tissues, and producing long chain fatty acids. SCFAs can act on leukocytes that modulate immune responses, beneficially impact hormones that influence satiety, and influence neural signals that modulate appetite and food intake via the gut-brain axis [3]. Acetate can influence hormones responsible for cholesterol production, and may be transported to the muscle and the brain; acetate may thus contribute to effects attributed to the "Gut-Brain Axis". Propionate can play a role in hepatic regulation of glucose and synthesis of cholesterol, while butyrate can promote growth of protective colonocytes via colonic epithelial cells.

Consequently, microbes fed directly or indirectly by prebiotics can influence health via SCFAs that act either locally within the digestive tract, or distally, mainly via the liver, and to a lesser extent, in the muscle, kidney, heart, and brain [3].

XOS/AXOS

As noted previously, XOS and AXOS are oligomers of 5-carbon sugars, connected by β-1,4 bonds. Fewer types of microbes contain the enzymes and transport systems necessary to utilize XOS and AXOS, thus conferring greater selectivity to beneficial bacteria (see **Figures 5** and **6**). Mannooligosaccharides (MOS), although based upon a 6-carbon sugar backbone, are connected via β-1,4 bonds that also render them less susceptible to microbial utilization. In this section, we summarize clinical trial results from these novel prebiotic oligosaccharides.

Tables 3 and **4** summarize clinical trial results with XOS and AXOS, respectively, outlining impacts on the microbiome and various clinical biomarkers. Clinical trials with XOS were conducted with doses ranging from 0.4 to 8 g per day; most were in the range from 1 to 3 g per day. A statistically significant increase in *Bifidobacteria* was observed in most trials with XOS [34–39]. Improvements in laxation were noted by Childs et al. [34], Chung et al. [35], Iino et al. [40], and Tateyama et al. [41]; the latter was a study in constipated pregnant women. Studies by Childs et al. [34], Na and Kim [39], and Sheu et al. [42] observed improvements in triglyceride and cholesterol levels, at doses as low as 2.8 g/d, whereas Yang et al. [43] did not observe changes in triglycerides when dosing XOS at 2 g/d (2.8 g/d of a 70% purity product). Yang et al. [43] observed a tendency towards a reduction in OGTT insulin in predia- betes patients dosed with 2.8 g/d of 70% XOS, but no effect on blood glucose, unlike Na and Kim [39], who noted a reduction in blood glucose at a XOS dose of 2.8 g/day. Sheu et al. [42], in a trial with diabetic patients receiving 4 g/d of XOS, observed a statistically significant reduction in blood glucose and HbA1C.

Studies with AXOS complement studies with XOS, since most AXOS products are comprised of at least ~50% XOS. All studies noted an increase in *Bifidobacteria* [44–48]. Francois et al. [45] and Walton et al. [48] observed increased fecal levels of SCFAs (acetate, propionate, and butyrate). Reductions in cresol, indicative of reduced protein fermentation, were observed by Cloetens et al. [44] and Francois et al. [45] with AXOS, and Lecerf et al. [38] with XOS.

The bifidogenic effect and health benefits observed in the various clinical trials with XOS and AXOS where generally observed at a dose less than 4 g/d. By com- parison, approximately 10–20 g/d of FOS is needed to trigger a bifidogenic effect, and the required inulin dose is stated to be at least 15 g/d [49, 50]. Alfa et al. [51] required a dose of 30 g/d of resistant starch to enhance *Bifidobacterium* levels.

Miremadi et al. [52] summarized clinical trials in which various prebiotics were evaluated to assess cardiovascular impacts, particularly reductions in cholesterol, lipids and blood pressure. No improvements to lipid profile were observed with 5.5 g of GOS or 20 g/d of FOS (type not specified), whereas triglyceride levels improved following consumption of 10 and 20 g/d of inulin. Similarly, Alles et al. were unable to detect changes in blood glucose and lipid profiles following adminis- tration of 15 g/d FOS [53].

The higher doses required for a bifidogenic effect or clinical efficacy compared to XOS and AXOS is likely due to the greater selectivity of XOS and AXOS for beneficial bacteria, versus the widespread utilization of FOS and GOS [11, 23, 31].

Study authors	XOS purity / dose	Study design and duration	Endpoints
Childs et al. [34]	8 g per day with or without *B. lactis*	Randomized, double-blind, placebo-controlled, factorial cross-over study with 41 healthy adults; 3 weeks dosing, with a 4 week washout between dosing periods	Increased HDL levels by 0.07 mM (P = 0.005) Increase of 0.10 mM is associated with 10% risk reduction in coronary artery disease Increased # of bowel movements per day (P = 0.009) Increased Bifidobacteria (P = 0.008) Reduced IL-10 production (P = 0.049)
Chung et al. [35]	95% pure XOS, 4 g per day	Randomized, placebo-controlled trial with 22 healthy adults (9 control, 13 XOS); includes a 1 week run-in, 3 weeks of dosing, and a 3 week washout	Statistically significant increase in fecal moisture (alleviates constipation) and fecal pH, with return to baseline post-washout (P < 0.05) Statistically significant increase in *Bifidobacteria* content of the stool (P < 0.05)
Finegold et al. [36]	70% pure; 1.4 or 2.8 g per day	Randomized double blind placebo-controlled trial with 32 healthy adults; 2 week run-in period, followed by 8 weeks of treatment and a 2 week washout	Statistically significant increase in *Bifidobacteria* (P = 0.007; 2.4 g per day dose) and *B. fragilis* (P = 0.001; 2.4 g per day dose) relative to placebo/baseline
Iino et al. [40]	0.40 g per day	4 Weeks with 40 healthy adults	Improved stool frequency (P < 0.05)
Kajihara et al. [37]	3 g per day	Baseline data in patient population comprised of 14 adults with cirrhosis, followed by 2 weeks of dosing with 3g per day of XOS.	Increased *Bifidobacteria* content Reduced *Bacteriodes* Reduced ammonia in serum (statistical data not presented)
Lecerf et al. [38]	5 g per day WitaXOS (80% Pure); 1 g per day XOS + 3 g per day inulin	Randomized, placebo-controlled, doubled blind trial with 60 healthy adults (20 in each of three groups: (i) XOS, (ii) XOS and inulin, (iii) placebo); 4 weeks	Increased *Bifidobacteria* (P = 0.002) Increased butyrate (P = 0.036) Reduced cresol (P = 0.020), acetate (P = 0.011) and fecal pH (P = 0.033)
Na and Kim [39]	1.4 and 2.8 g per day	Randomized dose-dependent trial without a placebo with 14 healthy adults (7 per test group); 4 weeks	Reduced triglycerides, cholesterol and glucose in 2.8 g per day group (all P < 0.05) Increased *Bifidobacteria* content after 14 days (P < 0.05) Increased lactic acid concentration (P < 0.05) While study lacked placebo, subjects saw effect of dose response over multiple doses
Sheu et al. [42]	4 g per day	Randomized, double blind placebo-controlled trial in 26 adults with type II diabetes; 8 weeks	Statistically significant reduction in: Blood glucose (P < 0.05) (described as a point measurement of blood glucose, with the sample obtained after fasting) LDL (P < 0.01) Total cholesterol (P < 0.01) HbA1c (P < 0.05) Apolipoprotein B (P < 0.05)
Tateyama et al. [41]	4.2 g per day in 8g syrup	29 Adult Women (pregnant with constipation); all received product, at the same dose, for 4 weeks	Statistically significant increase in stool frequency and reduction in constipation Improvement increased every week over the duration of treatment Normalized stool consistency
Yang et al. [43]	2.8 g per day of 70% XOS	8 week trial with 13 healthy adults and 16 prediabetic (increased risk of diabetes) adults	Decreased *Firmicutes* Attenuated changes in *Howardella, Slackia*, and *B. hydrogenotrophica* Tendency to reduce OGTT insulin No effect on blood glucose, triglycerides Limited by small study numbers and lack of statistical analysis

Table 3. *Summary of clinical trials with XOS.*

For example, Mao et al. [23] found that 237 out of 453 strains (114 genera) of gut bacteria contained the necessary transporters and enzymes for some type/degree of FOS utilization, suggesting fairly widespread utilization of FOS, which will impact the effective dose required to grow the targeted beneficial bacteria. Similarly, Goh et al. [11] note the widespread ability of microbes to utilize lactose, suggesting the presence of LacS (or similar) transporters and enzymes in many microbes that

Study Authors	AXOS Product / Purity / Dose	Study Design and Duration	Endpoints
Cloetens, L. et al. [44]	AXOS, 10 grams per day (63% XOS, 17% arabinan, 12% glucan)	Randomized, placebo-controlled cross-over study with 20 healthy adults; 3 weeks with AXOS or placebo, with 4-week washout between treatments	Increased *Bifidobacteria* (P = 0.012) Increased *B. adolescentis* (P = 0.013) After 3 weeks, reduced urinary p-cresol (P = 0.011)
Francois, I. et al. [45]	Fugeia wheat bran extract, 3 grams per day, 10 grams per day (comprised of 79% AXOS, which, in turn, consists of 49% XOS and 10% glucan) [1]	Double blind randomized placebo controlled cross-over trial with 20 healthy adults; 3-week treatment periods interspersed with 2-week washout periods	At 10 grams per day: Increased acetate, propionate, and butyrate (P = 0.009, P = 0.05, P = 0.001, respectively) Reduced p-cresol (P = 0.039) Lower fecal pH (P = 0.039) Increased *Bifidobacteria* (P < 0.001) Increased stool frequency (P = 0.258) Reduced LDL (P = 0.168)
Francois, I. et al. [46]	Fugeia Wheat Bran Extract, 5 grams per day (comprised of 79% AXOS, which, in turn, consists of 49% XOS and 10% glucan)	Double blind randomized placebo controlled cross-over trial with 29 healthy children; 3 weeks	Increased *Bifidobacteria* (P = 0.069) Reduced isovaleric and isobutyric acid (P < 0.01) No impact on stool frequency or consistency
Maki, K.C., et al. [47]	2.2 grams or 4.8 grams AXOS per day added to cereal	Randomized, double blind, placebo-controlled cross-over trial with 65 healthy adults; 3 weeks treatment with 2 weeks washout between treatments	Dose-dependent increase in Bifidobacteria observed; statistically significant increase n the 4.8 grams per day AXOS group (P<0.001) No statistically significant impact on fasting blood glucose, total cholesterol, HDL, or triglycerides
Walton, G. et al. [48]	180 grams per day of Bread, with or without enrichment with 2.2g AXOS	Double-blind, randomized, placebo-controlled cross-over trial with 40 healthy adults; five periods of 3 weeks each	Increased *Bifidobacteria* relative to baseline (P = 0.0011) Increased butyrate (P = 0.041) and propionate (P = 0.045) Other results confounded by XOS/fructans in the placebo bread, which increased *Bifidobacteria*, short chain fatty acids, and bowel frequency relative to the baseline. The trend to further improve these characteristics via the AXOS-enriched bread were thus masked.

Table 4. *Summary of clinical trials with AXOS.*

would allow use of the GGan type of GOS. This is consistent with the observations of Makelainen et al. [31], who also observed excellent growth of many species, ben- eficial and pathogenic, on GOS (see also **Figure 4**). Ultimately, increased sharing of prebiotic substrates among bacteria means that a higher dose is needed to trigger sufficient growth of the targeted beneficial bacteria that produce the desired health benefits.

MOS

Manno-oligosaccharides (MOS), extracted from yeast cell walls, coffee or biomass rich in mannan, are typically comprised of up to 6 mannose subunits con- nected by β-1,4 bonds. MOS have not been as extensively evaluated in humans, but have been recognized as a nutritional supplement for livestock.

Salinardi et al. [54] observed a reduction in body volume and adipose tissue in men consuming 4 g/d MOS, attributed to an increase in fat excretion in the feces or inhibition of hepatic lipogenesis.

Kumao et al. [55] observed a reduction in fat utilization and an increase in fecal fat excretion in people consuming 3 g/d of MOS for 7 days. St.-Onge et al. [56] also observed a reduction in body weight and adipose tissue following consumption of 4 g/d MOS for 12 weeks. Umemura et al. evaluated the impact of 1 g/d of MOS consumption on fecal microbiota and laxation [57]. They noted an increased ratio of *Bifidobacterium* spp., and enhanced defecation frequency/ volume, reducing constipation. A study in mice by Zheng et al. [58] suggested that MOS acted synergistically with metformin, altering the gut microbiota in a manner that would decrease clinical diabetic parameters, including a reduction in blood glucose. This may have promise for future clinical trials and application of MOS plus metformin for management of type II diabetes.

Jahromi et al. [59] examined MOS supplementation (1 g/kg) to broiler chicks, and observed increased levels of *Lactobacillus spp*. and *Bifidobacteria spp*., while reducing levels of *E. coli* and *Enterobacter* by >50%. Navidshad et al. [60] compared MOS derived from yeast cell walls with MOS from palm kernel expeller, assessing their efficacy as a supplement (2 g/kg) to the diet of broiler chicks. The yeast- derived MOS improved weight gain, while reducing the intestinal percentage of *E. coli* in the birds. MOS has been reported to have receptors for fimbriae on *E. coli*, which can help to control or limit colonization within the digestive tract [61].

Jahromi et al. [62] also evaluated *In Ovo* injection of MOS, and feeding of MOS to chicks. The single *In Ovo* injection had some short term but limited long term effects, other than an increase in *Bifidobacterium* spp. at 14 days. Feeding MOS in the diet markedly improved levels of *Lactobacillus* and *Bifidobacteria*, while reducing levels of pathogenic strains of *Salmonella*, *E. coli*, and *Campylobacter*. Adding MOS to the diet also improved levels of serum immunoglobulins IgA, IgM, and IgG. The immunomodulatory effect of MOS in chicks was stated to be a significant benefit that could help improve productivity, and reduce disease (thereby reducing use of antibiotics). Zhao et al. [63] studied the impact of supplementing 0.1 wt% MOS or 0.1 wt% FOS in weanling pigs over 28 days. They observed greater average daily weight gain and average daily feed intake in pigs consuming MOS compared to controls. Nutrient digestibility also improved, along with diarrhea score (potentially by inhibiting *E. coli*). Collectively, the authors concluded that MOS could enhance piglet growth and health.

Summary

Increased understanding of enzyme and transporter expression in various microbes, and key differences between the various classes of enzymes and types of transporters, are enhancing our knowledge about microbial selectivity for sub- strates, including prebiotics. Differences in chemical structure, degree of polym- erization, and bonds between subunits affect microbial utilization of prebiotics.

Novel prebiotics derived from xylan, arabinan, and mannan are comprised of less common subunits based on 5 carbon sugars (xylose, arabinose), and/or are con- nected via β-bonds. The unique types of subunits and bond structures confer greater selectivity for beneficial bacteria, and health benefits at a lower dose compared to conventional prebiotics comprised of glucose, fructose and galactose subunits.

Clinical trials with XOS and AXOS led to improvements in laxation, triglyc- erides, cholesterol, and blood glucose, typically at doses from 1 to 3 g per day, far less than the 10–30 g per day required with FOS, GOS, inulin, and resistant starch. Preliminary clinical trials with MOS suggested the potential for weight management and reductions in adipose tissue at doses in the range of 4 g per day. Furthermore, MOS seems to have a unique capacity to inhibit proliferation of *E. coli*, and its addition to livestock feed has improved livestock health with a concur- rent reduction in antibiotic use. Ultimately, these novel prebiotics may usher in a new era of prebiotic utilization, driven by their greater selectivity for beneficial bacteria, and easier product formulation and efficacy at a lower dose.

Disclosures and conflicts of interest

The co-authors are both affiliated with Prenexus Health, a company that produces xylooligosaccharide prebiotics.

Author details

Bradley A. Saville[1,2*] and Sandra H. Saville[2,3]

1 Department of Chemical Engineering and Applied Chemistry, University of Toronto, Toronto, ON, Canada

2 Prenexus Health, Gilbert, AZ, United States

3 Saville Nutrition Consulting, Oakville, ON, Canada

*Address all correspondence to: bradley.saville@utoronto.ca

References

[1] Mayo Clinic. How long does it take to digest food—from the time you eat it to the time you excrete it? [Internet]. 2018. Available from: https://www. mayoclinic. org/digestive-system/expert- answers/faq-20058340 [Accessed: July 31, 2019]

[2] BC Open Textbook Project. The digestive system. In: Anatomy and Physiology, Ch 23. Pressbooks. Available from: https://opentextbc.ca/ anatomyandphysiology/ chapter/23-7- chemical-digestion-and-absorption-a-closer-look/ [Accessed: July 23, 2019]

[3] Tungland B. Short-chain fatty acid production and functional aspects on host metabolism. In: Human Microbiota in Health and Disease: From Pathogenesis to Therapy. London, UK: Academic Press; 2018. pp. 37-106. DOI: 10.1016/B978-0-12-814649-1.00002-8

[4] Hillman ET, Lu H, Yao T, Nakatsu CH. Microbial ecology along the gastrointestinal tract. Microbes and Environments. 2017;**32**(4):300-313. DOI: 10.1264/ jsme2.ME17017. Available from: . https://www.jstage.jst.go.jp/ browse/jsme2

[5] Zoetendal EG, Raes J, van den Bogert B, Arumugam M, Booijink CC, Troost FJ, et al. The human small intestinal microbiota is driven by rapid uptake and conversion of simple carbohydrates. The ISME Journal. 2012;**6**(7):1415-1426. DOI: 10.1038/ ismej.2011.212. ISSN 1751-7362

[6] Mohammed A, Guda C. Application of a hierarchical enzyme classification method reveals the role of gut microbiome in human metabolism. BMC Genomics. 2015;**16**(Suppl. 7):S16. DOI: 10.1186/1471-2164-16-S7-S16. Epub 2015 Jun 11

[7] Zimmermann M, Zimmermann- Kogadeeva M, Wegmann R, Goodman AL. Separating host and microbiome contributions to drug pharmacokinetics and toxicity. Science. 2019;**363**:eaat9931. DOI: 10.1126/ science.aat9931

[8] Fraenkel DG, Vinopal RT. Carbohydrate metabolism in bacteria. Annual Review of Microbiology. 1973;**27**:69-100. DOI: 10.1146/annurev. mi.27.100173.000441

[9] Gibson GR, Hutkins R, Sanders ME, Prescott SL, Reimer RA, Salminen SJ, et al. Expert consensus document: The International Scientific Association for Probiotics and Prebiotics (ISAPP) consensus statement on the definition and scope of prebiotics. Nature Reviews Gastroenterology & Hepatology. 2017;**14**:491-502. DOI: 10.1038/nrgastro.2017.75

[10] Makelainen H, Juntunen M, Hasselwander O. Prebiotic potential of Xylo-oligosaccharides. In: Charalampopoulos D, Rastall RA editors. Prebiotics and Probiotics Science and Technology. New York, NY: Springer; 2009. pp. 245-258. DOI: 10.1007/978-0-387-79058-9_8

[11] Goh YJ, Lu C, Klaenhammer TR. Genetic mechanisms of prebiotic oligosaccharide metabolism in probiotic microbes. Annual Review of Food Science and Technology. 2015;**6**:137-156. DOI: 10.1146/ annurev-food-022814-015706

[12] Franco-Robles E, López MG. Implication of fructans in health: Immunomodulatory and antioxidant mechanisms. The Scientific World Journal. 2015;**2015**:289267. DOI: 10.1155/2015/289267

[13] Gosling A, Stevens GW, Barber AR, Kentish SE, Gras SL. Recent advances refining galactooligosaccharide production from lactose. Food Chemistry. 2010;**121**(2):307-318. DOI: 10.1016/j.foodchem.2009.12.063

[14] Otieno DO, Ahring BK. The potential for oligosaccharide production from the hemicellulose fraction of biomasses through pretreatment processes: Xylooligosaccharides (XOS), arabinooligosaccharides (AOS), and mannooligosaccharides (MOS). Carbohydrate Research. 2012;**360**(Complete):84-92. DOI: 10.1016/j.carres.2012.07.017

[15] Singh S, Ghosh A, Goyal A. Manno- oligosaccharides as prebiotic-valued products from agro-waste. In: Varjani SJ et al., editors. Biosynthetic Technology and Environmental Challenges, Energy, Environment, and Sustainability. Singapore: Springer; 2018. DOI: 10.1007/978-981-10-7434-9_12

[16] Raigond P, Singh B, Joshi A, Ezekiel R, Raigond B. Resistant starch in food: A review. Journal of the Science of Food and Agriculture. 2015;**95**:1968-1978. DOI: 10.1002/jsfa.6966

[17] Ai Y. Structures, properties, and digestibility of resistant starch [thesis]. Iowa State University; 2013. 13558. https://lib.dr.iastate.edu/etd/13558

[18] Sanders ME, Merenstein D, Reid G, Gibson GR, Rastall RA. Probiotics and prebiotics in intestinal health and disease: From biology to the clinic. Nature Reviews. Gastroenterology & Hepatology. 2019;**11**:3838-3850. DOI: 10.1038/s41575-019-0173-3

[19] Hamaker B, Tuncil YE. A perspective on the complexity of dietary fiber structures and their potential effect on the gut microbiota. Journal of Molecular Biology. 2014;**426**(23):3838- 3850. DOI: 10.1016/j.jmb.2014.07.028

[20] Schneider E. ABC transporters catalyzing carbohydrate uptake. Research in Microbiology. 2001;**152**:303-310. DOI: 10.1016/ S0923-2508(01)01201-3

[21] Saier MH. Families of transmembrane sugar transport proteins. Molecular Microbiology. 2000;**35**:699-710. DOI: 10.1046/j.1365-2958.2000.01759.x

[22] Francl AL, Hoeflinger JL, Miller MJ. Identification of lactose phosphotransferase systems in *Lactobacillus gasseri* ATCC 33323 required for lactose utilization. Microbiology. 2012;**158**:944-952. DOI: 10.1099/mic.0.052928-0

[23] Mao B, Li D, Zhao J, Liu X, Gu X, Chen YQ , et al. In vitro fermentation of fructooligosaccharides with human gut bacteria. Food & Function. 2015;**6**:947. DOI: 10.1039/c4fo01082e

[24] Rossi M, Corradini C, Amaretti A, Nicolini M, Pompei A, Zanoni S. Fermentation of fructooligosaccharides and inulin by bifidobacteria: A comparative study of pure and fecal cultures. Applied and Environmental Microbiology. 2005;**71**(10):6150-6158. DOI: 10.1128/AEM.71.10.6150-6158.2005

[25] Scott K, Martin JC, Duncan SH, Flint HJ. "Prebiotic stimulation of human colonic butyrate-producing bacteria and bifidobacteria", in vitro. FEMS Microbiology Ecology. 2014;**87**:30-40. DOI: 10.1111/1574-6941.12186

[26] Garrido D, Ruiz-Moyano S, Jimenez-Espinoza R, Eom HJ, Block DE, Mills DA. Utilization of galactooligosaccharides by *Bifidobacterium longum* subsp. infantis isolates. Food Microbiology. 2013;**33**:262-270. DOI: 10.1016/j. fm.2012.10.003

[27] Hinz SW, Pastink MI, van den Broek LA, Vincken JP, Voragen AG. *Bifidobacterium longum* endogalactanase liberates galactotriose from type I galactans. Applied and Environmental Microbiology. 2005;**71**:5501-5510. DOI: 10.1128/AEM.71.9.5501-5510.2005

[28] Andersen JM, Barrangou R, Abou Hachem M, Lahtinen SJ, Goh YJ, et al. Transcriptional analysis of oligosaccharide utilization by *Bifidobacterium* lactis Bl-04. BMC Genomics. 2013;**14**:312. DOI: 10.1186/1471-2164-14-312

[29] Ryan SM, Fitzgerald GF, van Sinderen D. Screening for and identification of starch-, amylopectin-, and pullulan-degrading activities in bifidobacterial strains. Applied and Environmental Microbiology. 2006;**72**:5289-5296. DOI: 10.1128/ AEM.00257-06

[30] EjbyM, Fredslund F,Vujicic-ZagarA, Svensson B, Slotboom DJ, Abou Hachem M. Arabinoxylo-oligosaccharide capture in *Bifidobacterium*. Molecular Microbiology. 2013;**90**:1100-1112. DOI: 10.1111/mmi.12419

[31] Makelainen H, Saarinen M, Stowell J, Rautonen N, Ouwehand AC. Xylo-oligosaccharides and lactitol promote the growth of *Bifidobacterium lactis* and *Lactobacillus* species in pure cultures. Beneficial Microbes. 2010;**1**:139-148. DOI: 10.3920/ BM2009.0029

[32] Li Z, Summanen PH, Komoriya T, Finegold SM. In vitro study of the prebiotic xylooligosaccharide (XOS) on the growth of *Bifidobacterium* spp and *Lactobacillus* spp. International Journal of Food Sciences and Nutrition. 2015;**66**:919-922. DOI: 10.3109/09637486.2015.1064869

[33] Topping D, Clifton P. Short-chain fatty acids and human colonic function: Roles of resistant starch and nonstarch polysaccharides. Physiological Reviews. 2001;**81**(3):1031-1063. DOI: 10.1152/ physrev.2001.81.3.1031

[34] Childs CE, Roytio H, Alhoniemi E, Fekete AA, Forssten SD, Hudjec N, et al. Xylo-oligosaccharides alone or in synbiotic combination with *Bifidobacterium animalis subsp. lactis* induce bifidogenesis and modulate markers of immune function in healthy adults: A double-blind, placebo-controlled, randomised, factorial cross-over study. The British Journal of Nutrition. 2014;**111**:1945-1956. DOI: 10.1017/ S0007114513004261

[35] Chung Y-C, Hsu C-K, Ko C-Y, Chan Y-C. Dietary intake of xylooligosaccharides improves the intestinal microbiota, fecal moisture, and pH value in the elderly. Nutrition Research. 2007;**27**:756-761. DOI: 10.1016/j. nutres.2007.09.014

[36] Finegold SM, Li Z, Summanen PH, Downes J, Thames G, Corbett K, et al. Xylooligosaccharide increases bifidobacteria but not lactobacilli in human gut microbiota. Food & Function. 2014;**5**:436-445. DOI: 10.1039/ c3fo60348b

[37] Kajihara M, Kato S, Konishi M, Yamagishi Y, Horie Y, Ishii H. Xylooligosaccharide decreases blood ammonia levels in patients with liver cirrhosis. The American Journal of Gastroenterology. 2000;**95**:2514. DOI: 10.1111/j.1572-0241.2000.02712.x

[38] Lecerf J-M, Depeint F, Clerc E, Dugenet Y, Niamba CN, Rhazi L, et al. Xylo-oligosaccharide (XOS) in combination with inulin modulates both the intestinal environment and immune status in healthy subjects, while XOS alone only shows prebiotic properties. The British Journal of Nutrition. 2012;**108**:1847-1858. DOI: 10.1017/ S0007114511007252

[39] Na MH, Kim WK. Effects of xylooligosaccharide intake on fecal Bifidobacteria, lactic acid and lipid metabolism in Korean young women. Korean Journal of Nutrition. 2007;**40**:154-161. DOI: 10.1097/01. NPT.0000288316.43613.63

[40] Iino T, Nishijima Y, Sawada S, Sasaki H, Harada H, Suwa Y, et al. Improvement of constipation by a small amount of Xylooligosaccharides ingestion in adult women. Journal of Japanese Association for Dietary Fiber Research. 1997;**1**:19-24. DOI: 10.11217/jjdf1997.1.19

[41] Tateyama I, Hashii K, Johno I, Iino T, Hirai K, Suwa Y, et al. Effect of xylooligosaccharide intake on severe constipation in pregnant women. Journal of Nutritional Science and Vitaminology. 2005;**51**:445-448. DOI: 10.3177/jnsv.51.445

[42] Sheu WH-H, Lee I-T, Chen W, Chan Y-C. Effects of xylooligosaccharides in type 2 diabetes mellitus. Journal of Nutritional Science and Vitaminology. 2008;**54**:396-401. DOI: 10.3177/jnsv.54.396

[43] Yang J, Summanen PH, Henning SM, Hsu M, Lam H, Huang J, et al. Xylooligosaccharide supplementation alters gut bacteria in both healthy and prediabetic adults: A pilot study. Frontiers in Physiology. 2015;**6**:11. DOI: 10.3389/ fphys.2015.00216

[44] Cloetens L, Broekaert WF, Delaedt Y, Ollevier F, Courtin CM, Delcour JA, et al. Tolerance of arabinoxylan-oligosaccharides and their prebiotic activity in healthy subjects: A randomised, placebo-controlled cross-over study. The British Journal of Nutrition. 2010;**103**:703-713. DOI: 10.1017/S0007114509992248

[45] Francois IEJA, Lescroart O, Veraverbeke WS, Marzorati M, Possemiers S, Evenepoel P, et al. Effects of a wheat bran extract containing arabinoxylan oligosaccharides on gastrointestinal health parameters in healthy adult human volunteers: A double-blind, randomised, placebo- controlled, cross-over trial. The British Journal of Nutrition. 2012;**108**:2229- 2242. DOI: 10.1017/ S0007114512000372

[46] Francois IEJA, Lescroart O, Veraverbeke WS, Marzorati M, Possemiers S, Hamer H, et al. Effects of wheat bran extract containing arabinoxylan oligosaccharides on gastrointestinal parameters in healthy preadolescent children. Journal of Pediatric Gastroenterology and Nutrition. 2014;**58**:647-653. DOI: 10.1097/ MPG.0000000000000285

[47] Maki KC, Gibson GR, Dickmann RS, Kendall CWC, Chen C-YO, Costabile A, et al. Digestive and physiologic effects of a wheat bran extract, arabino-xylan-oligosaccharide, in breakfast cereal. Nutrition. 2012;**28**:1115-1121. DOI: 10.1016/j. nut.2012.02.010

[48] Walton GE, Lu C, Trogh I, Arnaut F, Gibson GR. A randomised, double-blind, placebo controlled cross-over study to determine the gastrointestinal effects of consumption of arabinoxylan-oligosaccharides enriched bread in healthy volunteers. Nutrition Journal. 2012;**11**:36. DOI: 10.1186/1475-2891-11-36

[49] Whelan K. Mechanisms and effectiveness of prebiotics in modifying the gastrointestinal microbiota for the management of digestive disorders. The Proceedings of the Nutrition Society. 2013;**72**:288-298. DOI: 10.1017/ S0029665113001262

[50] Bouhnik Y, Vahedi K, Achour L, Attar A, Salfati J, Pochart P, et al. Short-chain fructo-oligosaccharide administration dose-dependently increases fecal bifidobacteria in healthy humans. The Journal of Nutrition. 1999;**129**:113-116. DOI: 10.1093/ jn/129.1.113

[51] Alfa MJ, Strang D, Tappia PS, Graham M, Van Domselaar G, Forbes JD, et al. A randomized trial to determine the impact of a digestion resistant starch composition on the gut microbiome in older and mid-age adults. Clinical Nutrition. 2018;**37**(3):797-807. DOI: 10.1016/j. clnu.2017.03.025

[52] Miremadi F, Sherkat F, Stojanovska L. Hypocholesterolaemic effect and anti-hypertensive properties of probiotics and prebiotics: A review. Food & Function. 2016;**25**:497-510. DOI: 10.1016/j.jff.2016.06.016

[53] Alles MS, Roos NM, Bakx JC, van de Lisdonk E, Zock PL, Hautvast J. Consumption of fructooligosaccharides does not favorably affect blood glucose and serum lipid concentrations in patients with type II diabetes. The American Journal of Clinical Nutrition. 1999;**69**:64-69. DOI: 10.1093/ ajcn/69.1.64

[54] Salinardi TC, Rubin KH, Black RM, St-Onge M. Coffee mannooligosaccharides, consumed as part of a free-living, weight-maintaining diet, increase the proportional reduction in body volume in overweight men. The Journal of Nutrition. 2010;**140**(11):1943-1948. DOI: 10.3945/ jn.110.128207

[55] Kumao T, Fujii S, Asakawa A, Takehara I, Fukuhara I. Effect of coffee drink containing Mannooligosaccharides on Total amount of excreted fat in healthy adults. Journal of Health Science. 2006;**52**:482-485. DOI: 10.1248/ jhs.52.482

[56] St-Onge M, Salinardi T, Herron-Rubin K, Black RM. A weight- loss diet including coffee-derived Mannooligosaccharides enhances adipose tissue loss in overweight men but not women. Obesity. 2012;**20**:343- 348. DOI: 10.1038/oby.2011.289

[57] Uemara M, Fujii S, Asano I, Hoshino H. Effect of "coffee mix drink" containing Mannooligosaccharides from coffee Mannan on defecation and fecal microbiota in healthy volunteers. Food Science and Technology Research. 2004;**10**(2):195-198. DOI: 10.3136/ fstr.10.195

[58] Zheng J, Li H, Zhang X, Jiang M, Luo C, Lu Z, et al. Prebiotic Mannan- oligosaccharides augment the hypoglycemic effects of metformin in correlation with modulating gut microbiota. Journal of Agricultural and Food Chemistry. 2018;**66**(23):5821- 5831. DOI: 10.1021/acs. jafc.8b00829

[59] Faseleh Jahromi M, Liang JB, Abdullah N, Goh YM, Ebrahimi R, Shokryazdan P. Extraction and characterization of oligosaccharides from palm kernel cake as prebiotic. BioResources. 2016;**11**:674-695. DOI: 10.15376/ biores.11.1.674-695

[60] Navidshad B, Liang JB, Jahromi MF, Akhlaghi A, Abdullah N. A comparison between a yeast cell wall extract (bio- Mos®) and palm kernel expeller as mannan-oligosac-charides sources on the performance and ileal microbial population of broiler chickens. Italian Journal of Animal Science. 2015;**14**(1):3452. DOI: 10.4081/ ijas.2015.3452

[61] Yang Y, Iji P, Choct M. Dietary modulation of gut microflora in broiler chickens: A review of the role of six kinds of alternatives to in-feed antibiotics. World's Poultry Science Journal. 2009;**65**(1):97-114. DOI: 10.1017/S0043933909000087

[62] Faseleh Jahromi M, Shokryazdan P, Idrus Z, Ebrahimi R, Liang JB. In Ovo and dietary administration of oligosaccharides extracted from palm kernel cake influence general health of pre- and neonatal broiler chicks. PLoS One. 2017;**12**(9):e0184553. DOI: 10.1371/ journal.pone.0184553

[63] Zhao PY, Jung JH, Kim I-H. Effect of mannan oligosaccharides and fructan on growth performance, nutrient digestibility, blood profile, and diarrhea score in weanling pigs. Journal of Animal Science. 2012;**90**(3):833-839. DOI: 10.2527/jas.2011-3921

Permissions

All chapters in this book were first published in PPPBNH, by InTech Open; hereby published with permission under the Creative Commons Attribution License or equivalent. Every chapter published in this book has been scrutinized by our experts. Their significance has been extensively debated. The topics covered herein carry significant findings which will fuel the growth of the discipline. They may even be implemented as practical applications or may be referred to as a beginning point for another development.

The contributors of this book come from diverse backgrounds, making this book a truly international effort. This book will bring forth new frontiers with its revolutionizing research information and detailed analysis of the nascent developments around the world.

We would like to thank all the contributing authors for lending their expertise to make the book truly unique. They have played a crucial role in the development of this book. Without their invaluable contributions this book wouldn't have been possible. They have made vital efforts to compile up to date information on the varied aspects of this subject to make this book a valuable addition to the collection of many professionals and students.

This book was conceptualized with the vision of imparting up-to-date information and advanced data in this field. To ensure the same, a matchless editorial board was set up. Every individual on the board went through rigorous rounds of assessment to prove their worth. After which they invested a large part of their time researching and compiling the most relevant data for our readers.

The editorial board has been involved in producing this book since its inception. They have spent rigorous hours researching and exploring the diverse topics which have resulted in the successful publishing of this book. They have passed on their knowledge of decades through this book. To expedite this challenging task, the publisher supported the team at every step. A small team of assistant editors was also appointed to further simplify the editing procedure and attain best results for the readers.

Apart from the editorial board, the designing team has also invested a significant amount of their time in understanding the subject and creating the most relevant covers. They scrutinized every image to scout for the most suitable representation of the subject and create an appropriate cover for the book.

The publishing team has been an ardent support to the editorial, designing and production team. Their endless efforts to recruit the best for this project, has resulted in the accomplishment of this book. They are a veteran in the field of academics and their pool of knowledge is as vast as their experience in printing. Their expertise and guidance has proved useful at every step. Their uncompromising quality standards have made this book an exceptional effort. Their encouragement from time to time has been an inspiration for everyone.

The publisher and the editorial board hope that this book will prove to be a valuable piece of knowledge for researchers, students, practitioners and scholars across the globe.

List of Contributors

Tomoko Ohshima, Tomomi Kawai and Nobuko Maeda
Department of Oral Microbiology, School of Dental Medicine, Tsurumi University, Yokohama, Japan

José Maldonado
Unit of Pediatric Gastroenterology, Hepatology and Nutrition, Department of Pediatrics, Virgen de las Nieves University Hospital, University of Granada, Spain

Christina C. Tam and Luisa W. Cheng
Foodborne Toxins Detection and Prevention Research Unit, Agricultural Research Service, United States Department of Agriculture, Albany, California, United States

Kirkwood M. Land
Department of Biological Sciences, University of the Pacific, Stockton, California, United States

Małgorzata Ziarno
Division of Milk Biotechnology, Department of Biotechnology, Microbiology and Food Evaluation, Faculty of Food Sciences, Warsaw University of Life Sciences – SGGW (WULS-SGGW), Warsaw, Poland

Rouba El Khati and Dolla Karam-Sarkis
Microbiology Laboratory, Faculty of Pharmacy, Saint-Joseph University, Beirut, Lebanon

Anne-Judith Waligora-Dupriet and Marie-José Butel
UMR-S U 1139, Pathophysiology and Pharmacotoxicology of the Human Placenta, Pre- and Postnatal Microbiota, INSERM, Hospital University Department Risk in Pregnancy, Faculty of Pharmacy, Paris University, France

Rebeca Monroy-Torres, Marco Antonio Hernández-Luna and Sergio López-Briones
Cuerpo Academico de Medicina Traslacional, University of Guanajuato, Campus Leon, Guanajuato, Mexico

Xochitl Sofía Ramírez-Gómez
University of Guanajuato, Campus Celaya Salvatierra, Celaya, Mexico

Maria Inês Sucupira Maciel
Graduate Program in Food Science and Technology, Federal Rural University of Pernambuco, Recife, Brazil

Michelle Maria Barreto de Souza
Technology Center, Federal University of Paraíba, João Pessoa, Brazil

Jessica Dayara Álvarez-Rosales, César Ozuna, Rubén Salcedo-Hernández and Gabriela Rodríguez-Hernández
Division of Life Science, University of Guanajuato, Guanajuato, México

Daniel Hernandez-Patlan and Bruno Solis-Cruz
Laboratorio 5: LEDEFAR, Unidad de Investigacion Multidisciplinaria, Facultad de Estudios Superiores (FES) Cuautitlan, Universidad Nacional Autonoma de Mexico (UNAM), Cuautitlan Izcalli, Estado de Mexico, Mexico

Billy M. Hargis and Guillermo Tellez
Department of Poultry Science, University of Arkansas, Fayetteville, AR, USA

Elena Franco-Robles and Cristina Doriany Balcón-Pacheco
Department of Veterinary, University of Guanajuato, Irapuato, Guanajuato, Mexico

Joel Ramírez-Emiliano
Department of Medical Sciences, University of Guanajuato, Leon, Guanajuato, Mexico

José Sergio López-Briones
Department of Medicine and Nutrition, University of Guanajuato, Leon, Guanajuato, Mexico

Ciro Gargiulo Isacco
Human Stem Cells Research Center, Ho Chi Minh City, Vietnam
Department of Interdisciplinary Medicine (DIM), School of Medicine, University Aldo Moro, Bari, Italy
Pham Chau Trinh University of Medicine, Danang, Vietnam
International Institute of Gene and Immunology, Ho Chi Minh City, Vietnam

Kieu Cao Diem Nguyen
Human Stem Cells Research Center, Ho Chi Minh City, Vietnam
Department of Interdisciplinary Medicine (DIM), School of Medicine, University Aldo Moro, Bari, Italy

Andrea Ballini
Department of Biosciences, Biotechnology and Biopharmaceutics, University of Bari Aldo Moro, Bari, Italy
Department of Basic Medical Sciences, Neurosciences and Sense Organs, University of Bari Aldo Moro, Bari, Italy

Danila De Vito
Department of Basic Medical Sciences, Neurosciences and Sense Organs, University of Bari Aldo Moro, Bari, Italy

Angelo Michele Inchingolo, Stefania Cantore, Gregorio Paduanelli, Alessio Danilo Inchingolo, Gianna Dipalma and Francesco Inchingolo
Department of Interdisciplinary Medicine (DIM), School of Medicine, University Aldo Moro, Bari, Italy

Bruno Solis-Cruz and Daniel Hernandez-Patlan
Faculty of Superior Studies Cuautitlán, Multidisciplinary Research Unit L5, National Autonomous University of Mexico, Cuautitlán Izcalli, Mexico State, Mexico

Billy M. Hargis and Guillermo Tellez
Department of Poultry Science, University of Arkansas, Fayetteville, Arkansas, USA

Bradley A. Saville
Department of Chemical Engineering and Applied Chemistry, University of Toronto, Toronto, ON, Canada
Prenexus Health, Gilbert, AZ, United States

Sandra H. Saville
Prenexus Health, Gilbert, AZ, United States
Saville Nutrition Consulting, Oakville, ON, Canada

Index

A
Arabinose Side Group, 209
Arabinoxylanoligosaccharides, 205
Atopic Dermatitis, 14, 17, 19-20

B
Bacteria Strains, 11, 13, 15, 40-41, 50, 52, 92, 154
Bacterial Cell-free Probiotics, 1
Bacterial Colonization, 11-12
Bacterial Fermentation, 120, 203
Bacteriocins, 3-5, 8, 27, 61, 68, 131-132, 134, 143, 153, 157, 187, 193, 199, 201
Bifidobacteria, 2, 8, 11-13, 15, 17, 19, 21, 24, 27, 29, 32, 36, 43, 45, 50-51, 53, 55, 63, 69, 72, 152, 154, 157, 162, 170, 207-210, 215, 218, 220-221
Bifidobacterium Lactis, 13, 19, 221
Bifidogenic Effect, 17, 185, 195, 215
Bioactive Compounds, 11, 99
Biogenics, 1, 3, 5-6

C
Cecum, 91, 185, 195, 214
Clostridium, 26, 28-29, 36-38, 42, 61-63, 65, 70, 72, 74, 76, 80, 83, 91, 137-138, 140, 146-147, 157, 203-204
Commensal Bacteria, 12, 23, 151, 162
Cytokines, 11, 28, 61, 66-67, 69, 151, 156, 158-159, 168, 171, 173-174, 193-194

D
Dental Caries, 1-2, 7
Dysbiosis, 2, 12, 15, 23, 26, 60, 62, 64-69, 74, 76, 79-80, 86, 91, 138, 140, 149, 152-153, 155-156, 159, 161, 163, 166, 168-171, 174, 177-178

E
Eczema, 15, 17, 22
Enteropathogenic Bacteria, 13, 133
Exogenous Probiotic, 1
Extracellular Enzymes, 141, 207, 209

F
Firmicutes, 25, 29, 60, 63, 65, 70, 90-91, 152, 161, 170, 203
Flucosyllactose, 16
Fructofuranosidase, 207
Fructose Subunits, 205

G
Galactan, 207-208
Gastrointestinal Tract, 23, 29, 39, 41, 44, 49, 51, 91, 103, 131-132, 137, 141, 144-145, 150, 152, 157, 177, 182, 190, 195, 198, 202-203, 219
Gelatinization Temperature, 206
Gingival Sulcus, 1
Gut Microbiota, 8, 11-13, 15, 17-19, 23, 29, 38, 60, 62, 74, 76, 78-79, 81, 83, 86-87, 97-98, 110, 138, 148, 165, 167, 193, 198, 200, 203-204, 218,
Gut-brain Axis, 60-61, 76-78, 80, 148-149, 163-164, 214

H
Hepatology, 11, 13, 18, 31, 36, 81, 197, 220
Humoral, 16, 150, 170
Hydrolytic Enzymes, 206, 209

I
Ileum, 38, 135, 140, 185, 190, 195, 203, 214
Immunonutrients, 14
Infant Formulae, 11-17
Interferon, 66, 77, 151
Intestinal Microbiota, 11, 17-18, 22, 26, 81, 86, 88-93, 99-100, 102, 130, 138, 140, 152-153, 155-159, 162-164, 181, 183-185, 187, 192, 195
Isomaltooligosaccharides, 157, 204-205

J
Jejunum, 150, 185-186, 190, 195, 199, 203

K
Ketogenic Diet, 75, 77, 83

L

Lactobacilli, 2-5, 7, 11-13, 15, 17, 24-25, 27-28, 50, 56, 58-59, 69, 72, 98, 110, 126, 144-145, 157, 165, 189, 210, 221

Lactobacillus Fermentum, 4, 13-14, 19, 158

Lactobacillus Strains, 8, 10, 13, 33, 35, 43, 173-174, 199, 210

M

Makelainen, 209-212, 217, 220-221

Mannooligosaccharides, 202, 205, 214, 220, 222

Microbiota Composition, 15, 18, 22, 144-145, 152, 165, 178

Microbiota Transfer Therapy, 74, 77-78, 83

Microflora, 2, 7-8, 33, 35, 41-42, 44, 51, 53, 65, 79, 143, 160, 163, 179, 192, 195, 197-198, 200

Monomers, 202-203, 205, 207-208

Mucosal Cell, 12

N

Necrotizing Enterocolitis, 14, 29

O

Oligomers, 202-203, 205, 209, 214

Oligosaccharides, 11, 15-17, 21-22, 25, 30, 33, 36-37, 57, 157-158, 164, 167, 183, 199, 202, 204-205, 207-209, 214, 217, 220-223

Oral Candidiasis, 1-2

Oral Indigenous Bacteria, 1

P

Pathogenic Microorganisms, 1, 86, 90, 148, 190

Pediatric Gastroenterology, 11, 13, 18-19, 37, 79-80, 222

Periodontitis, 1-2, 5, 174

Potential Pathogenicity, 2

Prebiotic, 11, 16-17, 19, 21-22, 25, 28-29, 31, 33, 36-37, 74, 105, 163, 165, 182-186, 190-192, 194-195, 197, 199-202, 209, 213-214, 217, 219-222

Preterm, 14, 19-20, 29, 37, 63, 77-78

Prevotella, 2, 64-65, 75, 79, 91, 152-153, 168-169, 204, 209

Probiotic Bacteria, 1-6, 12, 14-15, 17, 26, 32, 56, 58-59, 86, 92, 95, 100-105, 109-111, 116, 133, 145-146, 148, 157, 162, 192

Proteobacteria, 60, 63, 90-91, 138, 152, 170, 203-204

R

Resistant Starch, 25, 33, 204-205, 215, 219-222

S

Saccharomyces Boulardii, 13, 19, 36, 68, 82

Streptococcus Species, 203-204

Symbiotic, 11, 60, 91, 152, 157, 161-162, 167, 182, 184

T

Therapeutics, 10, 23, 28, 32-33, 37, 97, 177, 179

Tongue Dorsum, 1

Tumor Necrosis Factor, 66, 77, 166

V

Verrucomicrobia, 60, 90, 203

v

X

Xylobiose, 207, 209

Xylooligosaccharides, 157, 183, 202, 204-205, 209, 221

Xylose, 25, 204-205, 207-209, 218

Z

Zoetendal, 18, 37, 203-204, 219

www.ingramcontent.com/pod-product-compliance
Lightning Source LLC
LaVergne TN
LVHW080610210926
848485LV00003B/135